PRAISE FOR *Integrated Forest Gardening*

❖ ❖ ❖

"*Integrated Forest Gardening* fills a major gap in the canon of permaculture books, giving us, at last, a detailed guide to guild and polyculture design. No longer is this subject mysterious and daunting; in this book we now have specific instructions for designing and installing multispecies plant groups. Chapter 7, which describes fifteen guilds and their plant members, is a golden nugget worth the price of the book alone. This is an essential book for all food foresters and ecological designers."

— Toby Hemenway, author of *Gaia's Garden: A Guide to Home-Scale Permaculture*

"This rich feast of nature love by three experienced and working permaculture designers pushes into the hard task of creating recombinant ecosystems, a field where few have gone before. The authors expose the logic and lore of working guilds, the symbiotic plant assemblies of productive landscapes. Full of design insight into the needs and opportunities of both plants and the people who live with them, *Integrated Forest Gardening* offers a panoply of example guilds, work procedures, and luscious images to inspire and guide the perennial food gardener onto a path of ecological renewal."

— Peter Bane, author of *The Permaculture Handbook* and publisher of *The Permaculture Activist* magazine

"This is an intimate insight into the world of plant guilds. The authors have taken the broad land-based overview and zoom the reader into the micro detail of these plant polycultures. Details of root structure, seeding patterns, and relationships with the surrounding environment have been carefully observed and are well laid out in the plant guild lists. This book is an important contribution to every permaculture designer's library and will appeal to all those wishing to grow sustainable polycultures, whether broadscale or in the garden."

— Ben Law, author of *The Woodland Way* and *Roundwood Timber Farming*

"*Integrated Forest Gardening* makes the process of creating complex agroecosystems more understandable and achievable. It is a fine guide to designing forest garden and polycultural systems using permaculture principles."

— Martin Crawford, author of *Creating a Forest Garden*

"We stand at a new threshold. The history of food production has tended evermore narrowly toward monoculture, whereas a sustainable future can be based only on polyculture. But we're desperately short of knowledge on polycultures. While mainstream research still chases the chimera of fossil-fueled monoculture, a small band of visionaries are working to develop the knowledge we need to carry us forward to the future. Just such are the authors of this book, and the wisdom it contains is part of that movement."

— PATRICK WHITEFIELD, permaculture teacher;
author of *The Earth Care Manual*

"For the design work we do at Midwest Permaculture, when we need experienced advice on planting systems we turn to the three gentlemen who collaborated on this impressive work. The book is thorough, accessible, and timely. So wish we had this insightful compilation when we first started. It's a gem!"

— BILL AND BECKY WILSON, Midwest Permaculture

"Reading *Integrated Forest Gardening* was like taking a walk through a well-orchestrated whole-systems design! This book spoke my language to me as a plant enthusiast and systems thinker. It is rare to find in one book such depth of user-friendly detail. It demystifies the mythical nature of the forest garden and brings its strategies to easy application. This book is a must for all plant lovers."

— JUDE HOBBS, Cascadia Permaculture

INTEGRATED FOREST GARDENING

The Complete Guide *to* **POLYCULTURES** AND **PLANT GUILDS** *in* Permaculture Systems

WAYNE WEISEMAN, DANIEL HALSEY, AND BRYCE RUDDOCK

CHELSEA GREEN PUBLISHING
WHITE RIVER JUNCTION, VERMONT

Copyright © 2014 by Wayne Weiseman, Daniel Halsey, and Bryce Ruddock.
All rights reserved.

Unless otherwise noted, all photographs copyright © 2014 by Wayne Weiseman, Daniel Halsey, and Bryce Ruddock. Unless otherwise noted, all illustrations copyright © 2014 by Daniel Halsey and Kellen Kirchberg.

No part of this book may be transmitted or reproduced in any form by any means without permission in writing from the publisher.

Developmental Editor: Makenna Goodman
Copy Editor: Laura Jorstad
Proofreader: Michelle Moran
Indexer: Linda Hallinger
Designer: Melissa Jacobson

Printed in the United States of America.
First printing July, 2014.
10 9 8 7 6 5 4 3 2 1   14 15 16 17 18

Chelsea Green Publishing is committed to preserving ancient forests and natural resources. We elected to print this title on paper containing at least 10% post-consumer recycled paper, processed chlorine-free. As a result, for this printing, we have saved:

18 Trees (40' tall and 6-8" diameter)
8,717 Gallons of Wastewater
9 million BTUs Total Energy
583 Pounds of Solid Waste
1,607 Pounds of Greenhouse Gases

Chelsea Green Publishing made this paper choice because we are a member of the Green Press Initiative, a nonprofit program dedicated to supporting authors, publishers, and suppliers in their efforts to reduce their use of fiber obtained from endangered forests. For more information, visit www.greenpressinitiative.org.

Environmental impact estimates were made using the Environmental Defense Paper Calculator. For more information visit: www.papercalculator.org.

Our Commitment to Green Publishing
Chelsea Green sees publishing as a tool for cultural change and ecological stewardship. We strive to align our book manufacturing practices with our editorial mission and to reduce the impact of our business enterprise in the environment. We print our books and catalogs on chlorine-free recycled paper, using vegetable-based inks whenever possible. This book may cost slightly more because it was printed on paper that contains recycled fiber, and we hope you'll agree that it's worth it. Chelsea Green is a member of the Green Press Initiative (www.greenpressinitiative.org), a nonprofit coalition of publishers, manufacturers, and authors working to protect the world's endangered forests and conserve natural resources. *Integrated Forest Gardening* was printed on paper supplied by RR Donnelley that contains at least 10% postconsumer recycled fiber.

Library of Congress Cataloging-in-Publication Data
Weiseman, Wayne.
  Integrated forest gardening: the complete guide to polycultures and plant guilds in permaculture systems / Wayne Weiseman, Daniel Halsey, and Bryce Ruddock.
      pages cm
  Complete guide to polycultures and plant guilds in permaculture systems
  Includes bibliographical references and index.
  ISBN 978-1-60358-497-5 (pbk.) — ISBN 978-1-60358-498-2 (ebook)
1. Edible forest gardens. 2. Permaculture. I. Halsey, Daniel. II. Ruddock, Bryce. III. Title. IV. Title: Complete guide to polycultures and plant guilds in permaculture systems.

SB454.3.E35W45 2014
  635 — dc23
              2014012290

Chelsea Green Publishing
85 North Main Street, Suite 120
White River Junction, VT 05001
(802) 295-6300
www.chelseagreen.com

♦ ♦ ♦

We would like to dedicate this book to the generations of conservationists and ecologists who strive to protect our natural systems, the indigenous cultures that built living systems without destroying their habitat, and the generations to come, which will surely depend on ecological solutions to restore our earth to its natural abundance.

CONTENTS

❖ ❖ ❖

CHAPTER ONE: WHAT IS INTEGRATED FOREST GARDENING?   1

The Permaculture Design System ❖ The Design Process ❖ Essential Templates of Good Design ❖ What Is Forest Gardening? ❖ Perennials and Polycultures ❖ Climate and Scale ❖ Climate Change and the Importance of Integrated Crop Production ❖ Capturing, Storing, Cycling, and the Sustainable Homestead ❖ A Word About Compost: Extending the Life of a Resource ❖ Water and Earthworks ❖ What Is a Plant Guild? ❖ The Scientific Basis for Plant Guilds ❖ The Importance of Plant Diversity ❖ Applications of Plant Guilds in Permaculture and Forest Gardening ❖ How Guilds Work ❖ Designing for the Niche ❖ Studying Guilds in Their Natural State

CHAPTER TWO: THE STRUCTURE OF A PLANT GUILD   37

Perennial Polycultures ❖ Defining Your Niche ❖ Nutrient Cycling ❖ Carrying Capacity ❖ Understanding the Context of Your Site ❖ Start from Scratch or Follow Nature's Lead? ❖ The Integrated Living System ❖ Niche Dynamics ❖ Designing for Cooperative Competition ❖ The Importance of Sunlight ❖ The Position of Plants ❖ Determining the Quality of Your Soil ❖ Building for Nutrient Cycling ❖ Needs of a Forest Garden ❖ Five Considerations for Sustainable Design ❖ Permaculture Principles to Apply to Guild Design ❖ Yeomans's Scale of Relative Permanence ❖ Constructing the Plant Guild ❖ Questions You May Be Asking

CHAPTER THREE: SELECTING PLANTS FOR GUILD DESIGN   93

Understanding the Biome ❖ Functions of Plant Guilds and Polycultures ❖ Covering the Soil with a Blanket of Vegetation ❖ The Soil Regime ❖ Soils and Salt Tolerance ❖ Catastrophic Occurrences ❖ Agricultural Toxins ❖ When the Wind Doth Blow ❖ Terra Preta: The Dark Earth ❖ Growing Zones ❖ Selecting Plants for Resilience ❖ Understanding Sun Exposure ❖ Determine Your Soil Types ❖ Understanding a Plant's Tolerances ❖ Guild Design Basics ❖ Roots: Anchors and So Much More ❖ The Fabulous Fungi ❖ Nitrogen-Fixing Plants ❖ Seasonal Considerations ❖ Bloom Times ❖ Fruit Set ❖ Patterns of Growth ❖ Populating the Guild ❖ The Natural Range of Plants ❖ The Importance of Diversity

CHAPTER FOUR:	TREES: THE ESSENCE OF THE PLANT GUILD	151

Duir: Opening the Door on the Oak Tree • The Precious Pine • Old Man Hickory • The Maple: Sugar in the Gourd

CHAPTER FIVE:	DESIGNING FOR OPTIMAL SPECIES INTEGRATION	173

Beneficial Behaviors in the Permaculture Guild • Using Plants in Functional Pest Strategies • Agroforestry Techniques • Specific Plant and Animal Interactions in the Plant Guild • Everything and Everyone Is Lunch

CHAPTER SIX:	GUILD PROJECT MANAGEMENT	201

Implementation Time Line • What Are the Broad Site Preparations? • What Is the Sequence of Implementation? • What Steps "Complete" the Design Implementation? • What Is Needed for Long-Term, or Protracted, Implementation? • Budgeting the Financing • Time Lines • Design Decisions and Checklist • The Budget • Implementation, Management, and Maintenance

CHAPTER SEVEN:	CASE STUDIES: FIFTEEN PLANT GUILDS	215

Fruit and Nut Guild • Pawpaw Delight Guild • Four Vines Guild • Annual–Perennial Guild • Poisonous Plant Guild • Asian Pear Polyculture Guild • Ginseng/Sugar Maple Polyculture Guild • Boreal Forest Berry Guild • Salsa Garden Guild • Dwarf Cherry Tree Polyculture Guild • Ruddock Guilds • Moving Forward . . .

ACKNOWLEDGMENTS	283
APPENDIX	285
NOTES	289
RESOURCES	293
INDEX	299

CHAPTER 1

What Is Integrated Forest Gardening?

Keep close to Nature's heart . . . and break clear away, once in a while, and climb a mountain or spend a week in the woods. Wash your spirit clean. The clearest way into the Universe is through a forest wilderness. The gross heathenism of civilization has generally destroyed nature, and poetry, and all that is spiritual.

— JOHN MUIR

This book is about integrated forest gardening, but it's not simply about plants or forests. It is about the integration of all aspects of a land base into the development of healthy food, medicinal, and utility landscapes. The built environment, the waste stream, animals, plants, stones: All hold equal importance. When we step out of the house into the backyard garden, it would be difficult to define where the house ends and the garden begins. There is an "edge" created between the two, and neither is any less or more significant than the other. What, if anything, makes the house any less plant than a plant itself? As much as a plant is a biological entity, rooted to the soil, a living, breathing entity, exchanging gases, energy, nutrients, materials with its environment and with other plants and animals, so is a house a living, breathing entity in its own right, exchanging gases, energy, materials, with plants and creatures large and small, rooted in the same soils as the plants that surround it. When we speak of integration, we speak of whole systems, where all is working as one, striving toward ecological harmony, balance, and health.

Is it possible to split the two sides of a piece of paper apart without generating yet another front and back? How could we separate the forest from the trees, the driveway from the embankment, human "waste" from the septic system that is broken down by the same microorganisms that break down all organic matter? One of the central dictums of Permaculture is integration rather than segregation. If this is true, then a book about integrated forest gardening cannot only be about plants that make up polycultures, plant guilds, and

INTEGRATED FOREST GARDENING

When starting on design and research for integration of a forest garden on your site, having a comfortable workspace and drawing desk helps to keep your vision accessible as you adapt and make changes. Give yourself lots of elbow room and begin collecting the drawing tools needed to create a clean and easily read design.

food forests; it must be about all that goes into the creation of our primary producers, the plants, and about all the consumers that find sustenance from them and give back to them for their own sustenance in a never-ending cycle.

What do we actually see, hear, smell when we step out that door into the backyard, or an agricultural field, or a forest? What is actually there? What was here before this? Can we take the time to fully see what we are looking at? What does the day, the climate, the topography, look like? What animals have crossed the property in the last twenty-four hours? How does the soil feel as we crush it between our fingers?

Yes, we have pondered this concept of Permaculture, and food forests, and possibly that obscure word *polyculture*. It all sounds exciting, and somehow it all rings true; it is the right thing to do.

So what is Permaculture, anyway, and what are these food forests and polycultures, and why do we need to rigorously observe, inventory, and assess the landscape — and ultimately, why integrate?

Permaculture is a word coined by Bill Mollison in Australia in the 1970s with his student David Holmgren. Permaculture begins and ends with keen and protracted observation of a piece of land embedded in the larger bioregion, and the design of an integrated landscape where all elements and patterns are in functional relationship with one another.

You might ask, what exactly is *functional relationship* and how do we go about the business of setting this up? How do we set up a system where all elements relate to one another functionally? Imagine stepping out that door and into the landscape to see patches of grass interspersed with patches of bare ground, a single Norway maple tree dominating the backyard, a toolshed, a garage, a cracked concrete driveway, and a chain-link fence defining the perimeter of the property. What do we see beyond the obvious components? What patterns are visible? What is this land telling us? Is it at a modicum of health and balance? Is it producing food, medicine, and utility in unlimited supply? Is it beautiful, the soil healthy with a high percentage of organic matter, soil life, and good tilth? Is the water from the roofs of the shed and garage leaking off the property into the nearest storm drain, the gray water from the house flowing away through the city mains? How can we make use of all this and how will these flows benefit the property and the yield that we take from it? What more can we observe and learn from this landscape?

If we have read about food forests and the benefits thereof, we may realize that this property does not even begin to meet the ideal, does not reflect the kind of diversity found in a polyculture of high-yielding fruit and nut trees, and herbaceous species. The patterns and elements observed here are not in the least integrated; there is little, if any, exchange among the elements, with very little support. We have yet to find the kind of integration we seek within our landscape. Something is amiss.

So what is the solution, and why bother? Before digging deeper into a delineation of integrated forest gardening, polycultures, and plant guilds, let's stop and find answers to that question: "Why bother?" Why do this forest gardening thing to begin with, and what are its benefits?

When you walk down your suburban street, you might be amazed at the amount of wasted space. There is so much unacknowledged potential for raising crops. Meanwhile, food prices climb as the corporate farm continues to eat away at the small farmer. Indeed, the corporate farm, based on an industrial model, forces the small farmer out of business, endlessly applies harmful chemicals to the soil and plants, raises monocrops of grains and beans, and perpetuates what has become the "green revolution" on a mass scale. Eight crops are offered on this planet for primary consumption, but everyone needs to eat, not only the soy and corn merchants. And as upheaval continues to erupt around the world, what are the outcomes? What are people putting in place to deal with a rampant greed economy and environmental decimation? And the forests? What gives us life — predominantly trees — is continually turned into sawdust and into enough two-by-fours each year that if they were placed end-to-end they'd reach from the earth to the moon and back. This can be remedied, of course, by changing the way we think about the tree's purpose. It does not exist for lumber alone, but is actually the very root of all our sustenance. This involves changing the way most of us think about gardening and farming in general.

Stated bluntly, the earth is under great duress. We are witnessing the destruction of the planet's life support systems: the arable soil, potable water, breathable air, and forests that provide shelter for the majority of the planet's species.

This destruction is making it impossible to meet human needs as we surpass the carrying capacity of the planet. Human beings need food, shelter, fresh water, clothing, health care, a safe environment, and instruction and education for our health and well-being. Without these basic necessities we die. With the destruction of life support systems, human life becomes more desperate and untenable.

The cause for planetary destruction is a political and economic system that operates as if the earth's life support systems and species are only raw materials upon which to profit, regardless of destructive consequences. This global system is reinforced by states and armies that fight one another for control of diminishing resources. This system and the psychology behind it cannot be reformed. It will destroy itself by virtue of its own trajectory.

Under these conditions, the task of well-intentioned human beings everywhere is to join together to design and put into practice projects that meet basic needs in harmony with the needs of the planet. Permaculture is a design science that can help us accomplish this. Integrated forest gardening is a key ingredient in generating the yields that we require for our sustenance and the protection of the earth's biological resources.

Today Planet Earth needs a revolution, a revolution rooted in Permaculture. Permaculture addresses the needs for food, health care, shelter, education, and security. Permaculture is based on a universal ethic that does not recognize the categorization of human beings into different camps for the purpose of competing for resources. The Permaculture ethic is based on care for the earth, care for all people, and sharing the wealth for the greater good.

The Permaculture Design System

Every thought, word, and action carries consequences. Each person shares responsibility for the development of others. Ethics and values grow naturally out of these practices: care of the earth, care of people, ethical surplus distribution, building an ecological–human support base. Care for and love of the environment are at the core of our Permaculture work. How we do this is best encouraged through love, nurturing, cooperation, and reverence for all life.

The Permaculture ethic seeks to create ecological and environmental harmony, stability for future generations, and appropriate technologies that sustain rather than harm. It gives us a method to immerse ourselves in the landscape and exhaust all possibilities of an understanding of place. The Permaculture way promotes true self-reliance through design and hands-on practice for farmers, gardeners, homeowners, urban dwellers, educators, administrators, communities, businesses, students, ranchers, landowners, architects, environmentalists, regional planners, builder-developers, and you. Ultimately we can achieve balance and health by synthesizing applied biology, eco-technology, and integrative architecture: the merging of renewable energies and biological-earth systems.

Permaculture strives for the harmonious integration of the built environment, access and circulation through a property, microclimate, plants, animals, soils, and water into secure, productive communities. The focus is on the relationships created among elements by the way we place them in the landscape. We do this by studying and mimicking patterns found in nature.

Permaculture entails the integration of energy-efficient buildings, wastewater treatment, recycling, and land stewardship, economic and social structures that support the evolution and development of more permanent communities. Permaculture design concepts are applicable to urban as well as suburban and rural settings. From households to bioregional planning, Permaculture design is not limited by scale.

Permaculture is about whole systems, not about separate components. Because each element in a landscape or the built environment affects every other element at a site, a complete, comprehensive assessment is paramount to develop healthy, productive, energy-efficient relationships among elements for the benefit of all. Paying attention to every detail—topography, climate, water, wind, sun, activity nodes and corridors, buildings, machinery and tools, the waste stream, plants, and animals—enables us to make best use of what is already on the ground, and what we intend to put there. With a dynamic interaction of elements in process, and an assessment of both spatial and temporal attributes and patterns, organized around sound ecological principles, yields are maximized and the landscape balanced.

The idea and practice of Permaculture has been expanding ever since its founder Bill Mollison visited the United States in the 1980s. But in the last ten years it has begun to go viral. Permaculture is now moving into the mainstream at a rapid rate. Because of its comprehensiveness, Permaculture's time has come. It addresses agriculture, animals, aquaculture, the built environment, the waste stream, and building a nonlinear, zero-waste system.

Basically, Permaculture is a complete life path where nothing is taken for granted and where everything that we place into the landscape sits in functional relationship to everything else. Given the state of the world, people are seeking ways to become more self-reliant, to know where their food comes from, to diminish their ecological footprint on this earth. Permaculture supports this revolution in terms of reintegrating with the natural cycles that have existed long before us and will exist long after we are gone. Permaculture, as stated in its ethics, is all about care of earth, care of people, fair share.

There are many examples of how, when planned accordingly, even the industrial machine becomes a worthy proponent of reducing and reusing, not to mention recycling. When design focuses specifically on the primary goal of "functional relationship," and pieces of the system circulate materials and energies back into the system, we inch closer and closer to a zero-waste environment. When the "waste" that flows from the industrial juggernaut is organic waste, then all the plants in the system are fed from it. We can use backward thinking to follow this waste stream to its origins and eliminate toxic and poisonous materials so that the end result is healthy food for plants and animals. It is not a stretch to realize that this way of backward thinking to the source can be accomplished on the thirtieth floor of a Manhattan apartment, on a postage-stamp-sized property in the suburbs, or on a thousand-acre rural spread. All the same principles hold wherever we may plant our lives. The structure of a plant guild, polyculture, or food forest is contingent upon how the supply of materials and energies flows into it and whether or not they are worthy as foodstuffs for the guild's sustenance.

The Design Process

The key to the process of Permaculture design lies in rigorous, thorough, and ongoing observation. Like detectives, we piece together, incrementally through our observations of a landscape, the resources available at a site, and the patterns already present. In this way we can then make the least change for the greatest effect. We may find that what is already present in our landscape is worth keeping and maintaining. The large sugar maple tree in the southeast corner is a source of summer shade, sugary sap, and syrup; the black walnut tree on the northwest corner, a source of nuts and medicine and beautifully figured wood;

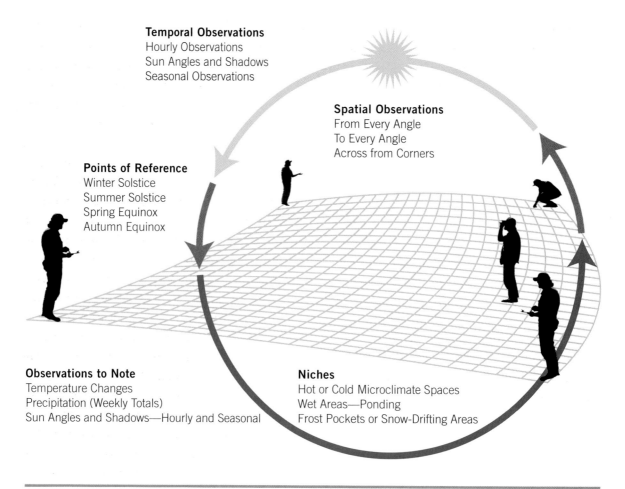

When assessing our properties for their resources and opportunities, it is important to look at them from all angles, during all seasons, and at all times of the day in order to gain a complete picture.

the daylilies in front of the house supply tubers, young stalks, flower buds, and flowers for the palate; the dandelions on the lawn provide food and medicine and act as "dynamic" accumulators of minerals from deep in the soil profile thanks to their subterranean taproots. As we design and imprint fresh patterns on the landscape, we augment what is already there and implement an integral and sound ecology into a property. All of the changes we make in the landscape refer back to care of earth, care of people, and fair share (benevolent distribution of goods and resources).

Let's look at the two words that Bill Mollison put together many years ago: *protracted observation*. Why these two words in particular? He originally stated that we should observe a property for one year before we start developing what could potentially be a high-yielding site, an integrated land base where the built environment, the waste stream, and the agricultural systems merge into a circular model that coalesces into an ecosystem of our own making, where nothing goes to waste, where everything in the landscape stands in functional relationship to everything else.

But why a year of protracted observation? Should we purchase a place and leave it untouched for a whole year? What if we need to grow food for our sustenance, or we need to construct shelter for our comfort, and we cannot wait a year? For how can we, as designers, do anything whatsoever without coming to some understanding of the climate, the landform, the movement of water, the plants and animals, and all the other natural elements on a piece of property? How can we utilize the biological intelligence that pervades a place if we have not completely immersed ourselves into that place and allowed it to speak to us, to tell us what it needs?

Inevitably we will impose patterns on the site, but what of the patterns that already exist? Are there plant guilds and polycultures already present? Have we taken the time to rid ourselves of concepts and simply see what we are looking at? Do we walk the land quietly with clipboard in hand and field guides in pockets? To put it simply: Design comes later. It always follows the inventory we take, and the assessment we make.

Essential Templates of Good Design

There are five essential templates to use as a model for designing and implementing integrated food forests:

1. The scale of permanence, initially described by P. A. Yeomans of Australia in the 1950s.
2. Sector analysis (delineating incoming and outgoing materials, energies, and forces coursing through a property) per Mollison in his book *Permaculture: A Designer's Manual*.
3. The zone system (zones 0 through 5, based on frequency and regularity of activity and visits to particular areas of a property).
4. Basic Permaculture principles and methodologies.
5. Needs, products, behaviors, and intrinsic characteristics of each element included in a comprehensive design. Wrapped around the design is a methodology based on the scale of permanence where we work from patterns to details.

These templates act as a way to organize and outline our observations and thoughts. Without templates, our accumulated notes and drawings, photos, maps — everything we use to understand a property — can get out of hand. What we are looking for is a cohesive pattern for all that we sense.

We can begin by asking, "What is most permanent in a land base, and how can we plan for the long term?" For instance, the scale of permanence begins with climate. Climate is something that we cannot escape. We are always in it. As we explore a property we make note of macro- and microclimate cycles: rain averages, prevailing winds, frost dates, snow and ice, weather extremes of hot and cold, available sunlight throughout the year. We can use what we know of the movements of these natural forces to develop water catchments, ventilation in our homes, windbreaks where needed, and planning for the life of our crops. After climate we look at landform and water; both are intrinsic components of any landscape. Water follows landform as gravity moves it through a property. Where there are depressions in the land and where the geology of the land has created slopes with ridges and valleys, we can take advantage of this natural movement of water by sculpting the land to move the water where it is needed most, or utilizing the preexisting landforms and the collection of water in various locations to place our plants for easy access, catch water for plants, animals, and people, and store water for future use. By understanding how water, people, and animals move about, access and circulation patterns through a property can

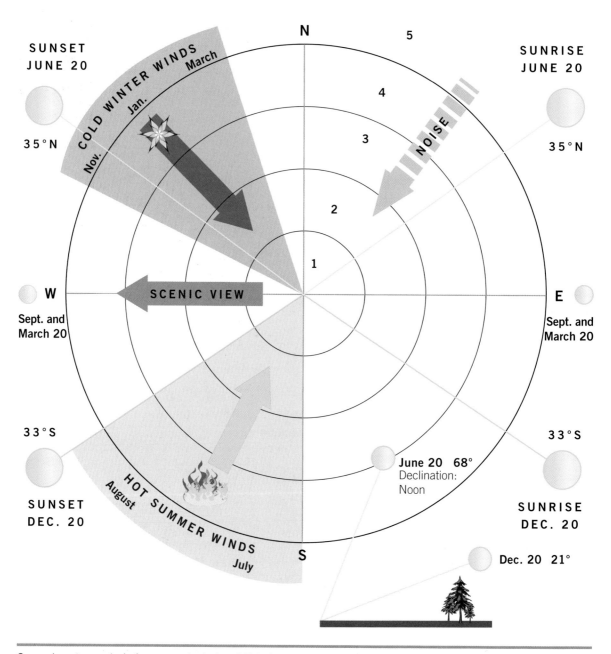

General sector analysis for sun and wind at 44° latitude.

WHAT IS INTEGRATED FOREST GARDENING?

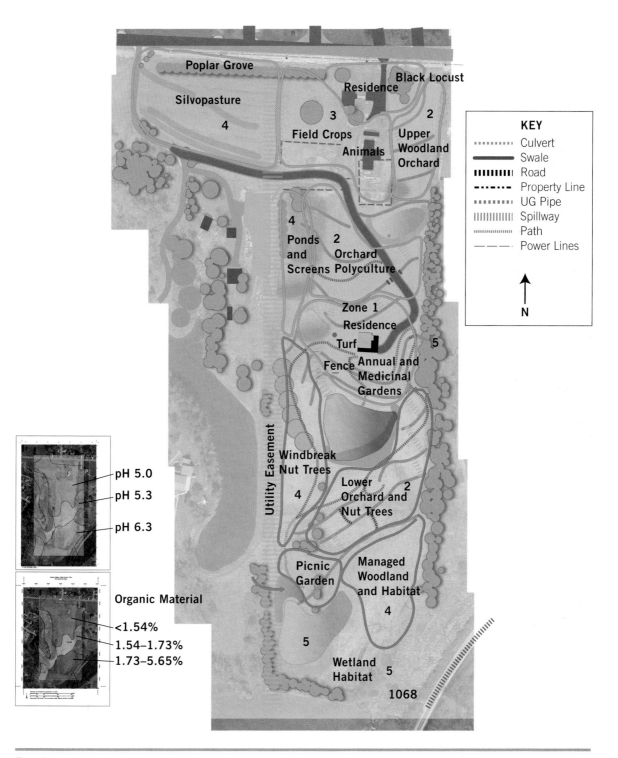

Functional spaces and zones of use.

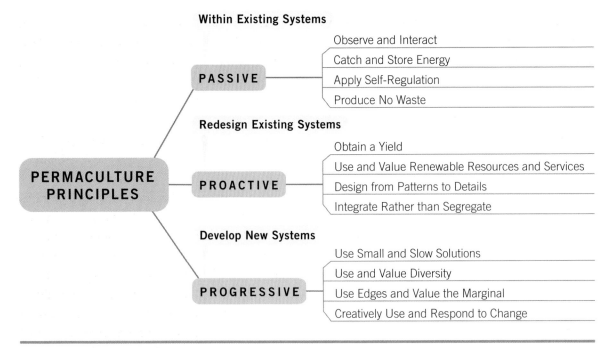

Permaculture principles can sometimes be arranged from passive to more aggressive actions. This is an adaptation by Daniel Halsey of David Holmgren's original list.

be used to keep the integrity of the land intact. For example, too much traffic in a sloped area will cause soil erosion. We can take note of this and plan our access routes accordingly.

Plants and animals are the bread and butter of our existence. These are our primary food sources. As we plan for vegetation, the need for high-yielding perennial species in integrated networks of food landscapes becomes central. The built environment is equally significant. This is where we spend most of our time, and it behooves us to design an abode that is healthy and clean, minimizes gadgets that use excessive fossil fuels, and takes full advantage of the sun, wind, and water that move about our property endlessly. Identifying preexisting microclimates and creating new ones to extend the growing season offers the possibility of producing food year-round. For instance, if we plant against a white wall of the house, the sun will be reflected onto our plants. The extra warmth this creates may extend the growing season a couple of weeks in the spring and fall. By planting sun traps with trees, and creating a protected pocket that faces the sun, we can lengthen the time that crops grown there will be free from the effects of climate extremes.

Healthy soil is all about fertility. In every form of agriculture and gardening, soil is the essential ingredient growing health-giving and vibrant plants. Healthy soil is created through the breakdown of organic matter, including the soil life itself. This soil life, this massive population of microorganisms, is our most valuable livestock asset. Without it the biomass that falls on the land would never turn into soil for our plants.

Aesthetics, culture, and social, economic, and political expression were not originally part of the scale of permanence, but these aspects of life are fundamental to a healthy and right livelihood.

How we go about supporting ourselves, how we organize as human beings, and how we color the world through our creativity, should possibly be at the top of the scale of permanence. All of these different facets on the scale — climate, landform, water, access and circulation, plants, animals, the built environment, microclimate, soil, and aesthetics — are part and parcel of the stew of all life. Using these categories as a template around which to organize our observations and thoughts is synonymous with assembling a fit and vigorous infrastructure on a property, providing a well-organized, structurally sound, effective base to build from.

Wayne Weiseman, Daniel Halsey, and Bryce Ruddock have been working as Permaculture design consultants and teachers of the Permaculture design certificate course, and advanced instruction, for many years with decades of experience among them. They have worked with countless clients developing master plans for diverse and sundry properties. Plants, and the combination thereof, are central to the development of any landbase where human sustenance and well-being are predicated upon sound design and implementation.

The flow of a consultation (whether with a client or personal) begins with observation and taking inventory of the property, then making a rigorous assessment, conceptualizing what is desired for the property, followed by design and finally implementation, in that order. Typical of architects and landscape architects, the flow from assessment to design leads us into the purview of many other professions as we work.

The Permaculture realm has become bogged down in building more gadgets, discussions about peak oil, facts and figures about the collapse of the world, and playing with the next new sun-capturing and fuel-saving devices. Is this the stuff that a comprehensive plan is about, or are these simply pieces of a deeper and more essential interlocking puzzle? What of the land itself? How would we really stay warm or dry or cool based on the movement of the wind and the configuration of the clouds and the feel of the soil? What about all that rain that falls to the ground and flows away, away, away?

What Is Forest Gardening?

Forest gardening is the act of creating and maintaining food forests. A food forest, in turn, is a deliberately designed, high-yielding, perennial plant system developed by human beings for their sustenance. We begin by observing and understanding the local ecosystem in which we are living. We utilize these ecosystem processes as our infrastructure and plug high-yielding food, medicinal, and utility plants into this framework.

Imagine an urban or suburban landscape whose every niche is filled with a diverse and healthy mix of flowers, fruits, herbs, and vegetables; where paths lead us to the next scent, the next fresh, juicy apple or pear or persimmon; where we can harvest a tomato right off the vine and bite into it on the spot. Imagine a food forest or, for that matter, a whole city of food forests, places of gathering, where people harvest nutrient-rich foods at their doorsteps.

There are many empty lots, former building sites, alleyways, parkways along street corridors, roofs, and back- and front yards that are ideal for planting, places where we can produce high yields for our sustenance. And in these places we can reestablish some semblance of the pristine wild places that were here, intact for millennia, before us. Nestled into these plant matrices, our buildings and homes are a few steps from tonight's salad or stir-fry. We are enmeshed in an effort to "relocalize," to bring the production of fresh food back into cities, towns, villages.

Now imagine a city as a garden city, teeming with the beauty and scent of many flowers and an unlimited selection of foods — foods that we choose to plant, foods that we carry a short distance to our kitchen and prepare, allowing us to sit down with the family to a good night's meal. There is no car going to the supermarket, no fuel being burned. We walk to the harvest; we are surrounded by health-giving plants. We meet others along the way and we talk, exchange ideas, find out from one another what fruit is coming ripe around the next bend, what herb can be used in what dish and how we might prepare it. We integrate.

Imagine cities, towns, and villages all over the world feeding a large majority of their population with food grown right at the doorstep. Imagine the urban gardening movement in America taking off in the next few years. It is incredible what can be grown on a postage-stamp-sized plot! The nutritional health of people in urban, suburban, and rural environments is equal to good health socially, politically, and economically.

Finally, imagine the ills of the "food desert" of endless chemically sustained monocrops eliminated. Instead we grow healthy produce within city, suburban, and rural limits, people gardening and farming together, evolving into tight-knit communities.

When people work the land, they benefit from the healthy physical practice of gardening and farming. When people grow their food close to home it eliminates all the supply-line issues of the corporate food scene and also the addictive use of fossil fuels for not only transport but also conventional farm practices: synthetic fertilizers, huge tractors and combines. When people gather to garden, the world's pressures and constraints go by the wayside and we breathe easily and eat proudly from the fruits of our own labors.

There is a strong movement using community gardening and food forest development to put communities back together. This typically revolves around growing food and helping local folks regain a sense of health and balance. Permaculture, which is the umbrella for all these practices, is certainly at the forefront of the movement to relocalize food resources and is having a profound effect.

Permaculture does not pigeonhole people into specific categories of race, religion, color, anything. It works for folks across the board. It is about earth repair and regeneration. Ultimately, it excludes no one because of the immanence of the practice, and the need for it at this time in history. There are hundreds of small farms, home utility gardens, community gardens, and naturally built structures popping up all over the world. Encouraging signs proliferate everywhere.

One of the central goals of Permaculture is to design and implement perennial food systems rather than simply continue with annual cropping. An ideal combination is a mixed culture of perennials and annuals.

Cities and suburbs are ripe for these kinds of systems: in abandoned lots, in city parks, along street corridors, in back- and front yards, and so on. When we supplant the typical annual-only culture and monoculture practices of the current agricultural realm, we help to reestablish ecosystem health through processes indigenous to the place in which we have settled. This can only bring renewed vigor and balance to what has been overwrought by the industrial mind-set, especially over the last two hundred years.

Just as mycorrhizae, a symbiotic association of mycelium (the vegetative part of a fungus, consisting of a network of fine white filaments) and plant roots, form a net in the forest soil that can extend for many miles and ties the forest together into a communication and nutrient exchange network par excellence, a plant guild, through its functional relationships, does the same. Root exudates, ingestion and excretion of organic matter by soil fauna, chemicals needed to support plant life, circulation of carbon, nitrogen, water, and oxygen: All these are tied to the production of biomass in the guise of root, stem, leaf, flower, and seed in a never-ending cycle that is always sharing, changing, metamorphosing, sprouting, expanding, contracting into seed, and recycling material into new life, until death gives birth to new life again.

In an integrated food forest, soil is covered by plants and organic litter. There is little room for "weeds." Indeed, when "weeds" do show up, they may prove useful in ways we haven't thought of before, as food, medicine, and ecosystem support. We can eliminate all the repetitive, backbreaking work of the annual garden by planting perennial systems and allowing the biomass produced to feed the forest and animals, and to create nutrient-rich soil that helps perpetuate the health of the system. Weeding, cultivation, and replanting go by the wayside. Perennial plants return every year with little intervention or maintenance. Soil is produced by the plants and animals that grow and live there. The perennial system follows the cycle of the seasons. A living mulch and the active soil biology create the ground that supports the forest.

In ecological design, the classic model is "the seven layers of the forest garden" formulated by Robert Hart, a forest garden enthusiast in England, in the 1980s. This model's partitioning structure allows for each plant to receive the sunlight it needs for photosynthesis. A canopy of trees is followed by an understory of plants suited to partial shade or full shade. Beneath the tall canopy may be multiple plant guilds partitioning the available sunlight and creating a nutritive environment.

As a model for designing and building the food forest, the seven-layer model works, but we do need to expand it to include what is happening underground in the soil and the possibility of inserting species into all the vertical layers between soil and canopy. For instance, plants roots are not limited to one level of the soil profile. There are taprooted, midlevel, and surface-feeding roots. And aboveground there are not only ground covers, herbaceous species, shrubs, trees, and vines, but also particular cultivars that grow at varying heights. The opportunities to fill every niche are endless. If our tallest tree is eighty feet in height, then spatially from zero to eighty we have occasion to choose from an inestimable variety of plant cultivars for food, medicine, utility, and the ecosystem services that these plants supply.

Perennials and Polycultures

The plants that persist in the landscape are the perennials. Year after year they return into spring sunlight or sprout fresh growth with no intervention and little maintenance. We humans may come and go, but the trees persist, the big bluestem persists, and the saguaro persists. The bur oak persists. The metabolic soil creatures at the feet of these great masters go on digesting and digesting and turn what falls into a smorgasbord of their own making, feeding animals, mycelia, flora and fauna. No two leaves the same, bark furrowing from freeze and thaw, birds snatching insects from their several niches, over decades, centuries. But the tree, as tree, goes on being tree, a cunning, brawling, persistent river of vascular madness and celebration. For those who witness and rigorously, patiently recreate the tree in perception and imagination, there is much to appreciate in the extraordinary spectacle of intergenerational wood.

Perennials are not limited to trees and woody species. There are also numerous herbaceous species that return as the spring sun rises above the horizon every year, or — in tropical environments — that

Big bluestem, a prairie grass. Photo by Matt Lavin

Saguaro cactus, a desert tree. Photo by Imagini Venske

WHAT IS INTEGRATED FOREST GARDENING?

Bur oak, a woodland tree.

Lush planted berms.

persist throughout the yearly cycle. In temperate regions you'll find perennial kale and broccoli, asparagus, bamboo shoots, bunching onion, garlic, horseradish, radicchio, rhubarb, Good King Henry, perilla, seakale, skirret, sorrel, land cress, groundnut, garlic chives, stinging nettle, and Jerusalem artichoke, to name just a few.

As we design and plan for perennial food systems, we plan for diversity of species in a balanced matrix. By establishing redundancy in the system, by including several species in the mix, should we lose ten species to a hard frost in early spring (for example) we may have another ten, twenty, or more species that survive and produce a yield for us as the year advances. Diversity and redundancy create resilience.

Polyculture means many, diverse plants. A polyculture is simply the cultivation of multiple crops or animals. It arises from two Latin roots: *poly,* which means "much" or "many," and *culture,* which means "to tend, guard, cultivate, till." How did the idea of "many plants" evolve from these two root words? It may be based on the word *agriculture,* from the Latin *ager,* field. By substituting *poly* for *agri* we come up with the idea that many agricultures means many plants, so to speak.

In essence, an integrated forest garden is one that is predominantly perennial, diverse, and polycultural in design. The word *integrated* is synonymous with *merged, fused, mingled, unified*. These words depict exactly what the Latin root of the word *integrate* describes: *integrare,* to make whole. As gardeners and farmers we strive to develop and implement perennial cultures that subsist through many generations by combining a diverse array of plant species that support one another; all these plants' needs are met by the plants, animals, and humans that are part of an integrated system. This is whole-systems thinking and designing par excellence.

The taprooted chicory brings up minerals from deep in the soil and barely casts a shadow.

Climate and Scale

The practice of integrated forest gardening is not limited by scale or climate. The authors of this book all reside and practice forest gardening in temperate climates: Wisconsin, Minnesota, and Illinois. This does not mean that healthy forest gardens are limited to temperate or more northerly climates. Integrated forest gardens are creatures of any and all climates. Savvy forest gardeners work within the constraints of a particular climate, basing their designs on the particular ecosystemic processes that are predominant in their particular bioregion. You might ask, "If I live in a prairie environment that has minimal woody species, if any, and where the tallest plant is the big bluestem at fourteen feet, how can I develop a forest garden? There is no forest here." No, but there is a canopy of grasses and many niches filled with a strikingly wide variety of species that form a kind of forest in their own right. We now have a template for further development.

Scale is not a limiting factor in the development of an integrated forest garden. Imagine a two-foot-diameter pot with several edible species working together to support one another in an ecosystem of their own making. Or imagine a two-thousand-acre "forest" garden, or an agroforestry system where animals and plants (especially trees) mingle in health-giving ways. We can take a yield from either scenario, and everything in between. When we integrate many plant guilds into a polyculture of diverse and assorted plant species, the number of possible yields is enhanced immensely.

If you reside in a temperate deciduous forest biome, you can mimic the ecosystem processes that prevail there and then work from this infrastructural template. A tropical rain forest? Ten thousand feet up in the Rocky Mountains? The Sonoran Desert? We have mentioned in this text that Permaculture begins and ends with astute and protracted observation and research. If we take our lead from this, we have a guide to the design and implementation of an integrated forest garden regardless of locale. We are not attempting to re-create a perfect replica of what came before settlement on our land. What we are doing is analyzing what was here, what has happened since settlement, and what this particular part of the country, this local ecosystem, can support in a balanced fashion.

Is there anything in this landscape worth saving and integrating into what we foresee? What plants and animals will thrive here based on climate, water, soil? Every landscape is unique; this cannot be overemphasized: *Observe, research, observe, research!*

An integrated forest garden is a creation of human imagination, and, as we say in Permaculture, the only limitations to any design are the limitations of the designer's imagination. Take your time. Walk the land. Observe. Write all your observations down. Hold off on conceptualization until you feel that you have thoroughly examined your place from every possible direction and facet. Then have at it — brainstorm to your heart's delight! Doodle! Draw, take out the colored pencils or the pastels and go for it! Create an Eden of your liking.

Climate Change and the Importance of Integrated Crop Production

Permaculture design must address changes in our climate over years or decades. We cannot resist the change or facilitate these changes, at least in our environment concerning plants and animals. Still, we can build more resilient ecosystems that have the ability to adapt over time. The speed at which climate change happens will fluctuate with increases and decreases in temperature. So if you consider that water freezes at 32°F (0°C) and condenses in the air at the dew point, that insects need a certain number of degree-days to emerge, and that plants need a certain amount of light and warmth for maturity, you see that 1°F (0.5°C) can make the difference between snow and rain and the loss of a crop. Our design work must reflect change over time. This can be best accomplished by incorporating as diverse a palette of plants in the landscape as possible, sculpted to retain resources and buffer extremes.

Diversity buffers the extremes in long-term change. A diversity of plants will build a resilient ecosystem, though some plants respond more slowly than others. The canopy shields and takes the brunt of the external conditions, supported by the understory, but a few years of drought can decimate an unprotected stand of trees (one with very little understory). Should this happen in a well-designed plan, the canopy will open, other plants will advance, and the old trees will fall and become organic matter on the forest floor, thus increasing the available water capacity for the other plants shooting up in their wake. It becomes a self-supporting and adaptive system no matter what the future brings.

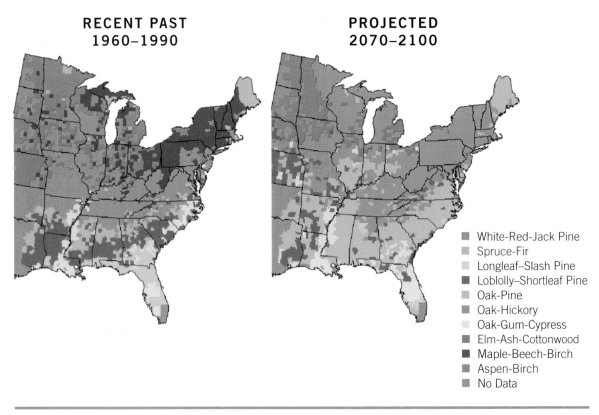

Shifting biomes: As conditions change, plants must change, too.

It would also be fruitful for us to adapt our expectations to changing resource availability, limiting our consumption and developing an attitude of intermittent abundance. For example, all summer our landscapes stockpile organic material and nutrients on and above the soil. While most of us are getting our food needs met based on seasonal availability, this abundance is coming to an end very quickly. Eventually we will have to change our diet and lifestyle to conform to a whole new set of environmental conditions. Perhaps we will need to become more like the plants and animals, less about on-demand resource availability and more about a catch-and-store, pick-and-preserve, cut-and-dry pattern of living.

We often get asked the question: How does the scale of climate change affect guild design? The answer is that it totally depends upon the design and its components. In one orchard, for example, you might have alpine strawberries, tiny little fruiting perennials that seem insignificant in comparison with the five apple trees and plethora of ground cover species. These strawberry plants — which have established themselves without any intervention on your part — make a lot of seed. You will undoubtedly notice over time that a small alpine strawberry is growing hundreds of feet away from the original planting, probably the result of a seed cast by some bird or chipmunk. Although many of the original strawberry plants may be gone as a result of the apple trees' crowding and shade, they continue to cast an influence throughout your property. In a similar fashion, it is also possible that a volunteer apple tree sprouts and sets fruit on

WHAT IS INTEGRATED FOREST GARDENING?

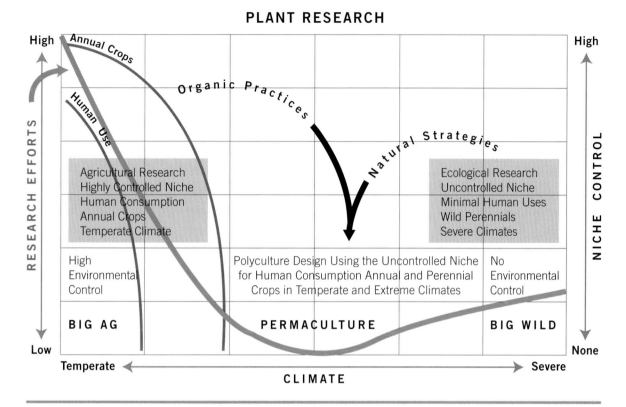

Building polycultures uses natural strategies and organic practices.

the opposite side of your house from the orchard. When you bring in any plant, if it can establish itself it has a one in ten chance of continued propagation.

Investment in research and development of ecological systems has focused primarily on highly controlled niches in the temperate climate and uncontrolled niches in severe climates. One of the goals of this book is to raise interest and research into ecologically sound and sustainable food production in the uncontrolled niche.

If we till up a backyard and plant a garden for one season and do not plant it again, it's pretty easy to see that the garden will be taken back by grass and weeds, and if mown will probably look like the rest of the yard within a few years.

When you think about it, all we are doing as designers is moving things around. It's much like redesigning our living room using the same furniture, or perhaps bringing some in from another room. By planting polycultures we are developing functional living systems that are self-supporting and enhance available ecological services. We influence the structure of the niche and the patch, allowing for the cycling of nutrients as we buffer environmental changes. Whether that landscape design will last for years or centuries is up to the next occupant, but for the most part, if the landscape design is ecologically sound, it's less likely to be disturbed by humans and more likely resilient enough to adapt to new circumstances.

Over time the species in our guilds may change — some come in, some go out — but the structure will still support whatever plants occupy the space, when it's designed well. Hopefully, with maintenance provided by animals as they harvest the nutrients and contribute their services, the design will expand and transition into other spaces, and additional plants and animals will benefit. The designed plot of land is influencing external ecological spaces as it is dependent upon them for genetic diversity and ecological integrity.

Climate disruption and climate change have accelerated to the point that a new issue has arisen: At this rate of change, plants and animals will not be able to reproduce and sustain the generations required in order to adapt. Although a great amount of genetic material exists in the world, with many possibilities for mutation, the process needs time. Intermittent disturbances that stress plants and ecological systems accelerate the genetic mutations that aid in this process. Fast and extreme changes, however, may not allow for natural adaptation by native plants over time; thus natural selection favors more resilient opportunistic species that may not be endemic to a particular bioregion.

As a result of rapid change, researchers are planting trees in the Boundary Waters Canoe Area of Minnesota to see what happens to hardwood trees over the next twenty years and how this affects the ecosystem compared with areas where sudden changes have not taken place. We need to continue such research to gain a macro perspective across a wide landscape. If you can imagine what the ecosystem in your region looked like five hundred years ago and then think of how it might have been able to adapt to changes in climate, you will realize that many of the related ecological services do not exist anymore — the grazers that controlled so-called invasive species and plant life have been exterminated; rivers have been moved and deepened; huge spaces of wetlands and trees have been leveled and filled. On the other hand, whatever happens, something will be here eventually that is adapted to the new conditions. There may be a time of great change, but if the soil is stable and organic material is present with its myriad microorganisms, a new ecosystem will emerge; over time, so will wildlife adapted to it.

We have seen areas where alien, non-native species seem to just pop up and take over. Are these invasive? What is an invasive species? It has been said that the only invasive species on earth is the human being. The dandelion has been in the United States for upward of eight hundred years. Is this thriving lawn plant any more or less "native" to this country than, say, a bur oak tree? The movement of peoples all over the face of the planet necessitates the movement of all species, plant, animal, and human. The horse was endemic to this country until it was made extinct in ancient times. It was reintroduced to this continent by the Spanish in the 1500s. Is it native or alien? This shift in biological populations is inevitable and has happened throughout time. Opportunistic species fill niches and thrive where others cannot. The natural succession of a forest may begin with "alien" plants and mature into a stately forest in the long term. Invasive? Good or bad? If we step back, take in a wider perspective, and attempt to understand the broader cycles of evolution and growth of the natural world, we may find that the so-called invasives are simply doing their job: setting up the infrastructure and building healthy soil for another ecosystem to come to fruition in all its richness and diversity.

Capturing, Storing, Cycling, and the Sustainable Homestead

A healthy and sustainable lifestyle is attained in a community that grows its own food, as long as the community supports the ecology of its surrounding landscape. Natural capital must be maintained and enhanced if the community is to be viable. Maintaining the health of the soil, water, and plants will increase the fertility and the possibility of growing more food in a smaller space.

Growing food more intensively, in smaller spaces, decreases the resources needed for food harvesting and maintenance. The amount of land needed in order to produce sustenance for human beings varies according to needs and preferences and the size of the population being fed. Is a city lot enough, an acre, five acres? When we move our food-producing plots closer to the home, these areas are no longer out of sight, out of mind. When we eliminate the need to grow in wide horizontal expanses, we allow many "wild" areas to reconstitute themselves into thriving ecologies. When we use vertical spaces on our properties, we open up more opportunities to pack in more utility plants per square foot while regenerating the soil and ecological support systems.

Soils covered by plants are less likely to heat up, fill with weeds, and erode. Plant cover keeps the soil moist, so it needs less water. Importing external resources depletes the areas from which they are imported and increases the costs and calories needed to grow food or maintain fertility. Using space on a property to grow plants for composting, green manure, and mulching decreases the need for external inputs and makes a property more sustainable. When we build richer soil through these means, we can grow more food in a smaller space.

Nutrient cycling on a property is of utmost importance in a sustainable system. Just *adding*, versus *building*, fertility increases dependence on outside sources, while growing nutrients on our own landscape is easier and more economical for us to manage. If we use the existing ecology as a guide, and limit our plant selections to species suited to the observed climate, soil, available sunlight, and microclimate, we reduce the energy, calories, and effort we must put in. The size of garden space limits the available nutrients and minerals to what is in the compost. Cycling nutrients from the kitchen and from harvested plants returns more nutrients to the soil, more organic material to the organisms that metabolize them, again making them available to plants. In order to grow healthy plant guilds and food forests, an understanding of nutrient cycling and the building of fertility into the land base are paramount.

Plants absorb nutrients, especially minerals, through their root and leaf systems. These minerals are cycled back into the soil and air as leaves breathe and roots slough off root hairs into the soil matrix. The creatures and fungi in the soil consume and break down this biomass, dispersing minerals back into the soil for uptake by plants again. This is the consummate cyclical process that enables plants to return year after year in a healthy and formidable manner. Without this buffet of minerals, plant life will suffer, growth will be stunted, and yields will decrease. Compost and mulches let the breakdown of organic matter continue unimpeded, and plants thrive.

Assessment of a property's ecological processes and natural capital allows us to create a list of plants directly suited to the existing conditions: planting zone, climate, soil profile, and more. Developing a plant system and lifestyle that is imprinted by the land reduces the stress on the grower, the livestock, and the landscape. The most efficient way to plant a landscape is to mimic the plant landscape that already exists. Using ecological analogues to the existing plant species will ensure a compatible plant system for the landscape.

Say you're living on a farm in Vermont on a hillside that is predominantly populated by sugar

> ## Nutrient Cycling
>
> **GIFTS TO THE VILLAGE**
>
> - Fruits and vegetables
> - Juices and pulp
> - Vitamins and minerals
> - Flowers and oils
> - Medicinal herbs
>
> - Storable foods
> - Spices
> - Honey, wax
> - Community activities
> - Animal feed
>
> **NUTRIENTS RETURNING TO THE GARDEN**
>
> - Soil from kitchen waste (worms)
> - Compost from the landscape
> - Mulch from grass clippings
>
> - Cardboard and paper
> - Water from roof gutters
> - Transplants from native lands

maples, and you want to raise sheep. Instead of clear-cutting your forests, ask yourself why you want sheep. If the answer is for meat, then suppose you raise pigs instead and graze them in the forest (silvopasture) while you also harvest the maple sap? Or you could raise the sheep in a silvopasture system. You could clear some of the less useful saplings or dying trees and plant some hazelnuts in the understory as well as the shrubs, herbaceous species, and grasses that the animals will consume readily. The animals deposit their manure and grind it into the topsoil with their hooves, thus improving the fertility of the land. Hazelnuts, maple syrup, and other fruits and nuts are harvested along with the animals. Diversification of yield is key here. In Permaculture systems this is known as stacking functions — not only using all available niches, but also promoting diversification of food, medicine, and utility.

A Word About Compost: Extending the Life of a Resource

Don't forget that the occupants of a property bring in a bounty of energy and resources (food and materials) through their purchases, and store this within the living space. Once these resources have been used, in many cases they are dispensed with via garbage removal and large trucks.

Giving a second life to all these purchased materials, however, may well increase the natural capital on your landscape. Carbon-rich and nutrient-rich foods and materials used in the kitchen can be put into a well-managed compost pile or can be reused as chicken feed or worm feed. The wastes produced by chickens and worms are then cycled back into the system and provide much-needed resources for plant guild health. The process also converts these scraps into eggs, meat, and ecological services. Allowing resources imported onto a property to travel down trophic (feeding) levels of organisms is much like the way we use water: Delaying, diverting, and creating new uses for refuse from each trophic level allows "waste" to become food. In a sustainable landscape there is no waste, only food.

In Permaculture the practice of catching and storing energy can be easily extended to nearly any resource and material. Whereas nature cycles all resources infinitely within its systems, we must consciously design ways to catch and store energy in our landscapes and living spaces.

The cycling of nutrients, of water — of any material brought onto the property, really — increases its usefulness and extends its life. This diminishes our need for new resources. The more frequently we can use the exact same drop of water in different situations, the less water we need to add to our living system. The more ways we can use the outputs of one element of our living systems as supporting inputs for other elements of the same systems, the more resilient our systems become.

Plants are the ideal catch-and-store tools. They store resources such as sunlight, water, and oxygen. Plants create natural capital. A plant guild or food forest is a sink of all things ecosystem. It reflects our living systems and the greater whole of the bioregion and habitat it roots its diverse feet in.

Water and Earthworks

As we design and install our plant guilds and polycultures, the guiding, storage, and smart use of water is tantamount to long guild life and the reduction of time and energy spent on irrigation and managing for water through mechanized gadgetry. Guilds and food forests are not designed as isolated entities in a landscape. They are part and parcel of the greater whole. We must consider not only the plants we are thinking about for our harvest, but all elements inserted into the land base. The discussion of water and earthworks here is but one of these elements, albeit a most important one.

In all systems great attention should be paid to conservation of the resources that exist on the land and pass through it. Any resource that can be slowed down and used on the property for an extended period increases the stores available. Say your hillside experiences a huge hurricane-like storm one summer. Instead of the water rushing right down the side of the hill, taking trees and topsoil with it, you want to trap that water and use it. You want earthworks such as terraces, ponds, berms, and swales to do this work for you.

Water may appear from the sky, on plants as condensation, or from the ground, such as via a spring or well. These are *sources* of water. But water is lost when it reaches the sink — the final location of the water, where it is out of reach for your use on your landscape. Water going down the valley off the property is out of reach. Water evaporating from uncovered soils or from dry spaces is out of reach, as is water that is contaminated after use. The goal is to use water as many times as possible before it reaches the sink. This technique and philosophy can also be applied to sunlight, wind, organic material, electricity, and human energy. This understanding and application supports the health of our guilds and polycultures and rewards us, if managed rigorously, with a future self-managed plant base.

Earthworks, such as swale-and-berm systems, terraces (shelves), or catchment ponds, collect water for use in integrated forest garden growing areas. Swales are ditches dug on contour. Contours appear on surveyors' topographic maps as curvilinear lines. These survey lines are all dead level in the landscape. Once the ditch is complete and the material scooped out of it used to build an earth mound, or berm, on the downslope side, all the water coursing downhill from higher ground will collect in the swale. Because the swale is dead level, the water will not flow on a gradient from side to side, but will percolate into the soil matrix, slowly soaking and sinking down and sideways through capillary action. Plants can quench their thirst readily with the slow-moving water, the groundwater supply is replenished, and topsoil is washed downhill but stays put in the landscape.

Imagine yourself as the rain landing on the landscape. As your drops accumulate into sheet flow, the force of gravity moves you down the slope, even when the slope is very slight. You flow over the landscape and shape yourself into whatever container you land in. All of a sudden you find yourself washing gently into a dead-level ditch. You cannot move left or right. You simply sit there and slowly seep into the soil at the bottom and sides of

Swales collect water on contour, creating increased soil moisture that is available to plants and trees with deep roots.

this ditch. You find yourself moving downward and sideways through all the nooks and crannies of the soil. You seep deeper until you reach an impervious layer of rock and compacted earth. From there you continue downslope — perhaps popping back out of the earth as a spring at the slope's base. In any case, some of you eventually arrives at the bottom of the watershed, enters a river, and then joins the ocean. Some of you may have been sipped by the searching roots and tongues of various plants and animals, only to be later transpired or excreted to join the journey again. The parts of you sitting on the surface evaporate into the atmosphere, and you form into clouds and, eventually, fall as rain again. You realize that all along you have been going in circles: rain, sheet flow, catchment, spring, runoff, river, ocean, evaporation, all over again in spades.

Using earthworks to guide and slow water is a major contributor to the health of living systems. Use of earthworks buffers weather change and climate conditions and creates a more suitable habitat for increased and predictable production. Depending on your type of soil, the shape, frequency, and size of earthworks you create will vary. Pay serious attention to the duration, frequency, and seasonality of rainfall in your area and how you might collect and disperse this precipitation. Undersizing catchment systems can cause serious erosion events. Oversizing a swale-and-berm system will increase the soil available to the plants on the berm and also increase the capacity of the swales to catch precipitation. The key to all this is to slow the water down so that the plants can drink readily and soil erosion becomes a non-issue. All earthworks that we create for the enhancement of plant habitats do the same thing: slow the water down. If we start slowing the water high in the landscape, by the time it moves downhill it is moving deliberately enough that its uptake by plants will be assured. If we allow water to move unimpeded downslope, with no earthworks in place, or without plants to drink it up, we encourage rapid sheet flow, erosion, and the loss of topsoil.

Some swale systems may seem oversized relative to the landscape yet may be required to handle extreme weather events. In sandy soils that drain the water from production areas quickly, extremely large swales are used since, frequently, the only time to catch water is during extremely high-precipitation periods, long rain events, or the spring melt when the ground is still frozen in northern areas. In clay soils large rain events will fill the swales quickly and take many more hours, possibly days, to drain into the subsoil. In either case the swale is quite large to retain the maximum amount of water within the landscape.

In smaller landscapes you can design for swales to be more frequent, allowing rainfall to spread over a broader area of the property. In a plant guild or food forest sitting on a hillside, you might install just one swale, or you might choose up to three or four swales spread apart, which allows space for orchard trees and fruiting shrubs.

Haitian farmers learn and teach skills in building swales. A young student uses an A-frame tool to level the berm along the swale.

Water lens: elevation cutaway. Illustration by Daniel Halsey and Kellen Kirchberg

In Ken and P. A. Yeomans's book *Water for Every Farm*,[1] keyline water-harvesting techniques are used in large pastures and properties. Beginning in the 1950s Yeomans resuscitated thousands of acres of land in his native Australia, the driest continent on the planet, by carving channels with a subsoiler or keyline plow shanks, six to eighteen inches deep, through the soil, more or less on contour. These channels readily absorb any water that falls on a property, percolating this rainwater to deeper levels of the soil profile where plants easily take it in through roots that move more freely through the soil profile because of the channels. After many years it is possible that springs will form at the base of large hills and in the valleys. Yeomans also designed water catchments high in the landscape and moved the water from these high catchments to several more catchments at lower spots on slopes, all of this water being utilized for gravity-fed irrigation. In all cases the water is caught, stored, diverted, and given many uses, for plants and animals, before it exits the property.

What Is a Plant Guild?

We frequently hear questions from Permaculture students and practitioners about plant guilds: how to configure them; why we select the plants we do; what functions the plants serve; why setting up particular relationships is beneficial to the guild; in other words: "How do you do it?" In Permaculture systems the plant guild is central to developing a diverse integrated forest garden. A permanent agriculture is an agriculture based on the planting of perennials. Perennial plant guilds are the foundation from which we begin. Think of an integrated forest garden as an assembly of plant guilds working together in functional relationship. A plant guild, put succinctly, is a beneficial grouping of plants that support one another in all their many functions: exchange of nutrients, water, and gases; physical buttressing; promoting diversification and redundancy in the system. Plant guilds also support animals and humans for all their food, medicine, and utility needs.

Toby Hemenway, author of *Gaia's Garden,* the influential 2009 book on home-scale Permaculture, says it nicely: "Guilds are an attempt to bridge the broad gap between conventional vegetable gardens and wildlife gardening by creating plant communities that act and feel like natural landscapes, but that include humans in their webwork. Vegetable gardens benefit only humans, while wildlife or natural gardens specifically exclude people from their ecological patterns. Ecological gardens . . . help our developed land to blossom into nourishing places for both humans and wildlife."

Clearly stated, there are no cookie-cutter guilds. What there is, is a scaffold, a structure, a set of diverse and multifarious functions that, when configured together, benefit groupings of plants that would not necessarily be there should these plants be growing in isolation. Functions such as nitrogen fixation, mineral and nutrient uptake and dispersal among the plants in the guild, the attraction of beneficial insects and pollinators, the filling of all niches from the soil profile to the canopy, and utility for animals and humans are how we define the structure of the plant guild. We can then select plants that deliver what we need. A plant guild is about shared functions. In essence, we are designing for these shared functions first. This is how the relationships that we need for optimum health and production in the plant guild come about.

The concept of a guild is old and did not originally refer to plants. Members of medieval craft guilds performed a variety of important functions in the local economy. They established a monopoly of trade in their locality or within a particular branch of industry or commerce. They set and maintained standards for the quality of goods and the integrity of trading practices in that industry. They worked to maintain stable prices for their goods and commodities, and they sought to control town or city governments in order to further the interests of guild members and achieve their economic objectives. The merchants and members of the craft guild worked together. This is what gave them the strength and perseverance to succeed. It is not difficult to apply this same idea to the plant guild: plants working together for the benefit of all. As stated in the *American Heritage Dictionary,* a biological guild is a group of organisms that use the same ecological resource in a similar way. In a plant guild the organisms are the plants and fauna, though many other resources come into play.

Bill Mollison, the creator of the Permaculture concept, initiated the use of the word *guild*. How did he decide on this word? Of course, this is only conjecture on our part, but craft guilds were occupational associations. In the same way, plant guilds can be considered occupational associations, providing each plant in the guild support, focus in the landscape, and the means to persist as a community.

In essence, plant guilds are not about the particular plants themselves; they're about the relationships among plants, and the biological processes, chemical processes, and shared functions that actively circulate among the plants selected. You don't have to be a horticulturalist to create a guild, though a basic understanding of plant functions goes a long way.

For example, nitrogen is what gives plants a kick start of growth in the spring. This very active element, found free ranging primarily in the atmosphere, is what greens up our spring fields and forests. In a plant guild nitrogen-fixing plants (plants that sop up free-ranging nitrogen from the atmosphere and infuse it through the root zone into plants) are required in order to store this element and supply it to all other plants to initiate the spring growth process. Legumes are prime examples of nitrogen-fixing plants: clovers, beans and peas, black locust trees. The alder tree, a non-legume, is a major nitrogen fixer, and is especially beneficial in a plant guild as a fast-growing tree species.

Another important function in a guild is the dynamic accumulator. Such a plant mines nutrients from different layers of the soil profile and brings them to the surface, where it drops them for other plants in the guild to ingest. Plants such as comfrey, dandelion, and other taprooted species are very efficient at doing just this. The dandelion absorbs calcium, iron, magnesium, phosphorus, potassium, sodium, zinc, copper, and selenium through its taproot and supplies other plants as its leaves die back and drop to the surface of the soil. There the soil biology, animals and fungi, break the leaves down and make their nutrients available for other plants.

Guilds or associations of living organisms are groupings of species that coexist in an environment, sharing resources or acting as resources for one another. Sometimes these relationships work to the benefit of the entire system and sometimes not. But whether an ecosystem is in balance is not readily apparent to the human observer. Often overall balance and shared functional relationships in a plant guild manifest over a length of time spanning years or even generations. This is probably one of the reasons that Bill Mollison recommends getting a good chair and sitting down with a notebook and pencil to record these relationships, spatial, temporal, and otherwise.

The Scientific Basis for Plant Guilds

In scientific literature the earliest mention of the guild as a grouping of species in an ecosystem was in the late 1800s. Plant guilds were originally discussed on the basis of similar plant lifestyles: vines and climbers, epiphytes, saprophytes, and parasites.

Another way the scientific community has defined guilds is as the functional structures of assemblages or groups of species. *Functional structures* are defined as the deliberate distribution of species in a functional space or habitat.

The actual word *guild* in the context of species relationships is the nearest translation of the German word *gennossenschaften*. The Permaculture definition of the word comes also from the idea that the boundaries of functional niches are set by competition for limited resources: sunlight, water, minerals, air. A stable guild community is based on complex trophic, or feeding, organization. By adding species to a system we enhance the number of niches or sites where more species can colonize, and doing so opens niches for species that previously could not survive at that location. A classic example of this niche enrichment concept is the planting of a third species between two incompatible ones. Think of a mulberry tree between a black walnut tree and an apple tree. The mulberry

WHAT IS INTEGRATED FOREST GARDENING?

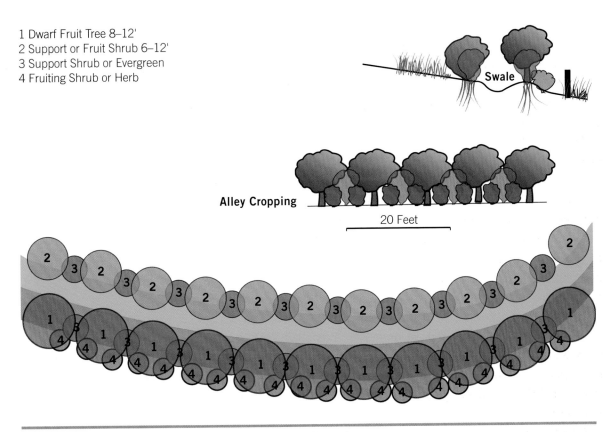

1 Dwarf Fruit Tree 8–12'
2 Support or Fruit Shrub 6–12'
3 Support Shrub or Evergreen
4 Fruiting Shrub or Herb

Neil's plant guild elevation. Illustration by Daniel Halsey and Neil Bertrando

neutralizes the allelopathic chemical exudate of the walnut, juglone — which inhibits other plants from growing near it — so that the apple can survive in closer proximity.

It has been established that it is actually the number of *species* in an ecosystem at any given time that determines the number of available niches in the system. In other words, it is the species and their distribution that allows the establishment of habitat for additional species. If the guild is dense with plants that are already competing for resources, then opportunities for colonization by further species will be limited. A guild assemblage, then, is an assortment of species in an ecosystem site that fill various trophic niches. We discuss this in greater detail in chapter 3.

The Importance of Plant Diversity

We are often asked the question: "If we plant in such a guild fashion, won't our gardens look untidy and messy? And do we want that?" The answer is simply — yes! But in reality, there are different levels of this kind of "messiness" that come in Permaculture design. How we define messy or tidy is, of course, quite personal. Sometimes we need to design for the suburbs, which are often encouraged by covenants and associations to be more linear and geometrically tidy. But the reasons we may not plant in straight, tidy rows is

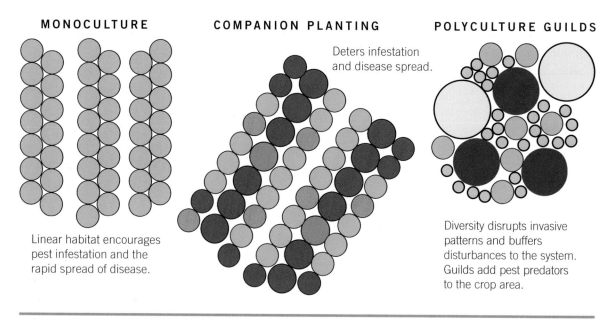

Progressive diversity benefits the ecology.

more than just an aesthetic preference of Permaculture designers or their clients. Row planting is an inefficient use of space, and it does not replicate patterns found in nature.

A diversity of species, whether plant or animal or fungi, affords an assortment of food, medicinal, and utility products in as little space as possible when we imitate the patterns found in nature. Nature strives to fill every niche in a landscape with some form of biological entity. Natural succession dictates that plants follow a more or less predetermined evolution toward a mature climax. Think of a blank slate of disturbed soil in a bioregion that supports a forest environment. Annual plants come in to colonize this site and set down roots. As the years fly by, biennials, shrubs, various-sized trees, and a mature canopy result in complex forest biomes. It is through thorough and ongoing investigation of our local region that we define how we will go about designing and patterning our gardens and farms, heavily based on the ecosystems that thrive naturally where we are.

Ecological benefits reduce pest control costs. An *Aleiodes indiscretus* wasp is parasitizing a gypsy moth caterpillar.
Photo by Scott Bauer, USDA Agricultural Research Service

On a small property this allows for a surplus that we can share with relatives, friends, and neighbors, or use in trade or barter. If we have too many tomatoes and apples, we can exchange them for the neighbors' extra black walnuts. We can use plants that we grow and harvest as medicines, thereby saving a portion of the costs of some medical care. Best of all, we enable new niche habitats for other species: birds, small mammals, and amphibians that will share the overall site with us, the designers, the forest gardeners. These other species will fill niche functions of pollination, pest control, food for humans, warning or alarm systems, cleanup, replanting of tree and fruit crops, and many, many more.

On a larger site, such as a small farm or market garden, you can design your guilds to enhance profit resilience by diversifying crops so that multiple revenue sources are available every year. The species attracted through guilds will also lower your pest control costs as the pests will never get the upper hand before they become lunch for another species.

Lastly, the careful selection of species for a guild group will allow you to bring in species that would normally be out of their growing area in your site. Select these species after you have delineated the functions you're designing for, rather than deciding on species first. If these "alien" species fit the function, and will work in combination with the other plants in the guild in a healthy and balanced fashion, then welcome them into your design.

However, you must also engage in ongoing maintenance of the plant guild containing any such species to monitor for invasiveness. It is a reality that even "native" plants may become invasive. With ongoing observation of the plant guilds within food forests, you can manage for diversity and restrain a species before it gets out of control. A diverse plant assemblage will act to eliminate possible invasiveness by encouraging all niches to be filled; invasive species thus won't have the opportunity to spread.

Permaculture's principles and methodologies apply across the board: in rural, suburban, and urban settings. These applied principles are not delimited by scale. What can be accomplished in a rural setting may also, based on principle, be accomplished on the thirtieth floor of an apartment dwelling in the middle of Manhattan. A plant guild can be devised in a container in an office courtyard, on a market farm in a rural setting, on a rooftop in a high-rise apartment building, on a suburban front yard, on the windowsill or under the kitchen sink in an apartment — anywhere. Remember that a plant guild is a beneficial grouping of plants that support one another; what would it matter if this is happening on ten square feet or thousands? The goal is healthy relationships among plants on any scale, where energy and materials are freely exchanged.

Applications of Plant Guilds in Permaculture and Forest Gardening

You are walking into your garden at the break of day. You hear birds beginning to chirp, listen to the bees awakening from their slumber, and see the diversity of plants lifting their heads out of the morning dew. It may feel like an exotic forest, but you're in your front yard in the middle of a subdivision in Detroit. The colors, scents, textures are exquisite. Before you go off to the office, you pluck some mint leaves from the herb garden and brew up a cup of tea. The four hens in the coop are producing four eggs a day this summer. You grab them out of the nest box, pick some strawberries and raspberries along the way, and go inside to cook breakfast. The bread that you add to the mix derives from the wheat and acorn flour that you harvested earlier in the spring. The jam you put on the bread you cooked up last autumn from a formidable harvest of various fruits: apples, figs, cornelian cherries, persimmons, pawpaws. Into the omelet goes spinach, dandelion leaves, Good King Henry, and amaranth greens. And lastly you pull out a handful of hazelnuts that have been supplying you with oil and protein since last fall. You grind the hazels in your hand-cranked oil press and use this oil to fry up the omelet. One more thing: dandelion, chicory, and beetroot coffee substitute!

Permaculture is a comprehensive system. What we mean by *yield* encompasses not only the plant world and forest gardens, but also the built environment, the waste stream, energy systems, animals, water, landforms — life in all its trappings. All these systems contribute to the development of a plant guild. The energy and materials embodied in structures, plants, animals, and humans placed in the landscape create dynamic relationships and functions of exchange and the instigation of a never-ending cycle of biological life. We might think of all these elements of life as a polyculture in itself. A polyculture of plants, numerous species intermixed into a green mesh, benefits us by ensuring that all organic matter is reincorporated into the soil where roots can find the nutrients needed for plant health and growth. What is "waste" in the biological realm but food for the millions of creatures that roam the "wilds" of the food forest both above- and belowground?

How Guilds Work

Plant roots are not simply absorbing nutrients from the soil; they also excrete minerals, vitamins, hormones, and the like back into the soil for uptake by other plants in the guild and food forest. These are known as root exudates; in guilds this constant sharing and exchange is key to balance and harmony among the plants. The ongoing ingestion and excretion of organic matter by soil fauna makes this possible. As the soil fauna ingest organic matter and digest and poop it out in a simpler form, plants promptly take it back into their structure. This consists of the chemicals needed to support plant life. The circulation of carbon, nitrogen, water, and oxygen is constant in a guild system. If these cycles are compromised, you may be in for trouble. It's important to understand how carbon cycles through the system and to set up the best possible scenario for this cycle to remain intact: lots of biomass and organic matter infused into the soil. This is food for the livestock in the soil, their breakfast, lunch, and dinner.

These exchanges are not simply happening belowground. High-quality, functional relationships set in place throughout the plant matrix create health for the plant guild and food forest. The oak tree offers shade to the burdock, which drills its taproot deep into the soil profile and pulls up the minerals that surface-feeding roots cannot reach. When the burdock in turn drops its mineral-rich leaves on the surface nearby, it helps satisfy the berries' mineral needs. When the berry bushes are flowering, the bees show up rather hastily to suck up the sweet nectar; unbeknownst to them, they are also gathering pollen and carrying it on to the next plant or back to the hive. Caterpillars somehow, seemingly spontaneously, generate on the leaves of these plants to eat, and are a harbinger of the colorful butterfly to come. We could go on and on drawing out these relationships.

By designing plant guilds that are thoroughly embedded in a bioregion and ecosystem; by configuring polycultures of plants that support one another; by paying attention to climate, landform, water, soil, historic, and ahistoric considerations; by understanding basic chemistry, biology, and physics; and by calculating the requirements of plants from sunlight to structure, we create the conditions for plants to thrive and supply us with sustenance.

Decades, or even centuries, are needed for natural ecological systems to go through levels of succession (the sequential development of plant or animal communities) until the plant species are balanced in number, position, and proximity. All throughout the process, multiple species may occupy the same space over time, and the soil and plant resources develop as the soil is modified by the various species that are passing through. But Permaculture designers and forest gardeners can accelerate the successional process by choosing plants for their designs that will fulfill all the needed ecological functions in a given niche — or, as we described earlier, the space that has specific characteristics and conditions that will only sustain specific plants. As the first pioneer plants add organic material and nutrients over seasons and generations in the niche, the niche will change.

Designing for the Niche

Designing for a niche allows you to customize the plant list you'll use for its specific ecological conditions. By using plants that naturally thrive in the site you've chosen, you set yourself up for the greatest success. You wouldn't want to be planting coconut trees in Wisconsin, for example. Site-specific plants use fewer resources, require less maintenance, and have a better chance of surviving unforeseen stressors of weather and predation. Plants in guilds and food forests are competing for space, light, water, and nutrients. Additional plants must fit into the remaining available niches. You aren't necessarily looking to enhance this competition for resources; you're looking to develop balanced systems where resources are shared and needs are taken care of for each plant.

Studying Guilds in Their Natural State

In an open field or pasture, full sun turns to partial shade under tall grasses, and established plants with wide fibrous surface roots share soil with deep taprooted plants. But a thin chicory plant can emerge from the dense shade of short broad leaves and climb into the sunlight, barely casting a shadow. The chicory reaches for sunlight above the grasses and legumes of the field or pasture. Plants fill every space and jostle for vital services until a balance is achieved — a balance that swings much like a pendulum as waves of species occupy the land.

Hills, once grassy, can become dense with years of Canada thistle; then all at once, in a season, they can be taken over by sweet clover. Maybe there is a shift in climate or rainfall averages. After years of thistles tapping deep into the soil and dying back — only to reemerge each following spring — the organic biomass left behind in their yearly wake and the drilling down of taproots prepares the ground for the future when sweet clover will germinate and thrive. The future plant residents of the hills wait in a seed bank far below, wait quietly as shaded understory, or are casually dropped by passing rodents and birds. Watching the succession of plants allows us to consider their purpose and effect. It helps us design plant systems that leapfrog the early successional stages and accelerate the process, positioning plants for their best advantage via ecological design.

Building a plant guild or polyculture requires us to follow the same rules nature has set for its players.

INTEGRATED FOREST GARDENING

Rules of plants: Plant systems throughout the world have a few things in common. Even though the conditions can be vastly different and the extremes of climate and disturbance create narrow niches of opportunity, basic needs must be met or the soil will be barren and lifeless. Fortunately, there seems to be a plant for practically every environmental condition on the planet. Plants need: sunlight, soil, nutrients, water, seeds, density, protection, harvesting, and integrated ecological services with other plants. When, how much, and the frequency of availability is adapted by each plant's physiology.

So what are these rules, and how can we use them to create abundant harvests, natural capital, and enhanced ecological health? We need to consider the structure of successful polycultures and plant guilds found in nature, and build the skills to mimic nature's momentum. Each niche, yard, lot, or field has a long list of potential plants that will thrive there, and it is our task to define them, find them, and design an ecosystem to support our and nature's abundance.

A tropical plant guild may not share the same cultivars as a temperate system, but in essence it shares the same functions: the need for nitrogen, minerals, hormones, vitamins, enzymes, oxygen, water, carbon, calcium, silica, a vast array of processes, mixing and matching, giving form to

Some Functions and Benefits of Plant Guilds

- Animal forage
- Air cleaner
- Carbon sequestration
- Fortress
- Flood management
- Wildlife habitat
- Insectary
- Mulch maker
- Nitrogen fixer
- Nurse
- Nitrogen scavenger
- Biomass
- Cleanser/scourer
- Compost
- Container garden
- Cut flower
- Dried flower
- Dye
- Food
- Fruit
- Protein
- Roughage
- Carbohydrates
- Vitamins, minerals
- Spice
- Storage, cache
- Essential oil
- Fiber
- Aromatics/fragrance
- Hanging basket
- Insect repellent
- Medicine
- Oil, wax, resin, or polish
- Ornamental
- Soap
- Wood
- Animal feed

the plants in the group. There are fermentative bacteria, fungi, microorganisms, earthworms, all bearing down on the organic materials sloughed off the living at death.

We are simply dealing with a community of beings that would be less without the ecological web of give-and-take, in the constant whirl of metabolic processes. It is as though the soil and the atmosphere just above it comprise the gut of this earth, and the plant is the beneficiary of a predigested food substrate bequeathed by sisters, brothers, aunts, and uncles in a magnificent potluck celebration. A divine marriage of sorts; a community of beings working together for the benefit of all life.

We, as human beings, have the uncanny ability to think, recall, conceptualize, and design a structural matrix, a foundational web that can support infinite plants with infinite functions. From the active whole to the active part, the great guild circle encompasses all plant life, and, through our ability to conceive functions and interrelationships, we can help create thriving ecosystems. We mimic local ecosystem patterns and functions and plug in high-yielding cultivars for human use. But it is about process, not the material-thing-in-itself.

In Chinese medicine, when, for example, the liver is mentioned, we are not speaking about the physical liver, per se. We are speaking about the liver process, a dynamic organ of transformation, an energetic basin of possibility, a multifunctional reservoir of life-giving potential. If we keep the energetic liver process in balance with all other energetic processes in the body, mind, and emotions, we maintain balance. The same holds true for the plants in relationship in a plant guild or food forest.

Think of a plant guild or food forest in this way: as an organism unto itself, yet embedded in a larger organism. Embedded within itself are organisms complete unto themselves, but the one thing that all larger and smaller organisms share is this one essential function: *relationship*. Because of the dynamic *nature of nature* we would be hard-pressed to design a system that is static — one that remains the same for all eternity, a pretty picture on a wall. The nature of nature is such that we never step into the same forest twice.

We cannot and must not separate our guilds or food forests from the place where the plants plant their feet and rise out of the rock, soil, and climate they find life in. And yet throughout the life of the guild or food forest, things will change. It behooves us to stay alert to our guilds through protracted observation, and allow the plants to suggest to our senses and perception what they need over time. If life is all creative process, what better way for us to merge with the creative genius than to become it: interactive, immersed in it, part and parcel of the community of plants we have so painstakingly designed, selected, and enjoyed the fruits and beauty of.

CHAPTER TWO

The Structure of a Plant Guild

What is a weed? A plant whose virtues have not been discovered.

— RALPH WALDO EMERSON

Give me a land of boughs in leaf,
A land of trees that stand;
Where trees are fallen there is grief;
I love no leafless land.

— A. E. HOUSMAN

Just as it is essential that a house has a strong and durable structural framework, so must a plant guild. By understanding how the plant guild extends into the space around it and how the members of the guild morph through time, the designer fuses function and structure, where form and function are not separate, but one and the same. A house must be built for the environment that it occupies. The foundation for a house on sand is different from a foundation for another house on clay. The roof on the house will be one type in Arizona and another in Alaska. Each of these adaptations to the environment ensures the house's resilience in weather extremes. These structural differences are natural responses to the environment intended to protect the structure itself. With plants, their structure or placement is made from the materials or species that are available. Thus the limiting factor to the existence of a plant in a particular place is its absolute dependence on the resources available. It is difficult for a plant to use anything but local materials.

The functions within the house are also different depending on the place. In one environment water needs to be added as humidity to keep the house from drying out; in another, water must be removed to prevent rot. Mechanisms that do this are built into the house's structure. Waste must be removed from the house so that it does not become a toxic environment. Similarly, plants expel their own waste to the environment via structural mechanisms. Fortunately, the immediate ecology of organisms around the plant will absorb those wastes as food, break them down, and return them to the soil as food for the plant. At this point the analogy of a house being like a plant begins to weaken: Everything in the house is dead matter. It has neither living tissue nor cells that grow and repair damage. Without constant maintenance a house will quickly degrade and become unlivable. Houses may have an environmental effect such as creating windbreaks on one side, or concentrating water with downspouts to form wet areas, but essentially they play a smaller biological role in the broader ecology that surrounds them.

Plants live by their relationships. As soon as a seed is dropped to the soil, the relationship begins. If the temperature is right and moisture is available, the seed will germinate and sprout. From this point a series of requirements for the plant to establish itself begins to unfold. The plant will send out roots looking for nourishment and a stem to the sky seeking sunlight and air. The seed has enough stored energy for this initial growth, but the plant will soon wither and die if it does not have a beneficial relationship with its environment. It must be in the right place at the right time and eventually have a functional role in that place. If in a sufficient amount of time the plant can have its needs met and begin a relationship with the other organisms in the local ecology, it has a good chance of fulfilling its life history. This is called the plant's *phenology* — the many stages of growth from seed to plant, followed by flowering, pollination, and eventually another generation of seed.

Communities of plants that have established themselves in close proximity to one another and share resources are called polycultures. Polyculture plants are integrated such that each plant has its needs met by the others and also delivers a beneficial function to the group. In the wild ecologists call them natural plant associates: plants that commonly grow together with no human intervention. Natural plant associates in the wild have very complex ecological relationships that involve not only the plants themselves but also dozens if not hundreds of other organisms, from bacteria to specific pollinators to animals. The basis for polyculture plant design by people comes from studying natural polycultures in the wild. These wild groups are composed of annual self-seeding plants, biennial plants that go to seed their second season, and perennial plants — the multiseasonal grasses, vines, shrubs, herbs, and trees. Natural plant associates are polycultures by nature.

The study in this book will focus on the structure to build these polyculture relationships and how to select plants for their position in a design that will deliver the functions needed, so that all plants will have their needs met. As a design progresses, it is important to understand that it's following nature's lead in all decisions, and mimicking the local plant patterns in each location. Just as you previously read about houses in different states, the structure of the plant polycultures will vary depending on where they are located. Generally, all plants will have the same ecological requirements in order to thrive, but by now it should be obvious that how those plants are placed and structured will differ based on whether it is a tropical environment, a woodland environment, or even the desert. Structure is the physical position of the plants in relative locations so that they can receive and deliver ecological services (to be listed shortly) to other plants.

Perennial Polycultures

For millennia there have been forms of annual crop agriculture seeking to enhance beneficial relationships among plants. Interplanting, or mixing up the plants out of rows, is used to take advantage of sunny space and soil resources in a garden and fill vacant spaces as early-season plants are harvested. Rotating crops is a type of interplanting, but done temporally over weeks as crops mature and are harvested. Companion planting, a beginning polyculture for pest control using natural plant chemicals, has had a wide following in the organic gardening movement. However, neither companion nor interplanting takes into consideration the use of perennial plants as major design components. Both interplanting and companion planting are forms of annual polycultures, but annual polycultures that benefit primarily humans, not the larger biome. They are seasonal and short-lived and still require annual tillage and disturbances to the soil that disrupt healthy soil composition. True polycultures are perennial and long standing.

A perennial polyculture is a resource-sharing plant guild. The plants are first placed to share by partitioning the resources of sunlight, water, soil, and nutrients. At this point the form and structure of each plant are important because they determine where it will be placed. Tall plants are placed away from the sun so shorter plants can be on their sunny side. Plants with long, narrow taproots are mixed into plants with wide, fibrous roots so they do not compete for soil resources. Beyond using the annual practices (changing plant species over time, rotating crops, and confusing insects with diversity), as each season passes a perennial plant guild leads to long-lasting and increased plant resources and a diverse polyculture that is of high value with low input. This also supports ecosystem services within and outside the planted area as pollinators and beneficial insects that are supported by the diversity of food sources and habitat extend the plant relationships beyond the immediate area. Birds will visit, get a few seeds, eat pests, and feed their young. In polycultures considerations are also made for the insect partners. Remember the relationships? They last all season long and into the next. When one plant stops flowering another is there to supply nectar for the beneficial wasps, bees, and butterflies, even beetles. The ecological functions (nitrogen fixing, nutrient accumulation, soil building, pollinator food, et cetera) within the polyculture are supplied by multiple species. This overlap, where many plants become redundant in their services, increases the resiliency of the plant system and ensures adequate resources when it is disturbed.

Later in this chapter we will offer a more detailed description of how to build a plant resource guild and polyculture. Plants are placed on the landscape based on their shape, size, and seasonal responses. Plants, of course, change throughout the year, and it's up to the designer/grower to pay special attention to these changes and plan for and take advantage of the resources that become available in each season. The choreography is dynamic as perennials pass through the season emerging, leafing out, and flowering at different times. Species grow at different rates over the years, eventually causing some to be shaded out and others to dominate. In time, initial plants that create soil structure and nutrient benefits may need to be moved to a new location. Trees and shrubs will begin to dominate the landscape as they expand. Understory plants, shaded out, will be outcompeted for resources. Unless they have partial- or full-shade tolerance, it is best to move them where they can be fruitful — perhaps a new planting bed. That, along with the many plant shapes and heights, offers natural patterns for placing the plants on the landscape. The design and site plan may seem frozen in time, though you can capture changes over time via your plant selections and notations on the design. The final master plan must be a flexible entity. It is finished to the point where nature takes over in some anticipated direction. The ecology surrounding the plants is a dance of cooperative competition[1] among minerals, plants, animals and humans. It is not forced into stasis like meticulously managed and groomed suburban yards adhering to imposed cultural standards.

Defining Your Niche

A niche is a habitat supplying the factors necessary for the existence of an organism or species. It's a relatively small space on the landscape that has specific environmental conditions. One niche might be the shade on the north side of a house; another might be a low spot in the yard that has full sun but is frequently flooded. Niche conditions such as precipitation, sunlight, and soil type, among others, define the niche and the plants that are appropriate for that space.

Biotic conditions involve the living components in a space. This includes the flora — all the plants living in the niche — and the fauna, or the living organisms such as worms, bacteria, fungi, and even mice. The flora and fauna in a niche are the producers and consumers of nutrients and organic material. The decomposers (fauna) are the tens of thousands, if not millions of organisms that break down old plant material to basic compounds to be absorbed by the plants, the flora.

Nutrient Cycling

Everything comes from the sun. Keep in mind the aspect of your land to the sun. In placement, partition plants in the forest layers so the light is shared by a diverse group of plants, and move the tallest plants away from the sun on the landscape so shorter plants get direct sunlight as needed.

Imagine the tightest circle of nutrients moving through a property. It could easily involve only a

Available sunlight is 40 times that used by plants.

Approximately 2.5% of available sunlight is used in photosynthesis.

Approximately 40% of photosynthesis results in biomass.

Approximately 30% of biomass is absorbed when eaten by herbivores.

1.5% of absorbed energy is turned into biomass.

Energy flow, photon to fork.

THE STRUCTURE OF A PLANT GUILD

> ## Nutrient Cycling
>
> **GIFTS TO THE VILLAGE**
>
> - Fruits and vegetables
> - Juices and pulp
> - Vitamins and minerals
> - Flowers and oils
> - Medicinal herbs
>
> - Storable foods
> - Spices
> - Honey, wax
> - Community activities
> - Animal feed
>
> **NUTRIENTS RETURNING TO THE GARDEN**
>
> - Soil from kitchen waste (worms)
> - Compost from the landscape
> - Mulch from grass clippings
>
> - Cardboard and paper
> - Water from roof gutters
> - Transplants from native lands

few components in a very small area. The producers, the plants, create the sugars in organic material using the sun's light, but they need the nutrients of the soil, many of which come from decaying plant and animal matter. The rain cycles through the sky to ground to pond and spring, passing through the plants as it goes. Gases are taken in by the plants while other gases are released. All of these cycle again through the sky and the ground, making their way perhaps through animals and organisms.

Each time nutrients, minerals, or gases pass through an organism or plant, they are metabolized, broken down, and changed. After being used they are expelled to the environment, where another organism or plant consumes the materials for its own purpose. This nutrient cycling continues at all levels of life. From bacteria to the largest mammals, each depends upon the trophic level of life above or beneath it to sanitize its waste as new compounds nourishing other organisms. Though slowed or accelerated by weather, this nutrient flow in healthy polycultures is a continuous process building soil resources for further growth.

So here's what it all boils down to. In nature and in every ecological system, nutrients continually cycle through elements. When the system is disturbed, for a short amount of time the cycling of nutrients may be diverted, but essentially they continue to cycle through organisms, plants, and organic material. Everything needed on a trophic level originates from the levels beneath it. Literally, waste is released to the organisms that decompose and metabolize the compounds to nutrients, which are then recycled into the system once again. All there is for humans to do is develop ecological cycles by placing the components in a relative location and create a healthy ecology.

Integrating ourselves into the landscape means making sure that each component of the design contributes to that landscape and increases its capacity to sustain us. Let's work to fit ourselves into the local ecological system that already exists and use our skills to enhance it and restore it to optimum health and yield. Nutrient cycling increases the capacity of the landscape to sustain life. The limit of the system is the carrying capacity.

Carrying Capacity

There has always been and will always be an eventual balance in nature. If a fox is to survive within its range, it needs a certain number of rabbits, mice, squirrels, or other prey. Each of the prey animals also needs a certain amount of resources to maintain its health and propagate a new brood. The range of the fox is dependent upon the food available to the prey. The prey also depends on a certain amount of predation to keep numbers in check so as not to overrun the available resources and face starvation.

A niche or landscape can only carry a certain number of any one species at any trophic level. In the plant world a diverse list of species growing in the landscape ensures that resources will not be depleted. The natural partitioning of roots and sunlight by the plants that are established creates a balance, for each participant is meeting all needs. If one invasive plant or animal overruns an area, it is only a matter of time before the resources specific to that organism are depleted and other organisms will take its place. Ecological designs repartition the resources and use a wide variety of plants for their ecological functions. This makes a resilient landscape capable of adapting to environmental extremes and disturbances, keeping the carrying capacity (sometimes referred to as "K" in research equations) high and in balance. The carrying capacity of a plant or organism's niche will constantly fluctuate within an acceptable range when in balance. Out of balance the capacity will rise sharply and/or collapse as resources are depleted for supporting or carrying a species. It can take many months or even years for a niche's carrying capacity for a given species to return, if at all.

Understanding the Context of Your Site

In the practice of design, use the ethics and principles of Permaculture and sustainability. This begins with the first principle of Permaculture, observation and integration. Listening to the land. Observe with every possible sensibility available. We use our eyes to see the existing plants and structures of nature, where resources collect in every season of the year, where the snow drifts, where fog collects on cold mornings, and the site conditions as they change through the day, the seasons, and the year. Look at the land from 360 degrees, from every corner and from every angle. Move throughout the property in the morning, in the evening, and late at night. Listening to the wind and the rain helps us understand where precipitation is being deposited. Feeling the ground with our feet helps us understand compaction in different areas while walking. Listening for frogs and birds is a clue to the ecological conditions of the property. The more you can observe your property in every season, the better you will understand its responses to the ever-changing conditions in its climate.

Ecological Permaculture designers create plant guilds in future tense, because site conditions change by the hour, and during the seasons. The design will change as plants mature and as tree canopies close. Some plants will be temporary as they build soil or nurse other plants while they establish themselves. The layers of the plant guilds will move

THE STRUCTURE OF A PLANT GUILD

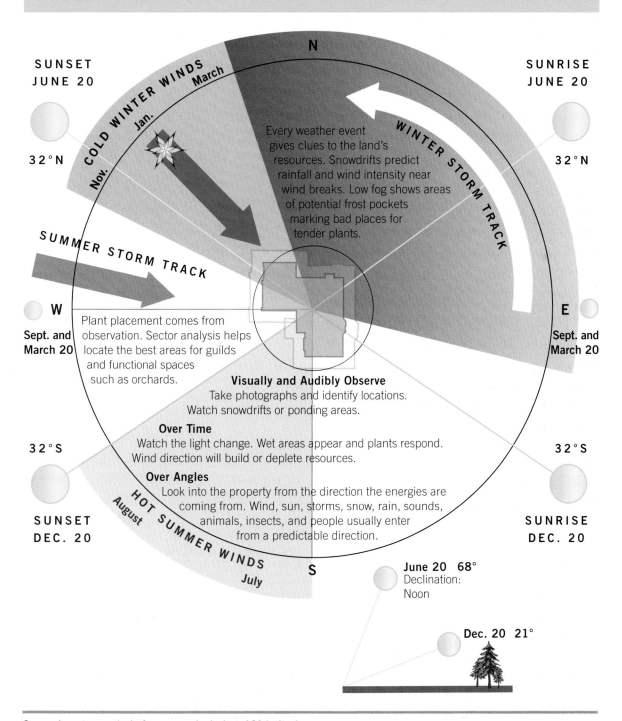

General sector analysis for sun and wind at 40° latitude.

Plant, Observe, Plant, Observe

When spring blooms, there is something that pulls us out of the house and onto the land; something that says, *Plant, observe, plant, observe*. The harvest is written into the planting, and the planting will bear fruit for us if we stay the course. A plant guild is a template for us to hone our efforts into an integrated unit of our design. The plant guild grid helps us stay focused and small and seek solutions that are small and slow. As we develop several guilds and place them into relationship with others, the dream of a food forest comes into being. We have had the luxury of designing and implementing all of this in small, incremental units, and we all know that there are many logical (ugh!) steps to be taken in order "to get it right the first time." Step by step. Assess, inventory, design, implement, manage, maintain, and interactively observe throughout.

— WAYNE WEISEMAN

---

vertically and horizontally, causing some plants to diminish while others flourish. Accelerated succession is built into the designs to work with nature. You need not spend hours every weekend trying to restrain the ecology in some ornamental or cultural aesthetic. Natural capital is built when the plants over-yield organic material and soil nutrients, along with harvest for human uses.

Designs follow the natural patterns that already exist and use nature's momentum to enhance the ecological services that are available. Along with following the seasons and the phenology of the plants as they sprout, bloom, and produce fruit, the design strategy also follows the shape of the land, where the water flows and the snow drifts. We work with the contours and plant across hills to increase the deposition of organic material and precipitation that may otherwise flow away from the system. Observing the natural patterns on the property allows us to work with nature.

All of this results in accepting the land's imprint in initial goals and designs. Designs do not imprint on the land some dominating structure, thinking we will force from it long-term benefits. By working with the land and accepting its ecological imprint, energy and resources are saved, natural capital is increased, and ecological services already available are supported. By doing this we as occupants adapt ourselves to what the land can provide and integrate ourselves into its ecological system. As a result the designs will intuitively reflect the patterns in nature that are present and appear to be a natural occurrence while being ecologically self-sustaining.

Assessing components in a niche helps us define its richness, its fertility, and its diversity. Walking into a backyard or field, the eyes see the richness of the property even if the plants and organisms cannot be named. Seeing a single tree surrounded by grass would denote little diversity and very little richness in the niche. On the other hand, seeing countless perennial plants, watching birds landing in shrubs and trees, hearing the croaking of frogs or the chirping of crickets, and smelling the earthy scents makes it obvious that the site has a rich ecology. Walking and looking confers a slope or angle to the land and the direction the slope faces, the aspect. The aspect is the compass direction (north, south, east, west). At this point start documenting the niche conditions in a quantifiable form, making observations and sketching the small spaces. Photographs are invaluable to this process. Features of the landscape can be documented from many angles and at different times of the year. Video recordings give a perspective that cannot be achieved with still photographs. Three-hundred-sixty-degree panoramas from the corners, high points, low points, and random places across the landscape are very helpful later while you are drawing out the design or explaining the positions of features to another person. Doing this guides future decisions so that the design aligns with the ecological conditions and resources available.

Site assessment calculates ecological conditions in the two categories already mentioned: abiotic (non-living, minerals, and so forth) and biotic (living cells).

ABIOTIC

- Geological
  - Soil types
  - Bedrock or base material
  - Topography
  - Slope and aspect
- Hydrological
  - Soil water and water features
- Meteorological
  - Climate
  - Precipitation
  - Sunlight
  - Temperature range
  - Prevailing winds
- Chemical
  - Soil pH
  - Soil nutrients and organic material
- Location
  - Elevation, latitude, and longitude

BIOTIC

- Biological and ecological
  - Soil organisms
  - Wildlife
  - Plants and fungi
- Cultural
  - Rural, urban, commercial, residential, industrial
- Social
  - Community, family, cooperative, public, private, homestead
- Economic
  - Resource-distressed, rich, adequate
- Historical
  - Previous use(s): fallow, livestock, crops
  - Previous owner(s)
  - Disturbances or contamination
  - Original condition: forest, woodland, prairie?

Once these site conditions are understood, it is much easier to adapt the design goals to the existing conditions. This allows us to begin understanding the factors limiting what we can do and the design drivers that guide us.

Start from Scratch or Follow Nature's Lead?

The next question is: What will the land support? Based on the site assessment, how can we improve or expand upon what is already there? What small changes will enhance health and resources on the site? The niche may be an empty lot or it may be an established woodland or prairie with preexisting native plants. Each offers opportunities for growing food. An empty lot probably offers little competition to new plants, but the soil may be degraded and have little nutrient value. Feeding the soil and building raised beds or berms for planting will take some initial effort. If the site has an established ecology, there are two choices: either tear out all the plants and trees that are inconvenient, or follow the existing plant structure and use it to support new plants that will fit in the system. The latter means you are adding to the system plants suited to the existing ecology. It is a lot easier to adapt the design to the existing ecology than to replace that ecology with a new plant system.

There are a number of good reasons to follow nature's lead and work with its momentum in the growing space. Established plant systems in woodlands, prairies, or old pastures have been setting down roots for many years. The plants have been spreading seed and building a seed bank in the soil. The soil itself has been modified by the plants to suit their needs and is filled with organisms and organic material tailored to the existing plant life. Even if you go in and cut down all the plants and till the soil, you are left with a massive seed bank in the soil that will now sprout and try to regenerate the original system. Small trees, shrubs, grasses, and weeds will continually emerge from the soil that you have disturbed. Grasses are relentless: Generations of seed are layered in the soil, just waiting for a little warmth and sunlight. For a small intensive garden it is expected that the soil will be disturbed while you plant crops and weed. If the niche is defined as woodland, woodland harvests should be the goal. If it is prairie, begin transitioning to a diversity of shrubs and small trees that are suited to the prairie environment. The variations of site and niche conditions are your guide to plant selection.

Now that the niche conditions have been defined, it's time to think about what plants have those conditions as requirements and what plants can add increased soil health and productivity. The purpose of site assessment is to understand the environmental conditions that exist so that you can align food production with them or add plants to the point that the site is productive for human use. This is not about extracting resources from the space. This is about enhancing and increasing its nutrient resources (natural capital) so that the harvest for human needs is increased and the soil resources are not depleted.

This is worth repeating. In Permaculture there are three important ethics: care for the earth, care for people, and fair share. So how are plant guilds created that reflect not only the three ethics stated here, but also the broader context of property, bioregion, continent, and world? This is done by acting locally with a worldview. Expect a majority of your sources to come from the ecosystem, and work to maintain a healthy ecosystem to hand down to the next generation. The more each of us relies on local resources, the less we pressure others to extract theirs. This extends from a personal to a federal level; at each level the ethics are the same. Building natural capital as individuals, communities, and a nation reduces dependence on others and their exported resources.

The polyculture model is self-repairing, self-sufficient, and cycles nutrients to a broad and diverse system of participants. In plant systems, strive to do the same and keep everything in the ecological loop. That is the "structure" we referred to — the ecological framework of plants and organisms that keeps nutrients moving through the system so nothing becomes toxic and everything has access to its daily requirements.

The Integrated Living System

Plant resources as well as Permaculture ethics and principles extend to our living space. The way we use energy, the waste stream, and countless other factors contribute to the health and well-being of

THE STRUCTURE OF A PLANT GUILD

Niche Dynamics

Nature fills every niche with varied plants for efficient use of sunlight, soil, and water. Plants naturally partition or divide up the spaces for those resources through a succession of species. After a disturbance, or as soil has become exposed to the environment through some geological event, plants begin to repopulate the new niche. At first the pioneer plants take hold. Dandelions, grasses, and an assortment of plants commonly called weeds begin to cover the soil and put down roots. These plants fill the initial niche and are tolerant of dry soils in full sun. Flat- and broad-leaved plants cover the soil and limit the growth of other plants.

Some root structures are fibrous, growing laterally, while others are deep and vertical. They may have taproots that dig deep for water, but they also bring minerals and nutrients up to the surface, changing the soil resource for the next set of plants. At this early stage the plants are partitioning the soil so that each gets its needs met. As soon as these plants are established they create a niche for other species, perhaps plants that need a little shade to get started. Gradually larger and larger plants fill in the spaces and shade out the pioneer plants, which are replaced by other shade-tolerant species. Soon nature has filled in all the spaces with plants with different types of roots and different tolerances so that the ground is completely covered. New plants of similar root structure are outcompeted by established plants. Only the plants that can find an unoccupied niche have a chance at survival.

As soon as these plants begin to sprout, the niche changes. The soil is covered, organic material enters, and soil moisture begins to rise. The original conditions that allowed for the pioneer plants to persist have changed, and new plants begin to appear. Each time a seed falls, a plant emerges;

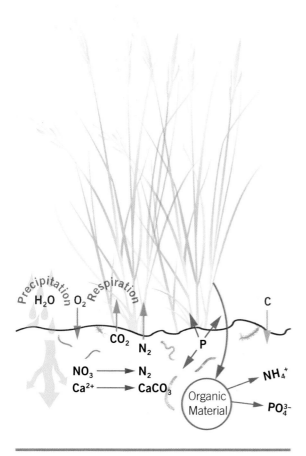

The cycle of soil organic matter. Illustration by Daniel Halsey and Kellen Kirchberg

plant systems by supplying a rich and plentiful stream of organic material and fertilizer. It is inevitable that whatever produce is trimmed or peeled away will be returned to the naturally recurring biological processes: Kitchen scraps, human waste, all "waste" can be reincorporated into the plant system through decomposers for the advancement, growth, and expansion of soil health. Building fertility into the system is key to our own health. Healthy minds and bodies depend on this. If the plant is the primary producer for all life, then our primary focus must be a conscious effort to supply what is needed in order to build a healthy and dynamic soil community.

Buckwheat soil cover.

it is in cooperative competition with the other plants. It must fit into the ecological system already present and compete with the other organisms for resources. This natural progression partitions the resources available as plants either succeed by finding the resources they need or die off for lack of an available resource niche. The new soil conditions may not be suitable for some seeds to germinate, so others will join the bank of seeds waiting for the required conditions: perhaps a wet season, warmer temperatures, or even fire, as some seedpods need exposure to fire in order to germinate.

Designing for Cooperative Competition

As you design planting beds for polycultures and guilds, take into consideration the cooperative

THE STRUCTURE OF A PLANT GUILD

Seven layers of the forest garden. Illustration by Daniel Halsey and Kellen Kirchberg

competition among plants. To build a successful guild, identify the resources available, examining the quality, quantity, and seasonality. Typically, these resources are limited to sunlight, soil, and water, but over time and seasons nutrients and minerals may also be partitioned.

The slope of the planting bed and its aspect toward the sun. This forms the basis for its partitioning sunlight. In the Northern Hemisphere the sun appears deeper in the southern sky as the latitude increases. While the sun at the equator is primarily directly overhead, northern latitudes have a sun angle that creates longer shadows. Partitioning the sunlight means placing plants relative to the sun angle. If the sun is generally in the southern sky, site tall trees on the northern edge of the planting area. Shorter trees are placed to the south, followed by shrubs and perennials.

As explained in chapter 1, seven layers is the classic model for forest garden structure. Formulated by

Robert Hart, a forest garden enthusiast in England in the 1980s, this sunlight partitioning structure allows for each plant to receive the sunlight it needs. The initial canopy is followed by an understory of plants suited to partial shade or full shade. Beneath the tall tree canopy there may be multiple plant guilds partitioning the available sunlight and creating a nutritive environment. Over time the seven-layer form has expanded vertically to include fungi, soil organisms, and aquatic plants. The main point is to think in layers, dividing up the resources according to their availability.

The Importance of Sunlight

The aspect of the sun and its angle on the landscape guide the initial placement of the overstory and subsequent understory layers. During the day the sun rises in the east, to reach its highest point at midday; eventually it sets in the west, its light changing with the shadows. An eastern forest edge may receive cool morning light as the sun rises to its midday peak and then be cast into the deep shadow for the rest of the day. The same area might have dappled sunlight depending on the density of the tree canopy. This effect might also be present on the east side of a building or wall. The sunlight available is limited to a few hours, requiring that the plants are especially suited to that environment. If the site enjoys full sun for a few hours, it may be sufficient for many plants. If the quality of the sun is dappled shade — patches of sunlight intermittently moving across the understory — a different set of plants may be needed.

In North America the sun shines laterally for most of the day. It falls at an angle between tree trunks and across the immediate landscape, allowing for sunlight space between plant guild members. This increases the availability of light to a larger area; it also adds to the ventilation and airflow within the guild.

Partition the sunlight to its maximum effect through placing full-sun plants according to their respective height and density. Once those full-sun positions are taken, you can position partial- and full-shade plants in the understory. As the plants are progressively closer to the ground, the final layer will be the ground cover itself. Depending on the sunlight available, ground cover plants can be selected for full sun, partial shade, or full shade. Building a dedicated plant list of trees, shrubs, perennials, grasses, and ground cover helps in the selection process.

At this point, keep in mind that with sunlight comes heat. As previously mentioned, the rising sun from the east is available during the coolest part of the day when moisture is present in the landscape. Later in the day, as the sun swings to the west, the morning condensation has evaporated and the air is much drier and warmer. Hardscapes such as buildings, pavement, large stones, and walls have now absorbed the day's heat and are radiating it back to the landscape.

Plants that receive the majority of their sunlight during this time of the day must endure the harshest heat, highest evapotranspiration, and least available moisture. Your guild plant list should include those that can tolerate or even require dry conditions for this niche.

The heat of the sun should not be seen as a limiting factor or negative attribute of the landscape. It is a condition that plants have adapted to. This also offers opportunities for microclimates where the sun's heat is absorbed by the thermal mass of stone or water, creating a heat island during colder parts of the day and moderating the heat of the day by cooling off at night. The radiant heat from the thermal mass buffers the temperature changes in the immediate surroundings. Plants in close proximity

THE STRUCTURE OF A PLANT GUILD

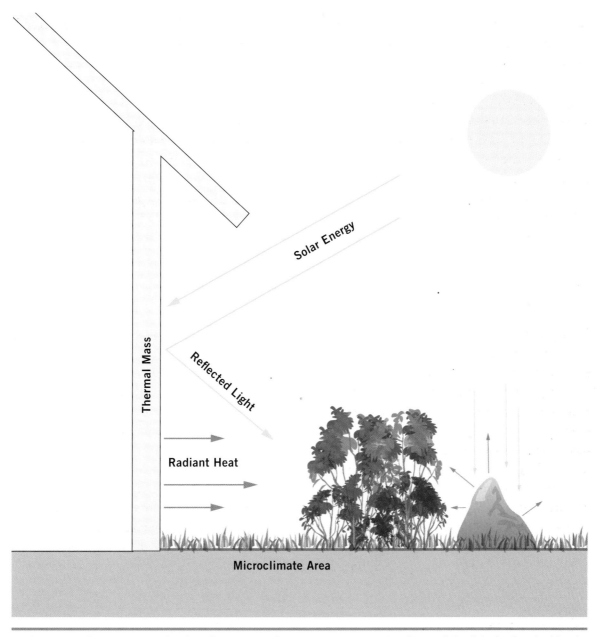

Radiant and reflected energy: Sunlight fills our planting areas on a sunny day. Some of the light is absorbed by the mass of the plants, the ground, and structures. Some is reflected, increasing the available light to plants close by. The light that is absorbed is then radiated as heat. This low-frequency radiated heat is emitted during the day and at night until temperatures equalize.

benefit from this service. This microclimate allows you to use species that may not otherwise be tolerant of the local conditions. Peaches, apricots, and figs that may not be suitable in USDA Zone 3b can be established within microclimates using the southern aspect. The addition of thermal mass causes a heat island effect. For example, the south side of a home with stucco or brick siding creates enough thermal mass and reflection to produce a microclimate within a few feet suitable for plants requiring a warmer growing zone and protection from freezing temperatures. Large stones in the landscape and water features can create similar conditions, additionally using reflection to increase the available sunlight for the plants.

The Position of Plants

The seven-layer forest garden pattern instituted by Robert Hart is a prime example of positioning plants based on height and the angle of the sun to the niche. Typically on flat ground the taller plants would be away from the sun and allow smaller plants access to the light on the sun side (the south side in the Northern Hemisphere, the north side in the Southern Hemisphere). A south-facing slope has the benefit of complementing the sun angle, catching more of the sun earlier in the season and more direct sun during the day. A north slope is hidden from the sun as it faces away and diminishes the sun's access to the slope. This has benefits for adapted plants in either case.

The slope and aspect of a hill have an effect on a plant's shadow and exposure to wind. Trees on the top of a south slope are overshot by the sun while those atop a north slope extend their shade far down the hill onto other plants. It's best to keep trees off the top of the slope, where prevailing winds may cause damage or an unwanted seasonal effect. Since the plant design is relative to available plants and desired harvests, the canopy or overstory may be sixty feet or six feet above the understory plants. The height of the tallest plants will have an effect on your choices of other plants to use. The trees and tall shrubs used in the overstory guide the choices of understory plants that will fit alongside or beneath.

An open canopy and forest edge may let the sun shine during different times of the day as it moves across the sky. Dappled or partial shade allows for an extended list of plants in a small area. Trees with open branching and minimal leaves help sun flow to the plants beneath.

At low angles the morning sun can extend briefly to plants deep under the canopy. The aspect of the sun can be managed for season, for time of day, and to increase infiltration to plants at all levels. A large pine tree with thick and dense foliage allows little sunlight to the ground beneath compared with a well-pruned deciduous tree.

Determining the Quality of Your Soil

When designing plant guilds, the tendency is to design for all vertical niches aboveground, from the herbaceous layer to the canopy. Below grade, within the soil profile, there are roots that seek nutrition in the upper layers of topsoil and taprooted plants that mine minerals, sometimes as deep as fifty feet. In Robert Hart's design, plant guilds are configured for seven layers aboveground, but nature is not limited to seven layers. The growth habits of divergent cultivars make even a single layer seem infinite in terms of height, spread, root, growth habits, sunlight needs, and so on. This is based on climate, landform, and soil fertility. For example, what the nursery catalog states is a fifteen-foot tree may only grow to ten feet — or if conditions are perfect, it may grow to twenty. With infinite possibilities, you are assured some semblance of order by delineating the plant regimes with average growth habits. Beneath the ground there is much

THE STRUCTURE OF A PLANT GUILD

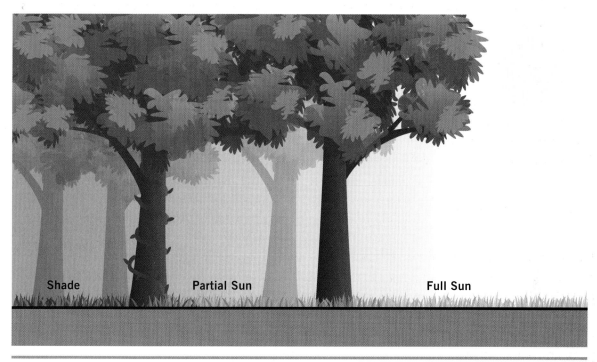

Edges are very important in ecological design, especially when deciding where to place plants. The edge of the forest is actually a transition zone from the adjacent niche's conditions to those of a woodland. Sunlight is filtered through the canopy and diminishes with the ever-increasing density of the overstory.

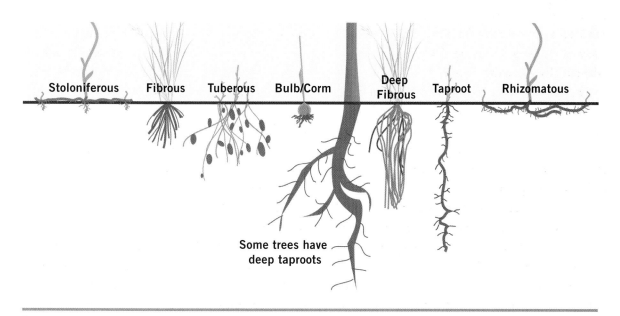

Root types: Planning plant placement by partitioning the soil resource requires a diverse selection of root types. Mixing root types allows for closer proximity among the plants. Illustration by Daniel Halsey and Kellen Kirchberg

Indicators of Soil Health

All of these conditions serve as indicators of soil health:

- Earthworms and the presence of organisms
- Color of organic matter
- Presence of plant residues
- Condition of plant roots
- Degree of subsurface compaction
- Soil tilth and friability
- Signs of erosion
- Water holding capacity
- Degree of water infiltration
- Movement of water in the soil profile
- Regeneration of groundwater reserves
- pH (acidity)
- Nutrient holding capacity

---

less certainty — yet another consideration for plant selection and placement.

Lateral, medium, and taproots penetrate different levels of the soil profile. There are niches aboveground, but just as many below grade. The spatial dimension in the soil profile is as complex as the densest of forest biomes, based on whether the soil is composed primarily of sand, silt, clay, abundant organic matter, or any of the many different combinations. A single wheat root cluster may travel deep below the surface in search of nutrients; the roots will travel different pathways depending on soil conditions. Carrots in heavy clay soils may be stunted, whereas those in silty soils may grow to the extreme. Differences in pH certainly affect what will grow where (blueberries love acidic soil, for example). With all of this potential variation, how can a plan be exact for the wide range of prospective root designs within the vertical and horizontal layers?

The simplest template is to divide the plant world into three levels of root growth: surface feeders, midrange, and taproot. By grouping plants according to these three levels in the soil matrix, you can place the plants closely and tap different portions of the soil, not only ensuring that they will avoid interfering with one another but also complementing their search for minerals, nutrients, and water.

Walking on the land, it is good to keep in mind the geological history of the region, its bedrock, and how bedrock and surface rock have been weathered through the centuries. With the advent of the Internet, this research is easily accessible. Substrate of rock (sandstone, limestone, granite?) colors the overall character of local soils, and depth of bedrock colors how roots will make their way through the soil.

Some of the other niche conditions that need attention for the building of soil fertility and access to plant nutrients through the root zone are physical, ecological, and historical. Abiotic physical effects such as drainage and climate are combined with species coexistence, competition, parasitism (the relation between two different kinds of organisms in which one receives benefits from the other

THE STRUCTURE OF A PLANT GUILD

Penetrometer and soil core.

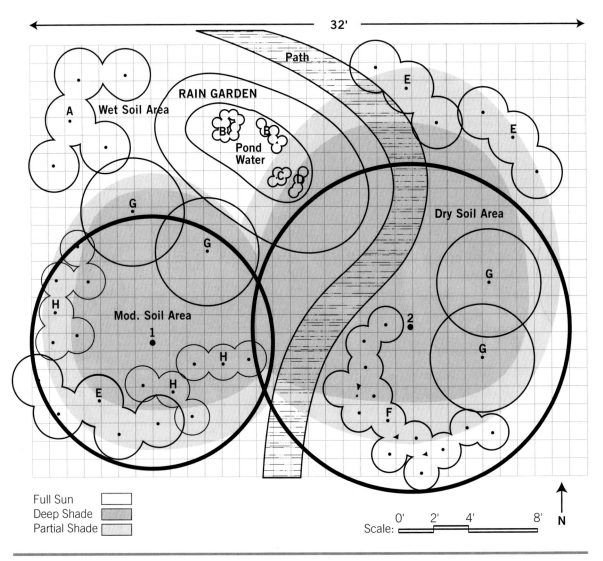

Pond niche exercise: Identify the plants by their ecological requirements.

by causing damage to it), mutualism (symbiosis that is beneficial to both organisms), microhabitat modification, and alternative food sources for beneficial insects. Due to limits of space in this book, we cannot spend much time defining all these terms. Think of them as keywords that may warrant ongoing investigation. Each research topic adds to your overall understanding of plant guilds and plant interactions with soil, water, warmth, and the need for nourishment. The more you investigate, the more deeply rooted you become in a place.

Building for Nutrient Cycling

All of us need food and water to prolong our existence. Plants are no different. The search for nutrients in the soil and from the air is a constant process affected by seasonal changes, species differentiation, and the location of the nutrients. Once the resources are used, the stored nutrients are returned and sent back into the local ecology, ensuring its longevity. How are plants organized in polycultures to ensure a constant flow of nutrients? Base your plant selections and placements on species needs and services. Perennial plants cannot be planted in seasonal rotations or relays like annual crops. Once a perennial plant is established, it will live for as long as the niche will provide for its needs or until it naturally dies back to soil. Forests use their own organic material to feed organisms that break it down and cycle nutrients for future production. The forest floor is also covered by a diverse set of plants that mine and collect nutrients and minerals in their biomass as wood, stems, roots, and leaves. The whole system of plants in the forest cycles the nutrients as fibers back to the soil. The forest (along with prairies, jungles, ponds, and wetlands) is an endless composting process breaking down and returning the extracted chemicals to soil. Watching the forest reveals how to grow and gather soil fertility by not only stockpiling and mulching material from different kinds of trees and plants, but also placing plants in positions where they will do the most good. Polyculture and guild design reflects this in accelerated succession.

Needs of a Forest Garden

Diversification

Plants mixed in the forest have differing strengths, needs, life histories, growth rates, heights, spreads, and tolerances. This allows for some plants to persist as the niche undergoes short-term disturbances, such as from weather or disease, or herbivory from animals. The system should have enough species to back up all its ecological functions, to create food for the organisms supporting the niche, and to buffer extreme conditions. In drought, those species that are drought-tolerant will flourish and cover the soil, whereas in wet weather another species will dominate. Differing shapes also affect the humidity and filtered sunlight.

Maintenance of Water Circulation and Total Recycling

Plant leaves collect morning dew, which rains down on the forest floor and can account for the majority of the plants' water needs. Air passing through the diverse canopy layers creates cool zones where water condenses. Each time, the water is distilled into its purest form. Deep and rich soil is like a sponge for water, keeping the root zone moist and slowing the loss of water to deeper soil profiles.

Maintaining this water and nutrient cycling is a product of not only deep and rich soils, but also the slope of the land. A perfectly flat — or even worse, flat and sloped — soil surface will cause considerable runoff during heavy rains. On the other hand, a pocked and uneven planting area filled with small depressions, berms, and swales will allow the water to stay in place and soak into the ground. The design process creates a landscape that collects as much water as possible to infiltrate and drought-proof planting areas. Many times this surface pattern may already exist; then it's just a matter of making some small berms to collect water and keep it from eroding the soil structure as it leaves.

Nondisturbance of the Soil and Continuous Soil Cover

One of the big problems with agriculture and even gardening is the fact that the soil is continually being disturbed during planting, weeding, and harvesting. It is very difficult for soil organisms to create a healthy soil structure when their processes are being disrupted by everything from trowels to plows. Leaving the soil undisturbed and covered with mulch and decomposing plant material protects it from heavy rains and hot sun. It also creates a cooling humidity that is best for slow discharge of nutrients to the root zone. Frequently, of course, animals are digging in the soil surface for roots or larger organisms to eat. Birds are constantly flipping through the leaves, and squirrels are burying and unearthing their foods. These disturbances involve what is more a mixing of materials than a disruption of the soil base. In any case a deep and dark mixed mulch will create the soil ecology best for the plants selected.

Plant Symbiosis

Most plants have a symbiotic relationship with other plants. After a number of seasons in which the plants have jockeyed for position in partitioning resources, they begin to put down broader roots and share the soil space. This can be in a passive or active role. A plant may play a passive role in the ecology by providing some moderate resources such as by offering pollen to insects, adding organic material to the soil, or simply serving as a nurse plant to protect a tiny seedling beneath its leaves. Other plants are more active in their ecological function, supplying nutrients to the niche. Nitrogen-fixing plants have a symbiotic relationship with mycorrhizal fungi and bacteria that nodulize nitrogen in the root zone. The plant trades its sugars for the nitrogen from the partner organism. When the plant dies back, this nitrogen is made available to broader ecology. Fungi in the soil also connect plants together, acting as conduits that move nutrients from one plant to another.[2]

Plants have symbiotic relationships with mycorrhizae. Inoculating the soil with the proper mycorrhizal associates will increase plant production and growth rate. The best way to find the associated mycorrhizae for a plant is to use its native soil to inoculate its new home in a forest garden. In the case of nut pine (*Pinus koraiensis*), for example, inoculating the soil will increase the growth rate and decrease the number of years until harvest. The mycorrhizal effect is impressive and should not be ignored. Happily, many plants are facilitated by the same mycorrhizal fungal species. Planting the associated symbiotic species, such as cow pea (*Vigna unguiculata*) (L.), can increase the likelihood of a beneficial effect for the primary plants. Again, using native soils will increase the availability of associated beneficial fungi.

Attracting and Providing Shelter for Wildlife

Attracting wildlife as part of the integrated forest garden creates an extended relationship with the surrounding environment. Providing shelter, habitat, and alternative food sources for wildlife is a trade-off for their ecological services. Birds will eat pest insects and a copious amount of seed produced in the forest. As previously mentioned, the bees, butterflies, parasitic wasps, and beetles that find alternative food sources within growing spaces will aid in pest control while plants benefit their populations. The presence of these beneficial animals and insects will stress the pests, reducing the need for additional biological controls. The larger space outside the property will benefit greatly from the island of nutrients for organisms that pass through. It is also a way to spread the beneficial plants to the surrounding community as they go to seed. Birds, animals, and insects carry genetic material across the horizon.[3]

Five Considerations for Sustainable Design

There are five major components found in a successful conventional landscape. As you can see from

Plant Placement Exercise

Neatly create an orchard landscape design with the list below. Include a path and use correct symbols and labels. Place the plants by their function. Trees have already been placed in the design. In the order of placement, consider each plant's function and need for sunlight.

Place the shrubs first, and then the nitrogen fixers near the canopy. There is a path that can be used for access and harvest. How does access to plants affect the placement? Strawberries need frequent picking, comfrey is chopped a couple of times a season, and apples may require a wide path for picking and moving baskets. What do the daffodils do? Place them so that their function can be best applied. White wild indigo fixes nitrogen in the soil. Placing these plants at varied distances from the trees will disperse services in a wider range. Creeping thyme is a spreading ground cover; disperse plants to converge over time. The hardest task may be to keep from placing the plants in symmetrical patterns. Be generally random, but allow the functions of the plants to be of service to the guild. There is no exact answer. Play with the plant positions and compare your results with the example provided.

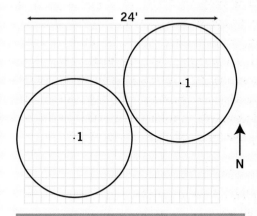

Two-tree exercise: Place the plants according to their ecological functions.

TABLE 2.1 PLANT LIST

ID	Species	Form	Light	Ht x W	Human Use	Ecological Function	Root	Spacing	Quantity of Plants
1	Honeycrisp	Globe	Full Sun	14 x 14'	Food	Soil Build	Fibrous Deep	14'	2
2	Alpine Strawberry	Mound	Partial Shade	12 x12"	Food	Mulch Build	Fibrous Shallow	12"	16
3	White Wild Indigo	Upright	Full Sun	36 x 48"	Flowers	Nitrogen Fixer	Fibrous Deep	4'	4
4	Daffodil	Upright	Sun or Shade	24 x 8"	Flowers	Rodent Deterrent	Bulb	8"	24
5	Comfrey	Upright	Sun or Shade	30 x 48"	Medicine	Green Manure	Tap	3'	8
6	Creeping Thyme	Low	Sun or Shade	Mass Planting	Path Borders	Ground Cover	Fibrous Shallow	6"	24

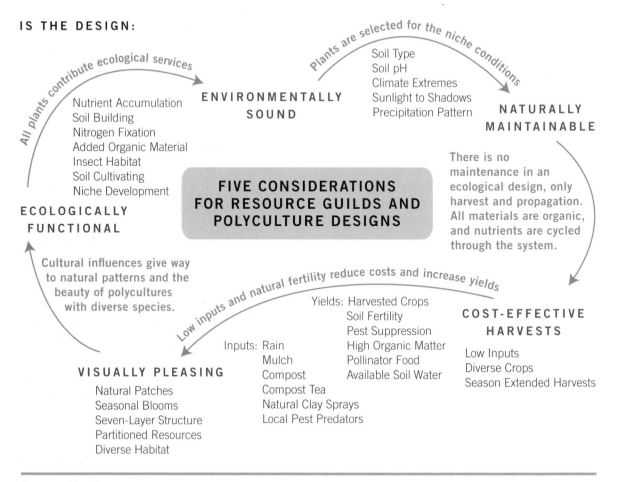

Five considerations.

figure 2.12, landscape that is visually pleasing is the first priority. The second is cost effectiveness. This order of priorities is obviously an economic model for the success of a landscape business driven by a cultural aesthetic and economic expectations. All of these considerations are also important in ecological design since they are a part of the design process and implementation. The paradigm for ecological design, however, redefines these considerations. What is visually pleasing is based on learned cultural values. Accepting a new ecological aesthetic standard diffuses the cultural influence. What is cost-effective depends upon the expected harvest, and for some it is merely a visual component; for others it's a matter of survival and independence. The beauty is in the resilience of the system and the freedom of lifestyle resources.

A landscape's function could be simple enjoyment by the property owner. However, the system within the land must be self-sustaining through its inherent ecological services and functions.

Maintainability in a conventional landscape setting refers only to the ability to move equipment and machinery. This is called access in the design process.

It is expected that all plants within a conventional landscape design will survive after its implementation. For some this is the definition of environmental soundness. Mostly, this refers to the plants being suitable to the environment and climate, while not dropping fruit on the turf or creating stains on the hardscapes.

Next, let's look at this a little differently within the framework of Permaculture and ecological design.

Permaculture Considerations

A Permaculture design should be:

- Ecologically functional
- Naturally maintainable
- Environmentally sound
- Cost-effective to harvest
- Visually pleasing

As you can see here "functional" has been replaced with "ecologically functional" to better describe the goal. With ecological function comes an expectation that the design will not conflict with existing systems. Within the framework of the ecologically functioning design, maintenance will be naturally supportive. One of the prime ideals of Permaculture is to design and develop self-managed systems.

All that is called maintenance to this point will later be seen as harvesting—ecological maintenance actually comprises harvesting ecological materials in the form of organic matter. In Permaculture all maintenance is harvesting; only the technology and mechanical systems need upkeep. The Permaculture view of the definition of *environmentally sound* is working with nature and increasing the ecological functions available to the food system or environments being developed.

Cost effectiveness is based on the yields, the harvests, and the harvest created for the organisms and plants that occupy the niches. Cost effectiveness is based on calories as much as initial investment of currency. Everything done will cause either an increase or a decrease in natural capital.

You may notice in the previous list that there are no numbers; this is because all are equal in priority. The systems are completely integrated; each consideration is a function of the others. A design that is visually pleasing is one that meets a new definition of beauty: In ecological systems beauty is in the taste of the harvested fruits, the enhanced ecological services that invite and sustain new participants on the landscape such as birds and other wildlife, and the many benefits of working with nature.

This is not to say that landscape design aesthetics aren't part of the design process. The structure of plant guilds begins with design practices that mimic nature. The order of placement, the patterns of the plants in the designs, and the curvilinear structure of the landscape create a visually pleasing design. For the most part you could follow a conventional landscape design process with all of its aesthetic drivers, simply choosing plants that are ecologically functional. This might create some difficulties in the beginning, but it allows for the adaptation of existing landscape design to an ecological system.

The goal through all these considerations is the enhancement of ecological services and the creation of natural capital. The designs must contribute to the ecological functionality of the property. It must be a beneficial change. Each component in the design is assessed as being beneficial or a pest that needs compensation from the system. Here are two examples:

- Either my washing machine has a beneficial effect on the environment or it's a pest. How can a washing machine be integrated into an ecological system?
- My lifestyle increases my natural capital or depletes it. How can my lifestyle create natural capital? How can I be a producer and not merely a consumer, including in my lifestyle the natural assets available to me?

Functional Spaces = Functional Design

- Allow for movement
  - Human access
  - Carts and equipment
- Collection of water, sun, and heat
  - Allow for use and banking of resources
  - Unrestricted growth space and succession for plants
- Supportive resources
- Provides a high net yield to all features
- Supports other plants, birds, and beneficial insects
- Buffers weather changes
- Needs no or minimal outside resources
- Uses low-level management (for times when humans are absent)
- Cycles all resources within the property

Make It Ecologically Functional

As functional spaces are developed in the design, remember to allow the movement of resources, wildlife, and humans. Access to the functional spaces on the property requires some forethought. Anticipate what will need to be moved and what will be moving it. As you develop bubble maps and transfer them to a draft design, you must include travel between the functional spaces. Allow access for humans and their equipment. Many times the size of the landscape dictates the size of the equipment used during implementation and harvest. A quarter-acre urban lot may need only a small wheelbarrow and a three- to four-foot-wide path in and around the production zones. A forty-acre property might require a large tractor with a wide turning radius. The access routes for this property will include major arteries for the tractor and trailer to move across the landscape. Paths are now the size of roads, and from these smaller roads may radiate paths where, again, wheelbarrows are used. Access is one of the first priorities in the process of design, since you must be able to move about the property without damaging the production areas. Placement of access roads and paths on contour is crucial to diminishing erosion and and can be augmented to create water catchment areas.

Once you have developed the access system, your next consideration is a system for collecting water and storing it in bulk or in the ground. Resources move across the property without restraint unless a managed system is in place. Water can cascade freely across the landscape and leave, creating a limited resource. Collecting water during weather events and storing that water on the property allows for increased resources for plants and animals. The availability of water defines the functional spaces, plant guilds, and their yields. Plan for multiple spaces to store water and organic material. Bank any resources that collect within the property.

Allow for movement of plants as they grow and as you accelerate the succession of the ecological system. Encourage unrestricted-growth space in order to increase harvest and ecological services. Ecological designs move and grow compared with conventional designs that are held in stasis and not allowed to expand.

A functional design has operative and supportive resources. Supportive resources banked in the soil, or on the soil, provide a higher net yield to all features. Ecologically functional plants provide services that support other plants. Deep collections of mulch on the forest floor beneath the plant guild later buffer changes in temperature or precipitation. Allowing for movement of resources across the property by human beings or nature diminishes the need to import outside resources, thus making the system more self-reliant. Ideally, a plant guild needs no imported resources and relies only on naturally supplied sun and rain. All other resources are cycled within the system of the guild. This is the goal.

Make It Maintainable

Like a forest ecosystem, the plant guild requires very little management. Human occupants vacate well-developed ecological systems for extended periods of time. These systems do not require frequent maintenance and can be harvested when humans return. This can be achieved through the passive and natural cycling of all resources within the property, which creates an increasingly self-supporting system.

As previously discussed, maintenance is harvest, but there are times when initial plantings are in their establishment phase and need some care. The goal is limited or no maintenance in the landscape. Prior to that there may be such tasks as dealing with drip irrigation for new plant systems while they establish themselves. Any mechanical gadget will need to be maintained on the property. Technology of any kind will at some point break down if it's not maintained. Well-designed and -implemented ecological systems are meant to be self-maintaining while harvested by humans or livestock, whether domestic or wild.

Access paths may need some maintenance if erosion occurs or heavy use (depending upon the frequency and equipment) causes them to break down. Make sure turf and borders around building foundations and perimeters are wide enough for equipment. A twenty-six-inch lawn mower may be the widest piece of equipment you need besides a wheelbarrow. In other cases a garden tractor may be used, which requires forty-eight to sixty inches.

Maintainable

Work toward limited or no maintenance:

- Access paths with usable width.
- Site defines usable width.
- Turf and borders wide enough for equipment.
- Multiple access routes.
- Space allowed for raw material and harvest.
- Mulching accessible.
- Slopes manageable by needs.

In order to efficiently harvest and/or maintain a property with mulch or compost, you need spaces to store the raw material before you place it in the landscape or after you harvest from growing areas. Pruning (harvest) and mulching around trees and shrubs require access to be maintained, as do seasonal plantings. Since the topography of the landscape may change, the slopes must be suitably planted and managed to prohibit erosion and take advantage of the aspect toward the sun. Steep slopes may not allow for any machinery access and only offer limited human access. These sites are thus limited to ecologically sound and appropriate plant species that are easily harvested relative to the difficulty and the fragility of the landscape.

As you work with nature on the property, sooner or later you will need access to each area for harvest or enhancement. Allowing for movement of equipment and material within the design spaces reduces the prospect of damage to those spaces and reduces the calories required to do the work.

Make It Environmentally Sound

The ultimate goal for ecological design is that it is environmentally sound, meaning it is self-supporting through the ecological systems and functions that have been included and implemented. It's important to keep Permaculture principles in mind when you make decisions in the design and in the materials you used to implement that design.

Make It Cost-Effective

Cost-effective design in ecological systems specifically refers to the yield and the harvest. The initial investment in an ecological system must pay out in dividends of nutrients to sustain itself. Nature does not skimp on resources; nor does it use cheap materials. Nature has high demands and takes little pity on plants or animals that are insufficiently prepared. During the implementation phase of ecological

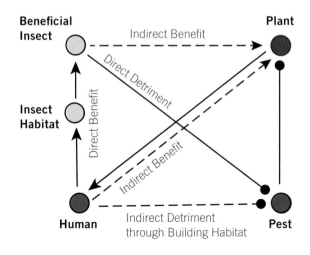

Place components in relative positions to build direct and indirect relationships. Let the plants do the work, giving nature an opportunity to respond.

design, quality materials and quality design are crucial; be sure to use healthy plants, viable seed, and rich soils (or develop these yourself).

To be cost-effective means to have an over-yielding of calories available from the final system. This requires simplicity in design and implementation. Hardscapes, or anything that needs to be engineered, must be well placed relative to all other components on the landscape. Hardscapes are expensive and difficult to move. Mechanical and technological components of the landscapes are expensive and easily reduced to rubbish. Working in the zone system[4] of placing materials and developing the resources from the center of activity to the outer regions is a small and slow solution that allows us to adapt to changes and make corrections. This is cost-effective in an ecological sense.

Make It Visually Pleasing

Visually pleasing as a phrase is loaded with interpretation. As mentioned previously, it is nearly completely culturally based. In the landscape there will be hard edges, usually around the built

Cost Effectiveness

- Quality design, plants, and materials.
- Low maintenance, self-maintaining.
- Simple in design and implementation.
- Hardscapes engineered and well placed.
- Small, staged, and slow solutions.
- Dedicated use of zones.
- Harvests of natural capital for all life.

Visually Pleasing

- Hard edges are softened with key plants.
- Curved edges are used for all spaces.
- Plants are placed in diverse patches.
- Sight-line "views" are maintained.
- Curved concept lines cross over hardscapes.
- Specimen and anchor plants are supported by understory plants and insectary islands.

environment. You can soften these edges the way it's done in conventional landscaping: with key plants. Key plants may be single plants, or a trio of plants placed to soften hard edges and add roundness. This is built into the structure of the landscape to hide many of the human-made elements or at least obscure them. Curved edges are used for most spaces. Avoid straight lines, since they add a mechanical and unnatural form. Plants are placed in diverse patches surrounded by various degrees of open space. This helps identify the different patches and makes them recognizable as people move about the property. These patches may also be similar to those found in nature by featuring similar species, thereby adding continuity in the landscape; think of an oak savanna or a mountainside of berry patches.

For security and to also observe the landscape, maintain sight lines. Vantage points from the windows of the living spaces or work spaces are delineated so that the occupants can observe activity in the landscape and enjoy the view. If there is a perspective to be maintained, it would be from these points. Another concept used from conventional landscaping is allowing the concept lines, for plant patches or planting beds, to cross over hardscapes. If a long curvilinear planting bed goes across a sidewalk or driveway, it continues on the

Aesthetics and the New Perennial Culture

Aesthetics: *a branch of philosophy dealing with the nature of beauty, art, and taste and with the creation and appreciation of beauty.*

— MERRIAM-WEBSTER'S
11TH COLLEGIATE DICTIONARY

Since the Permaculture design process, or any design process, is influenced by the designer's personal values, we need to assess our aesthetic defaults, critically examining and reevaluating why we act the way we do and how we make decisions and choices. We see more with our eyes through the aesthetic filter of our culture than through our innate organic vision. We appreciate great beauty in the natural world that we do not incorporate in controlled living spaces because we have learned that maintaining a separation between that world and personal space defines us as "superior" or "civilized." How we perceive beauty, which is a learned cultural aesthetic, causes conflicts and stress in the Permaculture design process. In serious ecological design, I believe, we need to redirect our vision to the functional aesthetic of natural systems. In the design process we must follow all the steps in the scale of permanence and apply the design process in a manner that redefines the personal aesthetic to a more natural expectation. This is done by using a structure of decision making, from large-scale patterns to details. Following an organic design process allows us to discover design solutions rather than impose them.

In order to do ecological design in a sustainable manner, we follow a functional aesthetic as nature does. What we perceive as beauty in nature is the functional aesthetic that is the result of billions of choices living organisms make in the name of efficiency and fitness. Those patterns that repeat in the vision show us the most efficient use of space and of resources, transportation of resources, and resilience. An organism's fitness is directly related to its ability to thrive in its niche. Each cell, each genetic mutation either increases or decreases the fitness of the affected organism. The simplicity of design for the most efficient use of space, resources, and structural resilience must also be sufficient to allow the system to survive changes in a niche. These changes in the world might be climatic, geological, or meteorological, or they might arise through competition with other systems.

When we as Permaculture designers begin to read the land, we want to make sure that we are being imprinted by the land so that the design is not imprinted (forced) on it. That is why the site-assessment process cannot be motivated by extraction of resources. We are only looking for available resources that we may enhance, restore, and integrate in a design that will increase the fitness of the land and its ability to buffer extreme events (which may deplete its resources). Within the boundaries of the property for which we are making a master plan, we are the new genetic code. We are like a virus in its most positive sense. We as ecological designers can supply a "new genetic code" bringing increased resilience to a property. We can help the land restore itself to natural fertility. We can assist the land in developing deeper

THE STRUCTURE OF A PLANT GUILD

> and richer organic material on its horizon. As stated in many Permaculture articles, we accelerate succession. We must see ourselves as stewards of the land, not its masters. By increasing the ecological services available to natural systems, we create increased resources for ourselves. We are the primary livestock in this natural system, yet also the stewards of this land. We know that if we were to vacate the property after the implementation, the natural systems would be left more resilient, deeper in organic materials, and at a higher state of natural restoration. There would be increased diversity of flora and fauna and increased levels of complexity in the ecology.
>
> The biggest challenge we have in ecological design is refraining from imprinting our personal, cultural, or economic aesthetic on natural systems that already exist. We ourselves are imprinted by our culture to create personal space that follows the cultural aesthetic. It isn't hard to imagine the design of a house built by a consumerist suburban American. It's also not hard to imagine how that design would change based on the native origin of the designer. That subconscious design driver, which infiltrates our ecological design, is most likely contrary to natural ecological design. This is the predetermined aesthetic that is subtly incorporated in the concepts. As we walk an undisturbed property, we can see the natural aesthetic of ecological systems, mostly in what we would call Permaculture zone 5. The natural aesthetic or beauty is the functional design of natural systems. As we move down the zones to zones 4, 3, 2, and 1, we can see the cultural design choices appear and begin to negatively affect ecological services. The greatest challenge we have is trying to adapt the predetermined economic and cultural design drivers (defaults) to the goal of a natural design (intention) based on enhancing ecological services rather than our economic and lifestyle preferences.
>
> — DANIEL HALSEY

other side as if it were never interrupted. This adds continuity and is also a familiar pattern in nature. It's as if the planting bed existed before the sidewalk, which is a disturbance. Erosion, a new creekbed, or even a falling tree sometimes splits large patches of planting beds in nature. The concept lines continue across any of these disturbances.

On any landscape, for as much homogeneity as there may seem to be, there are specimen and accent plants — the singular plants that pop up in a landscape seemingly all by themselves. They grew in a specific, open niche thanks to an opportunistic seed or root that fell there. Conditions were right for that single plant to appear. Specimen and accent plants may create a specialized niche that can only be occupied by a certain cadre of plants and provide shelter for a select species of birds. Following nature's patterns is an opportunity for diversity and creativity as opposed to symmetrical formulas or linear design. We will address each of these responses as we progress through the design process.

Permaculture Principles to Apply to Guild Design

The Permaculture principles of David Holmgren can be applied to design. This can help you make decisions about each functional space and plant system or guild.

Observe and Interact: Site Assessment

It takes time and interaction to wholly observe a site. The site-assessment process is part of this interactive observation. While assessing the property we are also assessing ourselves and learning about the components of each. Learn what each component on the site needs and what it supplies.

Catch and Store: Microclimates, Swales, and Harvests

The principle of catching and storing—in natural or human-made ways—applies to energy and natural resources. Many areas with microclimates might already exist on a property and can be used to extend the list of available plant species for plant guilds. Environmentally sound landscape design collects energy in all its forms and stores it for later use. All animals and plants do this. There is no waste—or rather, all waste is food and is cycled to the next trophic level where compounds are broken down to nutrients.

Food, Fodder, Fiber, F(ph)armaceuticals, Fun, and Soil Fertility

In order for a system to sustain itself it must have a yield of cycled and supportive resources. It must yield resources to the ecology around itself. For doing this, the feedback from the surrounding ecology includes the resources for the system to sustain itself. This may be a plant system or an animal. Each plant patch and each component of the ecological design must have a yield for nature to sustain itself and over-yield to sustain the human occupants so that resources will not be depleted.

Apply Self-Regulation and Accept Feedback: Take Notes, Photos, and Friends

Self-regulation is the ability to limit consumption, ensuring that resources will be consumed and cycled without excessive depletion. It allows for resources to renew at a natural rate and requires that consumers adapt to that rate, as that is the carrying capacity of the environment. Limiting consumption ensures a sustained supply of resources. Conservation measures immediately return an extended supply of resources. Reducing consumption extends this availability. It reduces the pressure on ecological systems and builds natural capital by the mere presence of excess organic material.

Accepting feedback with all our senses gives us information to make good decisions and to adapt to changing environments. All feedback is good. All feedback motivates change toward improved systems or corrective measures. Take notes, take photographs, and consider the observations of other individuals. Recording feedback over time helps you recognize trends and patterns in nature and in yourself. Some patterns take a very long time to establish or become visible. Keeping good records is a form of self-regulation that lets you accept feedback from nature and others. Starting with a little sketch on scrap paper or plant lists on a detailed spreadsheet, make a note to consider the feedback next time.

THE STRUCTURE OF A PLANT GUILD

Use and Value Renewable Resources and Services

A tendency to use ecological solutions reduces the need for external resources over time. Allowing renewable resources and services to develop on a property diminishes the inputs required. Technological solutions require the purchase of finished materials or equipment that will need increasingly more frequent maintenance and repair until they ultimately fail. Most equipment and technology puts toxic waste into the environment during its production, increases toxic waste while it is being used, and results in toxic waste upon its disposal.

Long-term ecological solutions can also be mitigated by cultural change. Changing default behavior can reduce many ecological problems and the damage they may incur. The three major solution categories employed are cultural change for an immediate benefit, ecological solutions that will build natural capital and restore diminished systems, and technological solutions that externalize their existence on distant environments and future disposal. Permaculture focuses on cultural change and ecological solutions, although it may use appropriate technology for certain applications as well as for observation, record keeping, and the facilitation of learning.

Produce No Waste: Compost and Cycle Nutrients

In ecological systems, some organism or other breaks down all natural compounds. The trophic levels in nature cascade the compounds and organic materials down through mega- and microfauna, in and above the soil. Each time the organic material is consumed it is broken down to nutrients, or simpler compounds. In nature there is no waste, there is only food, until the final product excreted by bacteria or fungi becomes the nutrients that are now available to plants.

In ecological systems, all the trophic levels occupy the plant niches so the organic material can be reduced to the nutrients required by plants. When organic material is removed from a plant niche, it must be returned sooner or later, through some mechanism. This is what makes a system self-sustaining. Livestock animals do this as they clean up the brush or fallen fruit from the harvest systems and then deliver the manure as fertilizer across the landscape. Worms do this in the ground and in container systems as we add excess organic material from the human living system to theirs. They break down the material using the bacteria in their guts; the resulting nutrients are then absorbed by the digestive system or excreted for other organisms and plants. A diverse selection of integrated wildlife processing organic material in a forest garden increases the availability of plant and soil nutrients.

Design from Patterns to Details: Use, Access, Patch, and Plant

In the landscape there are things that cannot be changed or that are slow to change, like climate and landform. These factors dictate much of what can and cannot be done in Permaculture. Weather extremes, cold or hot, humid or dry, define the types of plants that will survive. These patterns, being seasonal or daily, guide overall decisions when we're building an ecological system. These are large patterns that are for the most part predictable, including patterns of sun and shade, water movement, and soil type. These patterns guide us in decision making and in the devising of functional spaces. Working from these large patterns helps us make good broad-scale decisions that are then followed by more detailed decisions — for instance, plant species. The types of plants usable in the landscape may encompass a broad array of species, but do not select the species until you know which plant types will be used, and associated to the climate and landform. Thus, work from patterns to details. As mentioned previously, the design process includes access in the design space, a very large pattern that needs to be decided early on. We also learned about water movement and catchments, another large pattern that should be clarified early. Then we can move on to the details of ecological functions, the ecological roles of the plants, the goals that may be attained, and finally the plants that will play those roles and fulfill those goals in the guild. Start with the big patterns and then develop the little ones.

Integrate Rather than Segregate: Use Diversity and Plant Guilds

In many parts of our lives we segregate the components and isolate specific systems from other systems. Often the only thing that connects the systems in our lives is the electrical outlet into which they are all plugged. It is unfortunate that many of the living systems we use are not integrated, which would save space, materials, and energy. In Permaculture redundant systems are integrated. We want to make sure that all needs are being met from multiple sources and that those sources have multiple benefits in a redundant system. Nature has many redundant systems, many sources for one type of nutrient plants need, many types of organisms that will decompose organic material, both of which are interrelated and integrated into the ecological system. If one part of the system is under stress, another part can take up the slack. Each of these components is set in a relative location.

Relative location describes how important it is to place things in the landscape so that they can have the benefit of one another's ecological services. It also refers to placing structures near their resource base and where their waste can be used as food or compost, or cycled into an ecological system — for example, setting nitrogen-fixing plants near other plants that are heavy feeders on nitrogen. Or placing a chicken coop near the garden and the house, so that the birds can be fed scraps from the kitchen and their scraps can be fed to the garden. It's similar to a chess game where all pieces need to be placed in their most efficient position to benefit the other pieces. Anything harvested needs a place to be metabolized and returned to the system. This is integration.

Use Small and Slow Solutions: One Patch at a Time

Small and slow solutions are less likely to cause damage if there is an error. They also provide an opportunity to learn the principles and processes within a solution. In any landscape, dividing up the functional spaces into smaller spaces helps in implementation and does not require an initial large investment. It also allows for contemplation and the ability to develop skills along the way. Implementing a design over a year or longer reduces stress and lets you gather materials when they are at their cheapest. Buying everything possible on sale in the off season saves money.

There are all kinds of options. For example, you might create a small apple orchard of thirty to forty trees by preparing the soil, digging the holes, and planting all the trees in a matter of days, having purchased them at a local nursery for thirty to fifty dollars each. On the other hand, you might start with small guilds as a first step, planting five trees off season for about seventeen dollars and planning for a small grouping with the associated polyculture guild plants. In the meantime the larger orchards can be planted with rootstock in small holes in fairly short order. All trees are allowed to establish themselves over a season, and then the mature trees are pruned for scions to graft onto the rootstock. Using this technique the majority of the trees cost about seventy-three cents and have an opportunity to acclimate to the property much more quickly than the potted trees, even though they are much smaller. Within five years they will reach the height and productivity of the others. This is a small and slow solution, a patient result with little loss of production over time. Gardens, aquaponics, livestock systems, cheesemaking, and rainwater systems are similar: Start small and build them gradually as time allows and skill develops.

Use and Value Diversity: Remove Bias and Explore

We all have our favorite plants. We have our favorite resources and things that we tend to gravitate toward: comfort food, say, or a favorite sentimental blanket. Of course, things change and many times those resources no longer provide us with everything we need. If we are to enjoy a diverse diet, we need diverse sources of information to guide us in making good decisions. Valuing diversity in all things helps build resilience and buffer the extremes. A broad, diverse selection of plants on the landscape will offer options when drought hits or floods occur. Diverse plant species also feed a diverse group of insects, which then feeds a diverse group of birds.

Exploring new options in materials and new ways of thinking keeps systems moving forward, developing new solutions. Limiting ourselves to a select few options is very toxic. Much as if you were to eat only one specific food, or plant, things suffer immediately and slowly as the system begins to collapse.

Value the Marginal: Create Edge and Use Edge

An edge is where the transition takes place between two different ecological spaces. It can be as thin as a hair or a hundred yards wide. This is where the ocean meets the shore, where the prairie meets the woodland, or where a meandering creek snakes along its way with an edge on both shores that influences the plants and animals nearby. If we can create more edges, these transition zones collect energy and organic material. A forest edge slows the wind, collects dust and nutrients from the air, causes more rain to fall in that space, and collects and reflects light. Microclimates happen in the edges; these spaces are teeming with diversity of life multiplied by the organisms and plants from either side.

A single pond of one acre in a circle may have a thousand feet of shoreline. Taking that same area and making it the shape of a star, or dividing that pond into six ponds that equal the same area, multiplies the total amount of edge and the availability of resources in that niche. Even aquatic plants and fish benefit from the shallows and the transition areas of the shoreline. More shoreline means more feeding area and shelter, more area to spawn, and more interaction with the surrounding landscape. The small ponds have the same volume and can be supplied with water from the same source as one large pond. On a fence line, an edge between two fields, resting birds leave a huge amount of nutrients. Snowdrifts build up in the winter, depositing extra moisture to the soil below. Wind erosion of silt is slowed in the edges, which then gradually build up rich nutrients and fertile soils. Edges are good.

Margins are like edges. They are spaces ignored. Marginalized land is left alone, allowing it to develop its own soil ecology and resource cycling. It's a place where wildlife knows it is protected and has refuge. It is usually undisturbed, though it perhaps may have a few piles of rubbish or an occasional abandoned refrigerator. But for the most part things that are marginalized are left to fend for themselves and develop their own resources.

Marginalized people have done this for thousands of years and develop their own culture, diet, customs, and artwork. Marginalized individuals in society develop freethinking, unlimited by social pressures and academic controls. Jazz, a marginalized form of music, has been cited as a source of inspiration for many pop culture musicians. In the same way, resource materials that are left undervalued in the margins are the best resource for Permaculture and ecological design. Marginal land may contain the most valuable crops that can be grown in well-managed systems with sustainable practices. Marginally profitable plants ignored by nurseries and growers become hard to find and thus highly valued by the few that require them. Even the seasons have margins, sometimes called shoulder seasons; these transition times are valued for the small windows of moderate weather they provide to make changes prior to the new set of extremes. Missing, ignoring, and devaluing margins wastes resources otherwise integrated into the forest system.

Creatively Use and Respond to Change: Use Adversity

Change is a matter of fact and life. Just as things get comfortable, they change. Expect constant change — fast and slow — in everything from sunlight to temperature. Do not be afraid of change. Prepare for it. Take advantage of it. Changes are an opportunity to improve. The shift of energy can be used to create new energy and a new direction. In an ecological frame of mind, look forward to change — especially seasonal change, without which you would have little opportunity to improve systems or enjoy any harvest at all. Plants, animals, and trees base their lives on the living systems around them that have patterns of change; they respond to these changes and in many ways benefit.

After eons of time and repeated patterns of change, natural systems depend on cyclical changes. The ecology has come to depend on the changes that happen over seasons and cycles. Ecological functions depend on changes in sunlight and temperature. For them adversity may not be enjoyed, but for us it's an opportunity to reflect on a system's efficiency while there is an opportunity or a requirement that can be improved. Each time there is unexpected change, it is an opportunity for us to respond creatively and find the weaknesses in our systems. It is also an opportunity to test a new integration where one system might benefit or buffer change for another.

Many times creatively responding to change immediately puts us on the edge of culture and friendships. People who ride the edge are marginalized until others accept the new reality and the new adaptation. Those with diverse points of view who are willing to make creative changes in their living systems and risk loss of status are crucial pioneers. How else can solutions be adapted unless early adopters take the opportunity?

Yeomans's Scale of Relative Permanence

In the early 1950s P. A. Yeomans published a book called *Water for Every Farm*. In it he described his scale of permanence and its use in design. While working to save water for livestock on his vast acreage, he found there were things he could not change that required adaptation and other things he could change that required thoughtful planning. The scale of permanence starts with the overriding and mostly stable components of the environment and cascades to components that are more likely to change over time. Climate is the first, and it's somewhat predictable on an annual scale, whereas soil, the last item, changes by the hour as weather

THE STRUCTURE OF A PLANT GUILD

The Progression from Patterns to Details

The organic design process is as follows:

1. Observation (understanding)
2. Data collection
3. Site assessment
4. Design drivers and limiting factors
5. Human needs and resources
6. Site resource assessment
7. Systems and concepts (resource cycling)
8. Space uses and plant patches
9. Layers, plant selection, and polycultures
10. Anchor plant placement
11. Supporting plant placement
12. Ground cover

---

conditions and the functions of plants and animals modify its components of water, gases, temperature, and organic material. Imagine as you read the following list trying to force your imprint of the design on the landscape, rather than working with its momentum.

P. A. YEOMANS'S SCALE OF PERMANENCE

- Climate
- Land shape
- Water
- Roads (access and circulation)
- Plants, animals, and wildlife
- Buildings (the waste stream, energy usage)
- Fences and boundaries (zones of use)
- Soil

Climate

Climate is at the top of the scale of permanence. No matter where we are, we are embedded in climate. It is what dictates to us our growing season, how we configure our homes, and how we dress. Climate can be a constraint, or we can learn to work within these constraints and develop systems that are not limited by the patterns of weather, day-to-day, or by the macroclimate that dictates how the seasons play out, year-to-year.

Land Shape

Landform — the contours and the slope of the land's surfaces — is one of the more permanent structures and resources in an ecological landscape. It is its foundation. Most of the dynamics of a landscape begin and continue due to the shape of the land, the way it collects water, and the different aspects toward the sun that it may have. These things are difficult to change overall, although small earthworks projects can enhance or take advantage of the dynamics of the existing topography. Water may always continue to flow in a certain direction on the landscape, but if you can slow it down it will soak into the soil. The scale of change in landform, short of removing mountaintops as is done for mining operations, is beyond the scope and practicality of most ecological design. Use the natural momentum of the topography to collect and distribute organic and nutritive materials across the landscape. Given all the things that do change — many with some unpredictability — it is comforting to know that in most cases the landform will not.

Water

Water usually comes from a somewhat timed and predictable source, sometimes from a spring or an artesian well in a consistent supply. Water is a part of climate, and the collection of water is part of landforms, so it's fairly unchanging and easy to find on the landscape.

Roads (Access and Circulation)

Access in the infrastructure to move materials on the landscape is somewhat set relatively early on in the design process. This must be decided based on contours of the property and the different large-scale production zones that might exist. Paths and roads that make sense might already exist as well. Ridges and stable sections of land may be the only place to put a road or large access. A road should not cover the best, fertile soils. Wetlands are totally impractical for access. In the design process, and in the scale of permanence, roads and access are fixed, long-term components. The soil becomes compacted over time as gravel and other materials are applied, and the area becomes unusable for plant life.

Plants, Animals, and Wildlife

Trees are the long-term overstory, and once established can last many years, or even centuries. They provide a huge ecological service to the landscape and the surrounding ecology. Their long-term occupation of a niche makes them increasingly valuable as a resource, not to be disturbed. This includes the surrounding landscape within and beyond the drip line. Existing trees may be underplanted to enhance their health, but they are a design driver. This means that the needs of all the elements within the area of the tree are subordinate to the tree's needs, and must be selected carefully so that they do not compete with it.

Buildings (the Waste Stream, Energy Usage)

The structures on a property, once built, are generally of high value and unlikely to move for the purposes of landscape design. Especially when concrete foundations and structural engineering have been used, buildings are permanent fixtures that will most likely remain and/or be replaced in the future. Although some trees may not outlast a building, after it's built it is generally seen as subordinate to the existing trees. The trees themselves provide protection and cooling to the building, and possible materials for the future.

Fences and Boundaries (Zones of Use)

Perimeter boundaries, fences set in stone, wire, wood, or living hedges, establish the production zones. This built structure, for the most part, is long-term and is used as a guide for containing livestock or excluding wildlife from planting zones. The perimeter fence around a property is usually the strongest and longest-term, and since it often designates the property line, it is fairly permanent.

Soil

It would seem that soil would be part of the landform, but this thin surface atop the land changes constantly and is the most fragile of our ecological assets. It is easily eroded by wind and by rain if unprotected by plants. It freezes solid in the winter and dries out, in some cases to powder, in the summer. The nutrients in the soil change, based on their use by the plants above and the available moisture.

Organisms in the soil fluctuate greatly depending upon the season and the organic matter available. Soil types are also extremely variable. Thousands of combinations of soil types exist; defining these types makes it easier to select plants suitable for a niche. Although soil can be moved and can change in consistency, its pH is relatively stable since it comes from the base material below in the bedrock that has mixed with the organic material from above. The pH of the soil specifies which plants will thrive in the ecology and which organisms will be present. When working with plant lists and developing our plant guild polycultures, soil is the most significant influence on the plants we select. It is also the most fragile component, the most easily disturbed by our actions, yet the most adaptable for our needs.

The ever-present scale of permanence is a reminder of the larger patterns that shape the smaller ones. Though listed last, soil is the ultimate living feature, created by the climate and eroded bedrock. Over millennia, plants and animals came from and passed over the soil only to become soil once again. As it deepened the plants grew taller, animals more numerous, and species more diverse. Depletion of resources is the only factor limiting soil health. Polyculture designs that build soil and natural habitat also build diverse and resilient communities for abundant living.

Constructing the Plant Guild

Although we have given many plant guild examples in this book, the reality is that there are no cookie-cutter guilds. The examples presented simply operate as templates for further modification. As described throughout this book, guilds are subject to climate and topographic particulars. Soils, animal and insect interactions, food, utility needs and choices, social and cultural intricacies — all play significant roles in the construction of specific planting strategies.

As you create plant guilds, you may experience a number of paradigm shifts. Visualizing the landscape and understanding ecological systems at work requires a perspective of integration. Let's say, for example, that right now you're looking at a typical suburban lot designed with numerous plant types filling each ecological niche. You might look at this property and see a conventional landscape — one filled with plants from a local nursery, plants that have been developed and bred for aesthetics. Each plant symbol — the circles in a design — could be an ornamental tree, shrub, or perennial plant that is visually pleasing without any human harvest or ecological function.

As you begin to design your homestead or landscape, look at the plant circles differently. Each one of these plants must either provide a harvest for human uses — food, fiber, medicinals, fuel, or refuge — or at very least provide an ecological service such as nitrogen fixing or nutrient accumulation.

Now, the design of this landscape provides much more than a static matrix of colorful plants and seasonal interest. It provides for ecological roles that sustain the entire system of nutrient cycling among all the participating organisms. The plant guilds selected for this property sustain the human occupants and reduce their dependence on purchased resources. Each group of plants provides the human livestock (and participating wildlife) vitamins, nutrients, and minerals. The maintenance on this landscape, as opposed to conventional landscaping, is now harvesting. A yield is obtained from every plant. Each plant performs many functions and contributes to the redundant systems within a sustainable landscape.

Through our design comes the discovery of the many benefits of integrated ecological design, and new insight in developing solutions for the landscape. When designing a guild, stacking functions is one of the primary directives of a Permaculture system. Furthermore, stacking functions occurs not only in a physical space, but in a temporal one (through time) as well. In order to produce food, medicine, and utility year-round, pay specific attention to time as well as space.

Think of the guild structure as being on a grid. Looking down at a grid model, we clearly delineate vertical as well as horizontal direction and niches to insert plants specific to the local region. Looking vertically at a side view from deep in the soil horizons (layers) and moving skyward into daylight, extend your purview all the way up to the top of the canopy where climax species reside. Every level in between root zone and canopy is a niche for the stacking of plants. Looking horizontally shows you opportunities to insert plants lateral to a central focus.

From the perspective of stacking in time, the opportunity arises to select species that will produce yields across the entire year. By way of example, let's freeze time into a spatial construct in our minds. Observe time in a kind of physical manner: lettuce in spring, giving way to zucchini, giving way to four varieties of apples that mature over four months, giving way to parsnips in autumn, giving way to sprouts over the winter months. This "time guild" affords the chance to put fresh food on the table year-round. This is partitioning the resources in time.

Let's take this a step farther. Based on the site's geographic location, climate, topography, altitude, local ecosystem services, and habitat, select multifunctional plant species for the guilds. Think function, process, and action, rather than of the physical plant itself. This is how to create the relationships common to a Permaculture plant system. The goal is an active and integrated collection of selected species that create high yields ecologically and agriculturally.

You begin with a blank canvas, so to speak. The frame of the canvas acts as perimeter within which you can design and plot the guilds. Of course, those of us who have gardening and farming experience know that any perimeter drawn becomes rather fuzzy at the edges. There is no sharp delineation between one plant guild and the next. Nesting several guilds within the perimeter of a property creates a fully populated food forest.

As you configure and divide the canvas, whether blank or with some plants already present, proceed from both a top view (looking from above onto the property) and by way of elevation. Setting a grid down on the canvas gives a framework to work from, a physical framework to arrange the space of the guild. Insert functional species within the grid framework then proceed step by step to realize what the group of plants will look like physically and aesthetically, paying special attention to how all the plants of related species will survive, share resources, and react over time.

The mind map on the following page will help us organize and define the parameters in terms of plant selection.

The top plan grid will depict how the plant organization lays out horizontally on the ground. This is important for proper spacing in order to maximize sunlight and to allow for future growth in terms of proper sizing.

The elevation can be thought of in this way. Just as algebra has y- and x-axes plotted on a grid, plot the x-axis and make this the ground level at grade. Below this is the root zone; above the x-axis is all visible biomass. Climbing the ladder of the y-axis up and down the cumulative grid blocks gives a visual representation of the many niche spaces inserted for the selected species.

In the root zone there are flat-rooted, medium-rooted, and taprooted plants. Each of these root systems penetrates to a different layer of the

THE STRUCTURE OF A PLANT GUILD

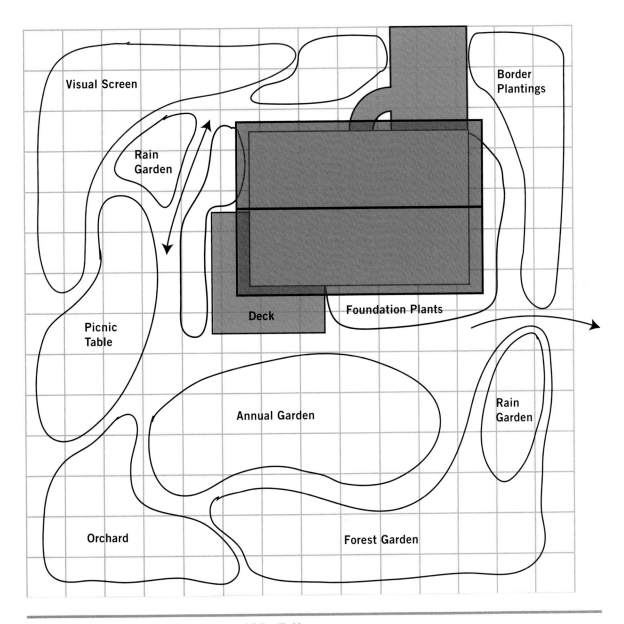

Grid template. Illustration by Wayne Weiseman and Kellen Kirchberg

soil horizon and seeks nutrients from levels that do not interfere with other roots in the vicinity.

Ground covers, herbaceous and shrub layers, small and large trees, and vines all afford us the opportunity to plug each gap in the matrix climbing above the forest floor. Doing this increases the possible actions and yields — which are theoretically unlimited. As Bill Mollison states, "The only limits to yield are the limitations in the designer's imagination."[5]

As a guide in creating guilds, plot a grid on a sheet of paper and work from there. Place plants so that their ecological function is shared. Since the landscape will be in continual change and expansion, place plants for growth as the niche changes.

INTEGRATED FOREST GARDENING

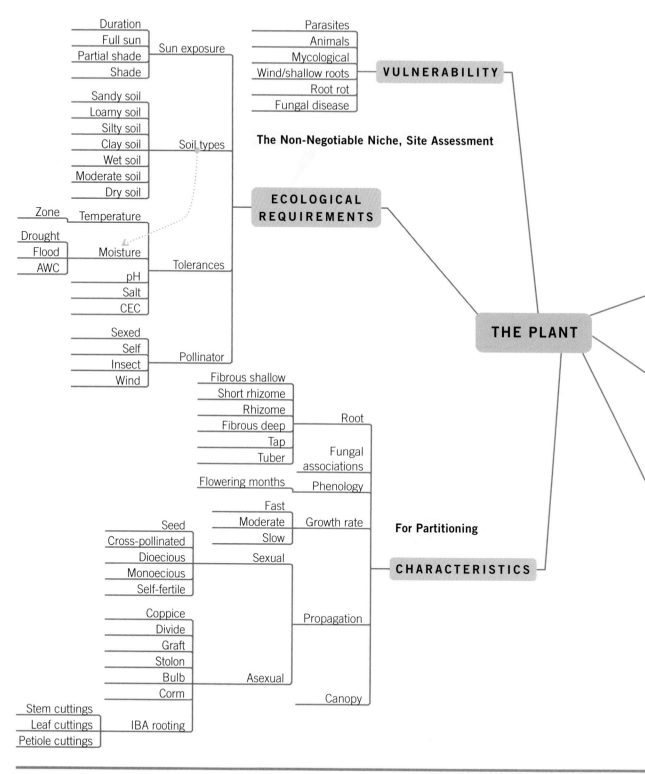

The plant mind map.

THE STRUCTURE OF A PLANT GUILD

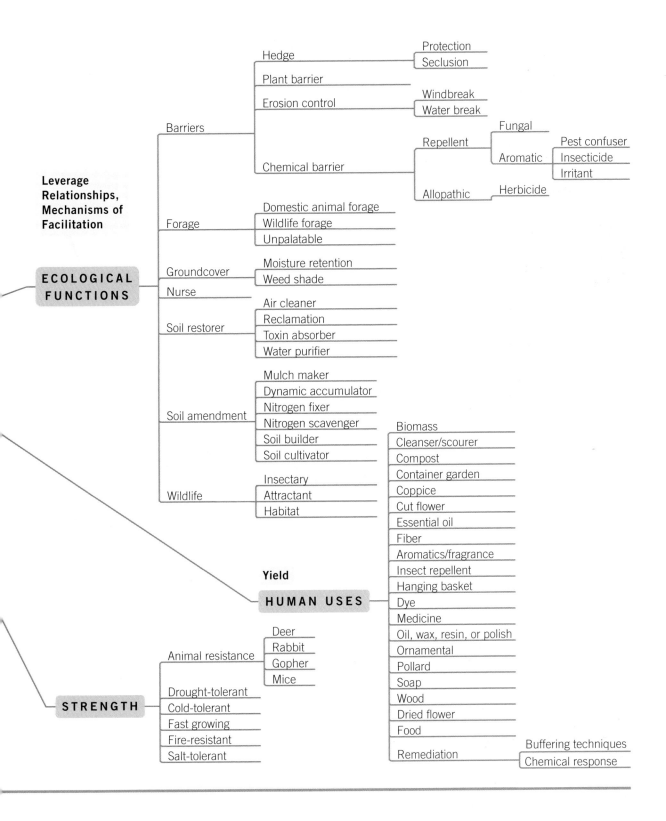

INTEGRATED FOREST GARDENING

Fast-growing and shading plants might be best in the rear, away from the sun so that as they grow they do not interfere with other plants' sunlight. For example, in the eastern woodlands biome, a forest environment, the ecosystem "climax" is a closed canopy of large deciduous and evergreen trees. An important aspect of designing for this is dynamically staging the guilds for various combinations of maturity on a scale from the smaller pioneering species to canopy plants. In terms of harvest, species differ greatly by when they are harvested; access to harvest each set of final fruits is needed. Make room for the human livestock to roam. The intermingling of time and space, form, and function in the plant realm is vast.

Structuring a Plant Guild: A Step-by-Step Process

As we create plant guilds, each niche is a functional space, an opportunity to plug in countless plants for high yields. Think of plant guild infrastructure as an outline yet to be filled with text. Each word and paragraph that we choose represents the idea that we wish to express. The guild is a combination of diverse phrases working together to form a story, an integrated and dynamic combination of plants, all nested together in space and time.

Analyzing niches and understanding seasonal effects are key to developing a plant guild. Seasonal effects must always be taken into consideration. Freezing temperatures and frost arise in autumn and diminish in spring; these limiting factors color design decisions. There are many guiding factors that dictate what and when to plant, as well as what USDA planting zone to use. USDA growing zones are shifting, allowing the planting of species that were formerly only possible in warmer climates. Every plant you select for a niche also needs to fit within the seasonal extremes. An understanding of seasonal and niche requirements is a good place to start in

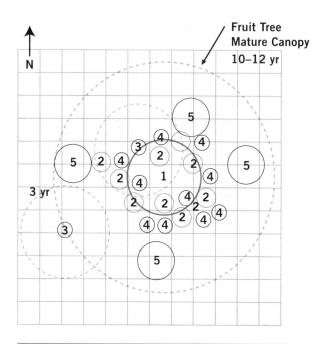
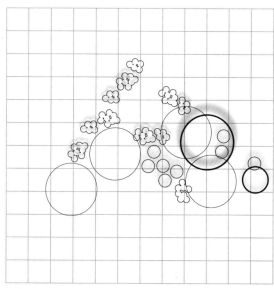

Scaling aerial and elevation. See the appendix for a full-page example.

the selection process. Buffer climatic extremes by creating microclimate niches on a property. Shelterbelts, windbreaks, and piles of stone contribute to season extension and the opportunities afforded us beyond the frost dates.

A niche is defined as a habitat supplying the requirements necessary for the existence of an organism or species, or the ecological role of an organism in a community, especially in regard to food consumption. Remember that plant guild members are in relationship with other plants in the guild space, and each choice you make will augment or diminish the ecological services supplied. It's up to you to research, understand the functions of the plants that you select for your needs, and link them up to the other plants available. For instance, we select nitrogen-fixing species in order to promote growth in the green parts of all guild members and to give an initial push to growth, especially in spring. Building a list of candidates for each function broadens the palette of choices for size and type.

Observe in detail the ecological dynamics of the site once the system is placed on the ground. Plan for succession (implementation, establishment, management, looking at change over a number of years). All layers can be placed at once or as succession develops through the years.

Developing the plant guilds for diversity uses an assorted pattern of plants and animals in a balanced matrix where interactions are as complex as an ecosystem in full bloom.

You can do this by mimicking local ecosystem services (the underlying templates of the system). Observing and exploring the local ecosystem helps you assess and inventory what the historical and a historical makeup of the land is and was. This supplies an infrastructure to plug the designs into. Each guild is like a template of the whole and vice versa. By doing this you are able, more easily, to find opportunistic plants that take advantage of the niches you create with the guilds in the context of local climatic and ecosystemic services.

Again, remember that there are no cookie-cutter guilds. There are only functional spaces within an ecosystem that are aligned with the local conditions. Keeping this in mind makes the design process much easier, since the focus is on only plants that are compatible.

Questions You May Be Asking

All throughout the process of designing a guild, questions will arise, such as:

- How does one plant affect the next, both positively and negatively? Are they compatible?
- What nutrients can the plants supply, individually and as a group?
- What plants will protect or "nurse" other plants?
- What animals will be attracted or repelled by the plants, and how can I use this to maximum effect?
- What is the relationship of the center to the edge of the niche space?
- How does the plant guild being designed affect the rest of the landscape, outside the immediate niche area?
- What human uses are supplied in the plant guild? Food, medicine, utility . . .
- In other words, what is my relationship with the landscape and am I part of the polyculture system? Who will use this plant (human, animal, another plant)?
- Why should I plant this particular cultivar when there are thousands of other choices?
- When do I implement the installation of the plants I have selected? How do I know if my selected plants are not genetically modified?
- Where do I put them? How do I maintain this landscape until it can manage itself?

Backyard Orchard and Design Process

My favorite polyculture designs have been backyard orchard guilds in small spaces, the first being a small corner orchard in the backyard of a suburban landscape design. Within the context of all the trees and shrubs in the yard, this little corner almost went unnoticed until it was apparent that all the plants were fruiting in some way. A small, roundish twenty-by-twenty-foot semicircle was packed with five fruit trees, strawberries, currants, serviceberries, and loads of fruiting ground cover. It became obvious to me many months later that this little corner was extracting huge amounts of nutrients from the soil and not really returning anything. Nitrogen fixers, dynamic accumulators, and plants for organic material were missing. I also found that most of these important and ecological functioning plants were not available at nurseries. It is no small wonder that they do not make it into many landscape designs, since they are relegated to the "native" and "rain garden" lists when offered.

Ecological functions need to be supplied by a majority of the plants in a design. I usually refer to my fruit trees as "diva" plants or anchor plants, since all we are really asking from them is to produce a harvest — and much of the time the other plants are there to support that end. Beneath and alongside those canopy producers are the hundreds of other smaller plants that feed the soil organisms and beneficial insects. In plant selection, pay attention to the major ecological functions that are required for a healthy agro-ecosystem (see chapter 5). There has been much discussion in books and on the Internet over the ideal proportion of ecological functioning plants relative to the canopy crops. Suffice it to say, put as many in as you can. The more nitrogen-fixing plants you have, the more nitrogen will be available for vegetative growth. The more dynamic accumulating taproot plants you have in your design, the more minerals and nutrients will be available for root and fruit production and can be composted or cut for mulch.

IMPORTANT ECOLOGICAL FUNCTIONS

- Nitrogen-fixing and scavenging plants (legumes, trees, shrubs, perennials, and grasses).
- Dynamic nutrient-accumulating plants (perennials such as comfrey, chicory, et cetera).
- Soil-building plants.
- Nectary plants (varieties blooming in all seasons).
- Insect habitat plants (for overwintering and protection).

These considerations are made during the plant-selection phase of the design, but you cannot pick plant species until you know the type of plant you are looking for. Using Robert Hart's seven forest garden layers helps us build the structure of the polyculture and define the plant types in the polyculture niche. I start this process by sketching the area I will be planting and building a small base map. Working with pencil and paper is the first step in the design work. Do not waste your time trying to jump right on the computer.

THE STRUCTURE OF A PLANT GUILD

Here are the steps I use:

- Use grid paper and make accurate measurements for the space.
- Create the base map to scale on the graph paper so that everything is proportional in size.
- Take a clean sheet of vellum or stiff paper and transpose the base map onto it, cleaning up the lines and making sure the shades are accurate.
- Remove the underlying original sketch and tape a new piece of tracing paper over the top of the refined base map.
- Roughly define the functional spaces on the base map. These are usually oblong bubbles curving on contour and/or surrounding existing trees with new planting beds.
- Name the functional spaces for what they will provide (annual garden, herb garden, orchard, compost pile, potting bench, and so on).
- Using a stencil, place large and small circles to represent the plants in the planting beds.
    a. Trees are placed first. Make sure they have room for airflow and sunlight.
    b. If the first trees are large and high-canopy, stencil in the smaller partial-shade trees beneath the canopy. (Some designs will not have a high canopy, and the smaller understory trees will be the top layer.)
    c. Draw in the shrub layer, usually groupings of three, five, seven, or more plants with widths of two to eight feet. Make sure the height of the shrub layer does not interfere with the canopy. Some shrubs are very tall.
    d. Place the perennial plants underneath and around the trees and shrubs.
        i. Perennial plants supply many of the long-term ecological functions.
        ii. Selecting perennials gives you your greatest opportunity for diversity in species.
        iii. Perennial plants can be extremely small or large. Some may be only a few inches tall and have a very narrow spread. Try to fill the landscape around the trees and shrubs with patches of different species of plants; repeat those patches throughout the landscape. Other areas may have a tapestry of many different plants intermixed. Use the stencil and the seven-layer technique to build a gradually sloping height of plants toward the edges of the planting area.
        iv. If the soil is not covered by plants, weeds will eventually do so. Fill that area and understory with perennial and ground cover plants. These can be curvilinear shapes, which will contain dozens of species or a single species.

Using the seven-layer process to develop the structure of your planting area will help you make sure the plants fit correctly around and under the canopy trees and shrubs. Trees first, then shrubs.

Neatness comes after experimentation. Use a few sheets of tracing paper as plants change positions and the design plays with the structure. Take your time and enjoy a game

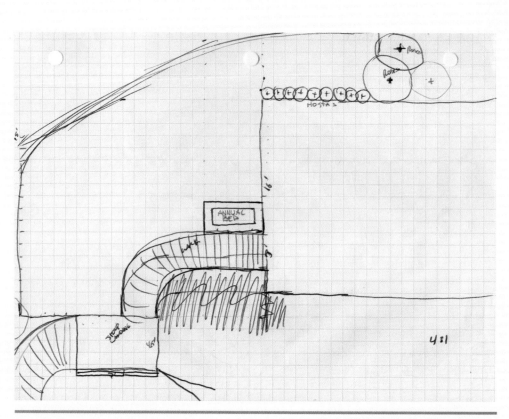

This design was started with a rough sketch based on measurements applied to graph paper.

of "Plant Chess" as you move the trees, shrubs, and perennials to their best advantage for sun and access.

Finally, after the positions of trees, shrubs, perennials, and ground cover are decided, redraw the design onto the final sheet of velum. This document needs to be clean and easy to read. Freehand drawing is not recommended unless you are adding texture and shadows. This is like an engineering plan with precise positions that you measure and transfer to the ground. You will thank yourself later for making neat lines and little visual noise (smudges, erasures, and scribbled notes). Make any notes off to one side, not in the planting area.

The final art has the placement confirmed and plant ID numbers on the symbols. These numbers align with the plant list you'll use to procure materials from the nursery. Your design should include the existing plants that will remain after the installation. Ground cover plants and a broad pattern of beneficials will fill spaces of sun and soil.

THE STRUCTURE OF A PLANT GUILD

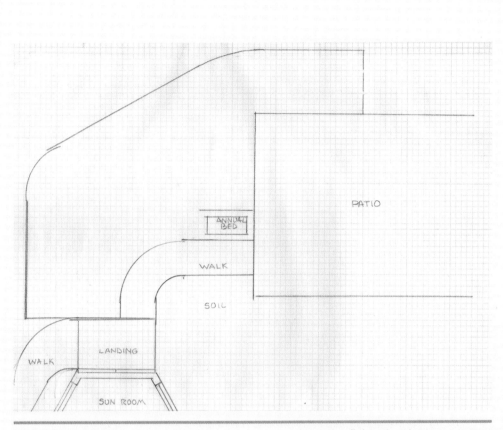

Then I transposed the shapes to vellum, tracing to scale, and refining the shapes to correct measurements. Clean vellum copies will make it easier to work, with little visual noise.

Table 2.2 (page 88) is the plant list for this design.

Building the plant list is more than just listing spatial considerations and food crops. Ecological function needs to be addressed. In the list for the orchard all plants have been defined for their ecological contributions, covering as many functions as possible. This adds diversity of services from many sources.

This design includes many plants for their ecological functions. The plant selection came after the plant types were placed. Although we always have some species of plants in mind while we design, we cannot let them determine the structure of the seven layers or the placement of other plants. Focus on the top-down design stages. Tree types first. Canopy spread is important, but you can estimate the canopy by type — such as dwarf, semi-dwarf, or standard — and by function: shade, windbreak, fruiting. Even though the species tree is not known, fill the space with one or more trees that complete the canopy area. You can pick which species later, and it does not limit you to a particular tree. I chose apples and cherries as my overstory. That leaves three or four feet for the shrubs underneath and to the side. I fill in the lateral spaces with shrubs and large perennials that have functions for the tree. If it's close to the tree or under its canopy, it should provide some benefit. At this point we are thinking size and function, not species.

After the shrubs and larger perennials are circled in, move to the smaller perennials and ground cover. These can be in patches, islands, and groupings. On the ends or side of your planting area is a patch that might be used for beneficial insects. Instead of making innumerable circles, draw in the area to be planted

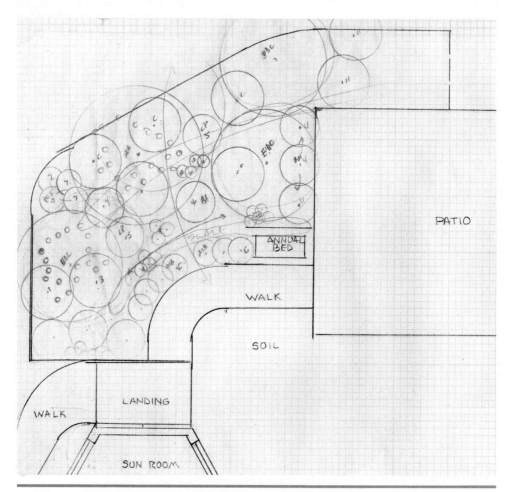

A piece of tracing paper is then used next as the plant positions and types are defined. Each step is done with tracing paper first to keep the process fluid.

THE STRUCTURE OF A PLANT GUILD

Final placed plants on clean vellum eight-scale grid paper.

TABLE 2.2 PLANT LIST

Quantity	ID	Name	Scientific Name	Plant Type	Height	Spread
3	1	Evans Bali	*Prunus cerasus* 'Mesabi'	Deciduous Tree	15'	10'
1	2	Honeycrisp	*Malus pumila*	Deciduous Tree	14'	10'
6	3	Comfrey	*Symphytum officinale*	Perennial	36"	48"
6	4	White Wild Indigo	*Baptisia alba*	Perennial	60"	48"
2	5	Leadplant	*Amorpha canescens*	Deciduous Shrub	4"	3"
7	6	Alpine Strawberry	*Fragaria vesca*	Perennial	12"	12"
8	7	Purple Coneflower	*Echinacea purpurea*	Perennial	36"	12"
3	8	Black Currant	*Ribes nigrum*	Deciduous Shrub	6'	3'
9	9	French Sorrel	*Rumex acetosa*	Perennial	24"	12"
6	10	Chicory	*Cichorium intybus*	Perennial	48"	24"
2	11	Serviceberry	*Amelanchier alnifolia*	Deciduous Shrub	10'	10'
1	12	Summercrisp Pear	*Pyrus communis*	Deciduous Tree	15'	12'
1	15	Groundnut	*Apios americana*	Vine	100"	36"
3	7a	Sky Blue Aster	*Aster azureus*	Perennial	36"	24"
3	7A	New England Aster	*Symphyotrichum novae-angliae*	Perennial	54"	36"
0	Exist	Rugosa Rose	*Rosa rugosa*	Deciduous Shrub	8"	6"
3	GC	Anise Hyssop	*Agastache foeniculum*	Perennial	48"	24"
3	GC	Lupine	*Lupinus* spp.	Perennial	48"	30"
48	GC	Daffodil	*Narcissus* spp.	Perennial	24"	8"
12	GC	Common Yarrow	*Achillea millefolium*	Perennial	36"	24"

THE STRUCTURE OF A PLANT GUILD

TABLE 2.3 ECOLOGICAL AND HUMAN USES

Name	Ecological Function	Human Use/Crop
Evans Bali Cherry	Spring Insectary, Wildlife Food	Food, Fiber, Wood
Honeycrisp Apple	Spring Insectary, Wildlife Food	Food, Fiber, Wood
Comfrey	Chemical Barrier, Domestic Animal Forage, Dynamic Accumulator, Insecticide, Insectary, Mulch Maker, Water Purifier	Biomass, Compost, Food, Medicine
White Wild Indigo	Dynamic Accumulator, Erosion Control, Insectary, Mulch Maker, Nitrogen Fixer, Soil Builder	Cut Flowers, Seasonal Decorations
Leadplant	Dynamic Accumulator, Erosion Control, Insecticide, Insectary, Nitrogen Fixer, Nurse, Soil Builder, Wildlife Food, Windbreak	Insect Repellent, Medicine
Alpine Strawberry	Mulch Maker	Compost, Container Garden, Food, Hanging Basket, Medicine
Purple Coneflower		Habitat, Nectary
Black Currant		Container Garden, Dye, Essential Oil, Food
French Sorrel	Insectary	Dye, Essential Oil, Food, Medicine
Chicory	Dynamic Accumulator, Insectary, Mulch Maker	Biomass, Compost, Dye, Food, Medicine
Saskatoon Serviceberry	Erosion Control, Wildlife Food	Food
Summercrisp Pear	Seasonal Insectary, Wildlife Food	Food
Sky Blue Aster	Insectary, Habitat	
New England Aster	Insectary, Soil Builder, Wildlife Food	Cut Flower, Medicine
Rugosa Rose	Nurse, Wildlife Food, Windbreak	Aromatics/Fragrance, Food, Medicine
Anise Hyssop	Aromatic Pest Confuser, Insectary, Water Purifier	Aromatics/Fragrance, Food, Medicine
Lupine	Domestic Animal Forage, Erosion Control, Nitrogen Fixer	Fiber, Food, Soap
Daffodil	Insecticide	Cut Flower, Dye
Common Yarrow	Aromatic Pest Confuser, Dynamic Accumulator, Erosion Control, Insectary, Mulch Maker, Nurse	Aromatics/Fragrance, Compost, Cut Flower, Dried Flower, Dye, Essential Oil, Food, Insect Repellent, Medicine

Initial plantings look sparse, but they will fill in and can be added to later.

and list the numbers of the plants to be used or a letter for the group of plants on the list. An insectary island could have aster, yarrow, chicory, echinacea, or many other plants, but for now just outline the space and give it a letter or number. Later you can make sure each letter/number corresponds to a plant on the list. Plants you choose that are not specifically related to another plant, such as a nitrogen fixer to a tree, can be placed in groupings and arranged as they fit when planted. Making a separate drawing helps plan the spacings and will make installation much faster. Once you get to the nursery or catalog, it's good to have a shopping list that you can stick to.

Groundnut (*Apios americana*) is the vine layer in this design. You will harvest the roots away from the trees, but allow the vines to travel up the tree's inner canopy. Winter pruning of the trees will include vine pruning as needed. Although the vine does not fruit, it could be the vine layer in the seven-layer strategy. The small fruit trees are not suited to heavy fruiting vines such as kiwi or grapes.

THE STRUCTURE OF A PLANT GUILD

> This polyculture models the seven-layer forest garden with its nutrient cycling patterns from nitrogen fixers to dynamic accumulators; ecological services are supplied within the planning bed and to the outer environment. Adding trees will add birds and their services to the space. Creating bloom time and nectar diversity within the niches will add ecological resources for beneficial insects and pollinators. For the steward, the garden will provide food, fiber, medicinal goods, mulching, and compostable material to build the soil and cycle nutrients. (See chapter 6 for details on installation, and pricing plants.)
>
> — DANIEL HALSEY

All these questions, and more, are a huge part of the process of selecting plants, drawing connections between functions and services, and implementing the design on the ground. We may spend the winter designing, looking through catalogs for species and cultivars, examining the property day in and day out for changes in sunlight angles as the year cycles through, the movement of water coursing through the property (where did it come from? where did it go?), how the wind blows through (or doesn't), and what animals traverse the property as the seasons change (who shows up when and why?). This questioning process is crucial to thorough design. It helps us configure in our minds and senses what the landscape can be, what it can produce, what it can do for the neighborhood or the countryside where we reside, and how it might bring health back to the land or serve some profound purpose that we cannot yet see, but know deeply is the right thing to do.

CHAPTER THREE

Selecting Plants for Guild Design

Even if I knew that the world would end tomorrow, I would still plant my apple tree.

— MARTIN LUTHER

Imagine the following scenario. It's winter and the holidays have ended. Snow lies blanketing the site that you have chosen for the gardens, planned for the following spring. Nursery and seed catalogs are beginning to arrive almost daily in the mail. This is a time for choices: What will be planted, where will it be sited, and which other plants and animals will share it? What do you want in your forest guild, and what do you need? Which plants will you make use of over a long time span? With observation and research there is plenty of room to uncover the many uses for plants, both the ones you know about and those you've yet to discover. Read this chapter to begin.

Previously you learned about the structure of plant guilds. This section describes the characteristics and functional niches of plants as well as their relationships with bacteria and fungi. A discussion of the larger environment of climate and soil ecosystems, or biomes, will set the plant guild in the context of its niche position. There are nearly infinite relationships in the global ecosystem dependent on plants. The diversity of these relationships and species in them is the key to exchanges of energies and materials within the ecosystem, from the guild level all the way to the global.

Trying to define a plant is like trying to define all of creation. Plants, as primary producers, are the basis for all life. They serve as sustenance for all creatures, carried through the food chain and transformed by secondary and tertiary producers. All of this metamorphosed substance goes into making soil for future plant growth, and the cycle repeats itself endlessly. Indeed, plants have the ability to manufacture their substance out of atmospheric carbon, moisture, and sunlight. From a seemingly insignificant seed, plants perform the magic of photosynthesis, send down roots, and build ring upon ring of growth into the atmosphere above ground. We can only marvel at the diversity of creative expression exhibited by the plant world. From the most meager algae or duckweed to the majesty of a redwood, plants occupy every possible niche on this planet, horizontally and vertically.

German botanist Gerbert Grohmann, in his book *The Plant,* calls the plant world the "light-sensitive organ of the earth." In other words, just as the human eye receives light and uses it to manufacture impressions of the external world, the plant is the receiver of the sun's rays and creates its substance, the image of the plant, through a synthesized mixture of sunlight, moisture, and carbon. This substance, the plant itself, is enabled by this photosynthesis, or light synthesis process of plant growth.

If plants are weak, mineral- and nutrient-deficient, then so will be the creatures that depend upon them for life. By increasing awareness of the issues that beset the modern world, by improving the fertility of the soil, and using the biological intelligence that surrounds and circulates through the ecosystem, gardeners can help fortify the plant world through regenerative practices. As soil is built into dynamic humus, plants' genetic stability will effectively be restored. There are no laboratory solutions that can do this, no test-tube experiments that can mimic what is already present in the ecosystem services that supply a plant's every need. Artificial fertilizers and manures create artificial plants. Something more comprehensive is needed. The "green revolution" is a revolution of quick fixes, and as the foundation of agriculture continues to fragment, the nutritional foundation of humanity and all life crumbles.

Plant guilds showcase that what we design into our plant matrices is a combination of species that thrive together, share resources, and contribute to the whole. When a community thrives, so do its individual members. The same is true with plant guilds.

Plant form is diverse to the point of infinity. Every geometrical configuration ever known is depicted in the way plants grow as they insert themselves, structurally, into the immediate environment. The design of a plant is predicated on efficiency, beauty, and interaction with the other kingdoms of nature: mineral, animal, human. These forms are based upon the species' adaptation to its surroundings.

Understanding the Biome

Easily recognizable units of climate, soil, and life that are contiguous across a large area are known as biomes. These are community units and can be thought of as guilds on a vast scale. Biomes are not just grouping of plants alone but consist of all the plant and animal life of the area. Global biomes have been classified by ecologists as the following:

- Tundra
- Northern coniferous forest
- Moist temperate coniferous forest
- Temperate deciduous forest
- Broadleaved evergreen forest
- Temperate grassland
- Tropical savanna
- Desert
- Chaparral
- Pinyon-juniper woodland
- Tropical rain forest
- Tropical scrub and deciduous forest

Nearly every one of these communities has a representative site within North America, allowing a highly diverse number of potential guild possibilities.

Planet Earth has its climate extremes, and within these extremes is the maximum tolerance of plants. The two major distinctions among the differing biomes are precipitation and temperature.

SELECTING PLANTS FOR GUILD DESIGN

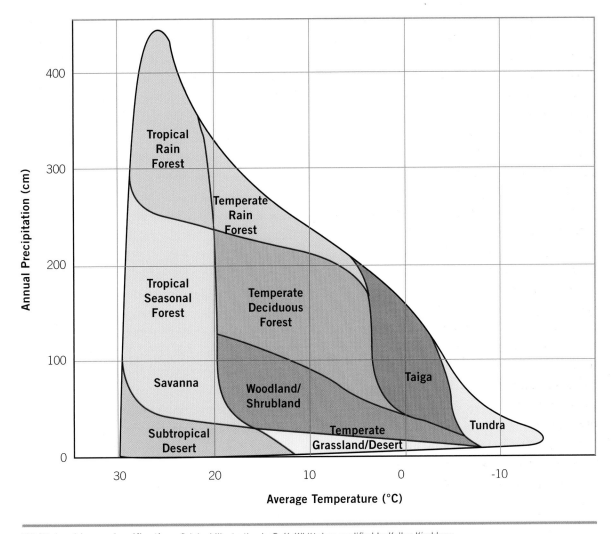

Whittaker biome classification. Original illustration by R. H. Whittaker, modified by Kellen Kirchberg

Plants can thrive in a range of extremes. Hot deserts can reach 109 to 120°F (43–49°C), and the Arctic can drop to -58°F (-50°C). Within this possible 178°F (100°C) range, plants have adapted either by dormancy during cold or hot periods, or by a life history that allows them to quickly emerge, pollinate, and disperse their seeds when conditions permit.

Along with temperature, it is the rate of precipitation that helps define a biome. The average annual precipitation within a biome can be close to zero, with brief weather events that may be years apart or provide constant moisture and occasionally intense rains as in a rain forest. Humid forests actually extract their own precipitation from the air on their leaves, dripping to the ground. In fact, a large percentage of moisture in a plant system may come from condensation. And while plants have adapted to all the tolerable extremes in our world, many have a broad range and can survive and even thrive in multiple biomes, but most are limited to a specific range within a specific biome.

Even within the biomes there are limiting factors for plants with a specific tolerance. Within the United States there are multiple biomes: grasslands, deciduous forest, desert scrub, alpine, and taiga. The Whittaker biome classification diagram depicts the various biomes via their combination of precipitation and temperature. Plant types are indicated by the biome name. Deciduous forests comprise trees that shed their leaves in autumn, while the coniferous forest biome consists primarily of tree species with evergreen needles that do not drop every year. Desert biomes will of course have a high number of dryland species such as cacti and other xeric or low-transpiration-rate plants. Understanding the biome and its climate conditions helps in plant selection. Knowing the climate extremes helps us make wise choices and reduce the labor required to establish and maintain plant health.

North American biomes range from the extreme cold of northern Canada to the deserts of Mexico and tropical forest in the Caribbean. Just like soil types, biomes are a pattern that help us make good design decisions. There may be exceptions, such as small niches and pockets of microclimate conditions that will allow some deciduous forest trees in grassland. These are often natural features of a biome. Along river systems in the North American Great Plains, cottonwood and willow trees are major landscape features and would have indicated to early travelers that water was nearby. We can expect that the number of these variances from biome norms will only increase as climate change alters traditional precipitation patterns and, with them, the boundaries of larger biotic communities. Look for these gray-area plants to add diversity and habitat that may be lacking in your broader landscape.

The transition zones between biomes might be miles wide or hundreds of miles wide depending upon the terrain. These transition zones or edges are highly productive since they involve the energies and resources of two biomes coming together, and often the extremes of both, adding a diversity of plants in that particular area. An example is a site in central Wisconsin known as the Tension Zone, just east of the city of Stevens Point. There the Arnott Ridge, a glacial moraine esker, is the westernmost boundary of the eastern woodland biome in the area. To its north a few miles begins the northern hardwood and pine forest biome, while to the west marks the beginning of the oak savanna biome. These are all natural boundaries that came about because of differences in soil formation and the resulting drainage patterns for soil moisture. Within a few miles in any direction can be seen major differences in the types of dominant vegetation as well as sites where many characteristic types from each of the three biomes come together at transition zones or edges.

It is important to adapt the structure of a successful polyculture design to the biome in which it is being placed by considering variables of climate, soil, and moisture availability.

Uses of plants in the context of their places in the global system of biomes can be referred to as functions. In other words, just what is a particular species doing that enables it to survive in its niche and contribute to the larger biome? All ecological functions need to be supplied by the plants involved. Although timing, life history, and soil play a huge role in plant selection and seasonal harvest, the proximity of the plants to one another is what allows these functions or services to be delivered. The diversity of plants allows for all functions to be provided even though extremes of temperature and precipitation may alter an ecosystem to another species' advantage. Years may pass as one species takes the dominant role until the climate adjusts to another pattern, and the key plant gives way to another. Nature has built in an adaptive resiliency based upon eons of experimentation with what

SELECTING PLANTS FOR GUILD DESIGN

North American biomes: Biomes are defined by their environmental conditions. The available plant species in the biome require those conditions. As climate and seasonal extremes shift, so does the biome and the plants it contains. Since the hot, dry desert biome is an extreme limit, it expands across the continent as temperatures rise, forcing an outward shift in all other surrounding biomes. Original illustration by K. A. Lemke will work across the edge transition areas between different ecosystems. Permaculture seeks to make use of that natural capital, or accumulated biologically based knowledge, and transform the human relationship with the natural world and thus our larger biome.

Critical ecosystem services provided by plants include clean air and water, erosion control, timber, wildlife habitat, upstream fisheries habitat, and food and medicines for all animals, not just humans. So in a natural environment or one that is carefully and minimally managed by humans, forest, field, and wetland plants moderate the extremes of weather, mitigating the impact of drought and floods. They shield animal species from the harmful effects of the sun's rays, cycle and move nutrients and water through the biome, anchor shorelines from collapse caused by erosion, detoxify and thereby neutralize toxins, contribute to climate stability through carbon sequestration, and provide a habitat for interspecies interactions.

Each species in the design will have its own unique life chronology. Germination, vegetative growth spurts, bloom time, fruit set, and maturity are events whose time line is known as phenology. Records of the first spring appearance of migrating birds have been kept for over a hundred years in England, as well as in North American locations. Farm and garden records detailing germination, flowering, and yields have been kept for hundreds of years — in the case of the French wine industry since the fourteenth century. These records are of great use in reconstructing an idea of what the past climate in a specific biome may have been like and how it has changed over time. Understanding each plant or animal's life history is a boon in designing an integrated polyculture plant guild with the adaptive resilience to survive long into the future.

In plant selection the goal is to determine the species most naturalized or adapted to a region and the niches that need filling in the guild design. In the process the individual seasonal growth and yield patterns of each species will become apparent. Ecological needs change as the plant develops through a growing season, and long-term needs change depending upon the age of the plant. Moisture and mineral requirements lessen as the plant reaches maturity and as the season of cooler temperatures nears. Individual species have evolved strategies allowing them to adapt to localized climate extremes; it is the variation among the plants that helps fill the seasons with harvestable crops, staggering the yields across weeks of crop maturity.

Annual crops such as bush snap beans and lettuce enjoy seasonal abundance and can be planted in relays and rotations to take advantage of seasonal preferences. In an area with cold winters, a shoulder season or edge interval between summer and autumn allows for the planting of a crop that may still mature before the first hard freeze. Fava beans (*Vicia faba*) and spinach (*Spinacia oleracea*) are two species that can thrive in late summer's cool nights and survive into late autumn until temperatures drop into the midteens Fahrenheit. Hot-weather crops such as tomatoes (*Lycopersicon esculentum*) can be started in a greenhouse or sunny window, then transplanted outdoors after danger of frost to mature in the warm season in mid- to late summer.

Domesticated annual crops are not considered native as they will only self-sow for a few generations before reverting to the characteristics of wild ancestors or failing to germinate. Plants that have retained a large portion of their wild genetic expression, such as cilantro (*Coriandrum sativum*), and borage (*Borago officinalis*), may continue to germinate and set fresh seed every year so long as suitable conditions on a site persist. Gardeners have developed numerous strategies based upon companion planting or mutualistic benefits among plant species based on phenology. An annual garden is a spatial and temporal microcosm of a model that can be adapted for long-term perennial vegetation design.

The life cycle of an integrated forest design can extend for decades and even centuries, provided no severe disturbance such as fire occurs. Within each phase of successional species maturity, the dominant species will change at each of the seven levels of a forest garden until a climax point dominated by a large overstory has been achieved. Taking notes regarding the guild characteristics over seasons and years and comparing that data will give a clear picture of the history or phenology of the site. For instance, thousands of plants may come and go beneath an apple tree over its life span. The progression of species and their decay products will build soil and enable a more stable array of understory and ground cover species to accelerate the natural process of successional change. A forest guild can be designed to place species in anticipation of the ecological services that will be needed over the years.

Functions of Plant Guilds and Polycultures

Like all living organisms, plants have certain functions or duties that they must accomplish in order to survive. In the process of doing so they are able to provide for the needs of other species: This is the basis of integration within a forest garden system. A plant's functions are those traits that maintain ecosystem resiliency. Functional roles in a guild are based on the site's conditions; some functions will take priority if a resource from another — say, nitrogen — is lacking. Permaculture design principles are a checklist for sustainable strategies. Each function of a species can be tagged with a corresponding principle to help define the time, frequency, and location for a functional plant type. Examples include the replacement of minerals such as nitrogen or phosphorus that are used up by subsoil taproots or seasonal bioaccumulation. This process pairs with a principle called the Law of Return wherein any nutrients lost must be provided for from the site if the system is to be maintained.

Ecological functions are processes inherent in a species of plant or animal that are used for its general well-being as well as serving as benefits to its niche or immediate environment. A species' benefits to the niche serve its own interests while helping other organisms at the same time. Examples include the ability of plant species to moderate pH, and the provision of waste nutrients by animal species that graze on the plant. Integration at its best. Be sure the plants you select as a part of the design for a particular guild have well-defined roles. If ecological functions are not in balance, then the overall survival of the guild may be in jeopardy.

When developing a plant community polyculture, functional diversity must be a priority if you are to provide for the needs of the various components. For example, balancing nutrient acquisition with a diversity of trophic feeding levels in the root zone by using plants with taproots, fibrous roots, bulbs, rhizomes, and tubers enables an abundance of species to coexist in a guild because they access different places in the soil zone for their needs. Another example is plants that provide shade or nutrients for others. Nitrogen-fixing symbionts contribute both protection from intense sunlight and leaf-building nutrients for other more demanding species. Locust trees provide the nitrogen as well as a dappled shade for strawberries, for instance, whereas alders do the same for nut tree species. Or a species may provide

TABLE 3.1 ENVIRONMENTALLY SOUND DESIGN: HOLMGREN'S PERMACULTURE PRINCIPLES AND THE DESIGN RESPONSE

Observe and Interact	Site Assessment, Look 360° for 365 Days, Occupy Patterns
Catch and Store	Microclimates, Swales, Harvests, Preservation
Obtain a Yield	Food, Fiber, Fodder, Fuel, Medicine, Skills, Fertile Soil
Apply Self-Regulation and Accept Feedback	Limit Consumption, Initiate Dialogue, Understand
Use and Value Renewable Resources and Services	Ecological Solutions, Build Natural Capital
Produce No Waste	Grow Resources, Cycle Nutrients
Design from Patterns to Details	Climate, Access, Functional Spaces, Patch, Niche
Integrate Rather than Segregate	Use Diversity, Systems Thinking, and Polycultures
Use Small and Slow Solutions	One Patch at a Time, Expand Proven Success
Use and Value Diversity	Remove Bias, Explore Origins and Alternatives
Use Edges and Value the Marginal	Create Edge and Use Edge, Find the Ignored Resources
Creatively Use and Respond to Change	Use Adversity to Motivate Lasting Change

protective cover, as well as nitrogen, to another plant during its early growth — the time when it's most sensitive to environmental extremes. The acacia acts as such a nurse plant for the young saguaro cactus, allowing its survival during severe droughts in the American Southwest deserts of Arizona and California.

In the natural environment, forest, field, and wetland species moderate weather extremes and their impacts. Within those ecosystems, guild components disperse seed, mitigate the effects of drought or flood, protect other species from an excess of ultraviolet light, cycle and move nutrients, stabilize river channels and streambanks, decompose waste products, control pest species, maintain biodiversity, and generate and preserve soils. They contribute greatly to stabilizing climate, purifying air and water, regulating disease-causing organisms, pollinating plants, producing biomass, and providing the interface for species interaction.

With integrated food forest design, an ecosystem that allows the buildup of natural capital is a major goal for a guild. *Natural capital* is a shorter way of referring to the sum total of ecosystem functions of the planet that sustain all life. Fundamental to the energy flows in an ecosystem are those functions that contribute to the visible resources of pollination, clean water and fisheries, and habitat. Central to these services are the benefits provided as food, medicines, and utility for human use.

In the natural environments of field, forest, and wetland, for example, weather extremes are moderated by the natural plant and animals present. These species serve as a carbon sequestration sink, mitigate drought and flood, disperse seeds, provide food to animal species and fungi, accumulate nutrients for later release to plants and animals, shield others from the sun's rays, protect streambanks and lake shorelines from erosion, and generate and preserve soils by decomposing wastes. They also control pest species and regulate disease-causing organisms, pollinate, produce biomass, and act as the interface for species interaction.

The species chosen for a polyculture guild design need to follow the same rules that nature has established for wild areas. Needs for all species must be provided for, with each providing an essential support for others, building resiliency into the system. Including multifunctional species as well as those that perform the same function will help the guild adapt to changes without major disturbance. Plants that provide the same process of nitrogen fixation can take over a larger part of that function, for instance, should disease or overgrazing by an animal reduce the effectiveness of one. Various species that have taproots bring up subsoil nutrients, and having several different plants that can do this enhances the availability and type of nutrients supplied to upper soil layers.

Animals that cycle through wide population swings due to disease and forage availability can be replaced with a similar species that functions similarly, although at a slightly different part of the food chain. Take the squirrel, which serves as a major mover of plant material and upper soil movement. Should squirrel numbers decline, then chipmunk numbers can increase and serve to move seeds and other plant material across the site.

Covering the Soil with a Blanket of Vegetation

Green manures are plants that have been grown for biomass and harvested before setting seed. They can be chopped and dropped on a site to enrich the soil, or plowed under to decompose there. The addition of the fresh vegetation provides a furnace for biological activity by soil organisms like insects, fungi, and bacteria. Mineral-accumulating plants such as burdock (*Arctium lappa*), and comfrey (*Symphytum officinale*) can be placed around shrubs and trees to provide nutrients and shield the upper soil layers from moisture evaporation. Woody plant debris decays more slowly and while doing so provides a substrate for the growth of insect and fungal species as well as a storage site for moisture.

Plants grow better when the soil is covered. Growing them in close proximity to sources of organic mulching materials allows for the breakdown of mulch into soil components. Annual crops perform this same function when they die to the ground at the end of the growing season; their biomass returns to the soil as decay progresses.

Specific plants can be used to seasonally collect and hold in the guild those nutrients that may

Chop-and-drop.

Hairy vetch. Photo by Paul Busselen (www.kuleuven-kortrijk.be/nl/Algemeen/Natuur)

Mustard. Photo by Paul Busselen (www.kuleuven-kortrijk.be/nl/Algemeen/Natuur)

otherwise wash or leach from the site. Dogtooth violets (*Erythronium americanum*) collect and hold phosphorous in their leaves in early spring so that it is held and not washed away in rains. When the leaves decay as the plant goes dormant in early summer, the phosphorous is released for use by later-blooming species. Similar functions are performed by spring bulbs such as tulips and daffodils. Late-season scavengers can perform a dual function both collecting a nutrient such as nitrogen or phosphorous — keeping it from washing into a watershed — and anchoring soil in place, preventing silt from running off into nearby waterways and thus depleting the soil. At maturity the plants can be cut for animal fodder, plowed under, or allowed to decay in place, adding carbon and humus to the upper soil layers.

In conventional agriculture this function is performed by seasonal crops such as vetches, clovers, and mixes of oats and field peas, all of which you can harvest either in late fall or in spring prior to seeding a main crop. In the garden the same technique, but on a smaller scale, is used. Plant species such as crimson clover, favas, peas, or buckwheat. Similarly, using annual legumes, grasses, or buckwheat as well as any quick-growing species to prepare the forest garden or vegetable plot allows you to increase organic matter and carbon-holding fungi in the soil prior to original plantings of any perennial species.

The workhorses of the plant world are dynamic accumulators, species that collect specific nutrients for later dispersal into the upper soil layers. These plants can be nitrogen fixers, nutrient scavengers that gather compounds discarded from a previous season's leaf drop, or simply particularly good at getting minerals like calcium, phosphorus, and potassium from the deeper soil layers. In a polyculture design such plants are valuable tools for nutrient cycling in the guild.

Using a soil test from a responsible lab can give you an idea of which nutrients may be lacking and the form that those foods need to take to become usable for the crop plants you'd like to include in a design. By adding carbon and fungi to the soil, decaying mineral accumulators will also be modifying the pH (acid–alkaline) balance in the accessible soil layers, which is an important determinant of which specific plant species are able to use mineral compounds. Other than dynamic accumulators, your only options are rotational grazing of livestock animals (with their droppings) and bringing in outside sources of plant nutrition — bagged fertilizers.

Nitrogen is the most common element other than carbon found in plants. It can be quickly added back to the soil by composting or even more simply by dropping cut leaf materials at the base of another species — say, trees or shrubs at the drip line. This allows soil microorganisms to make the delivery of nutrients to the root zone.

Phosphorus is necessary for flowering and thus the reproductive processes of all plants. Plants that can accumulate it in their tissues include legume species, mustards (*Brassica* species), and buckwheat. All three of these actually exude the mineral from their roots directly into the nearby feeding zones of other plants. Mustards are cool-season species primarily for early- and late-season use, with deep taproots. Buckwheat is an early-summer warm-season plant that suppresses growth of weed species with its shallow though wide root systems while loosening clay soils and building soil moisture levels when it decays. Buckwheat should be turned under before it has flowered or gone to seed to maximize these benefits (or the seed harvested for pancakes!).

Some plants collect potassium, a mineral essential for root growth. Comfrey (*Symphytum* species) and related plants such as viper's bugloss (*Echium vulgare*) draw up potassium from the deep soil where it either has washed down or exists in eroded form from rock. The cut leaves of these plants can be used as chop-and-drop fertilizer; or you can add them to root zones (where their nitrogen and potassium will boost growth) before you plant tuberous crops such as potatoes. Other plants in this group include most in the family Boraginaceae, including borage (*Borago officinalis*), forget-me-nots (*Myosotis* species), and the fragrant heliotrope (*Heliotropium* species). Additionally these borage family plants are sources of the element calcium from the carbonates stored in their coarse and hairy leaves.

Other sources of calcium include the leaves of oaks (*Quercus* species), black locust (*Robinia pseudoacacia*), and linden (*Tilia* species). Dandelions (*Taraxacum officinale*), watercress and nasturtiums (*Nasturtium* species), and chickweed (*Stellaria media*) are also excellent calcium sources. Some mineral elements are required by plants in minute quantities and are commonly available in the soil. These include iron, magnesium, boron, and copper. Selenium is essential in minute quantities for animal health, and most soils in eastern North America are deficient in it. However, in the American West the more alkaline soil types allow for an uptake of selenium in plants; this can reach toxic proportions for grazing species in areas of high soil selenium levels: the Dakotas, Montana, Colorado, Wyoming, and Utah. Certain plants act as indicator species for the farmer and grower in these areas. *Astragalus* species or milk vetches, certain woody asters, and prince's plume (*Stanleya pinnata*) accumulate up to three thousand parts per million of the mineral compared with fifty ppm in the region's barley, wheat, and alfalfa. Five ppm is considered a high level of selenium and is toxic to livestock. Many other minerals essential to plant life can be found within the soil; enabling healthy flora and fauna to exist in the soil food web allows access to them while holding them in molecular forms nontoxic to life.

The Soil Regime

The soil regime comprises geologic material — rocks and minerals essentially — that has been altered by climate and interactions with plants, animals, fungi, and bacteria.

The physical breakdown of this material by the freeze-and-thaw process or by wind and water erosion reduces larger particles to smaller ones; these in turn are reduced smaller yet by processes of chemical weathering by plant, soil fauna, fungal, and bacterial activity to use for their own growth. The plants, animals, and fungi are in turn broken down by decomposer species that return the nutrients to the soil.

Our ancestors noted that different soils have differing capacities to support life; some are more fertile than others. Important relationships among environmental factors and soil properties were noted and became parts of the farming traditions of various cultures. Lands in areas where limestone is present in the underlying bedrock, for instance, show a higher fertility than ones where quartz and sandstone are present instead. Low-lying areas where alluvial silt collected as rivers washed it down from higher elevations were also noted to enjoy high fertility — for example, the soils along the Nile and Mississippi Rivers.

There are five soil moisture regimes based on biomes and climate:

1. **Aquic** is wet soil characterized by an absence of dissolved oxygen. It has been saturated with water for at least a few days. Examples include

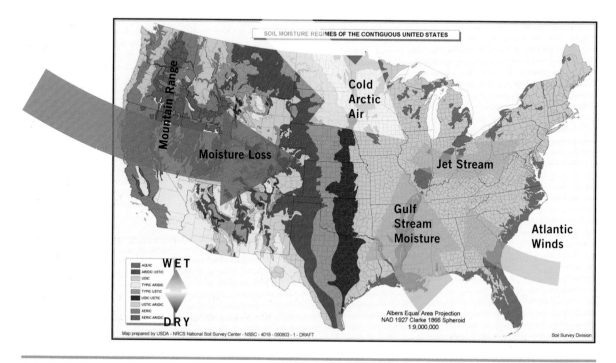

Soil regimes are formed by climate, plants, and erosion.

marshes, swamps, bogs, lakebeds, and even ephemeral ponds.
2. **Aridic** and **torric** are dry soils, such as those found in deserts. These soils have very little organic matter and subsequently feature low biomass or total soil life. Low precipitation and high temperatures result in a soil with a granular texture. The plants here spread themselves across a wide area so as to not compete for scarce resources.
3. **Udic** soils are generally found in the eastern United States, in high-humidity biomes. Plant diversity and interactions with bacteria and fungi create a large belowground soil structure teeming with life. Examples include the eastern woodland forests and oak savannas of North America. Because of the temperate climate, soil biodiversity in areas with udic soils has a high biomass.
4. **Ustic** soils are located in semi-arid grasslands. Seasonal rains cause a burst of plant growth during which life seems to literally spring from the ground. Plants grow from dormant roots or seed, and flower, fruit, and mature all within the time defined by the season of available moisture. The dry lands of the American Southwest are an example of such a soil moisture regime. Seasonal rains cause the dry grasslands to bloom with color. If you return during the summer months, though, you'll be greeted with an area of dry scrubby trees and shrubs along with a sparse ground cover of drought-tolerant plants.
5. **Xeric** soil is dry in winter and summer for forty-five or more days. This soil is typical of Mediterranean-type climates where winters are moist and cool and summers dry and warm. Much of Southern California is an example of a xeric climate.

A biome may have multiple soil regimes. For instance, along the Wisconsin River in the southwestern part of Wisconsin can be found areas of dry and sandy soils known as sand barrens characterized by species reminiscent of desert climates. Short-season annual plants make use of the limited available moisture while deeper-rooted species and those with leathery, waxy, or succulent foliage such as the prickly pear (*Opuntia* species) are able to survive daytime soil temperatures of 160°F (71°C).

Uphill from such a site may be black oak barrens, also hot and dry but with cooler soil temperatures and grassy soil cover over deeper soils and a diverse grouping of species such as hazel, wild roses, and strawberry. Downstream, above the eroded river channel before it enters the Mississippi River, are found all the major forest types of southern Wisconsin, including oak savanna, eastern woodlands, and floodplain forest. Oaks (*Quercus* species), maples (*Acer* species), sycamores (*Platanus occidentalis*), apples (*Malus* species), sweetgum (*Liquidambar styraciflua*), and ash (*Fraxinus* species) are all found here in an area of biological complexity and diversity where transition zones are sometimes only a few score yards apart. Few places on a property have a continuous soil type unchanged across the site. Move a few feet one way or the other, and the soil can exhibit different composition, texture, and drainage patterns. These differences can be due to deposition of silty sediments at different times or to human-caused disturbance.

The USDA Natural Resource Conservation Service Website includes a soil survey map for nearly every county in every state of the United States. These maps can also be found at local county extension offices or universities. The maps show locations of soil types and depths, and the accompanying texts give descriptive information regarding other soil characteristics, especially drainage patterns. For soil data specific to a potential integrated forest guild, we encourage you to take soil samples to a local extension lab for detailed analysis of nutrients, pH, and potential toxin contaminants.

Ratios of silt, sand, and clay define a soil's texture as well as drainage. Different plants have adapted to differing soil types during their evolution. Taprooted plants such as dandelion (*Taraxacum officinale*) and chicory (*Cichorium intybus*) have sturdy roots with the power to penetrate compacted clay loam soils. Plants with shorter fibrous roots require a lighter clay loam or even sandy loam soil that drains well to avoid rotting the roots and plant crown. Examples are strawberry (*Fragaria* species) and plantain (*Plantago* species). Many species of plant have optimal soils for best growth even though they may survive in a variety of different conditions. As a general rule, plants intolerant of saturated soils will not endure because the respiratory oxygen exchanges at root soil levels are insufficient to allow nutrients to be used by the plant or the soil organisms it depends on. Plants tolerant of wet soil may suffer from lack of sufficient moisture during a drought because of their higher water requirements and association with soil organisms that need a more saturated soil. Blueflag iris (*Iris versicolor*), for instance, is tolerant of wet conditions and able to grow in roadside ditches, but in the heat of summer the plant goes dormant as the water dries up.

Other species are tolerant of varying conditions, wet or dry, among them blackeyed Susan (*Rudbeckia* species), purple coneflower (*Echinacea* species), wild ginger (*Asarum canadense*), and chokeberry (*Aronia* species). These adaptable types of plants are used in rain gardens — temporary storage sites for runoff water that, on a larger scale, are known as bio swales.

The measure of alkalinity and acidity in a soil is known as pH, or potential hydrogen. This is the measurement of the activity of hydrogen ions or positively charged particles at the molecular level present in an aqueous or liquid solution of the material being tested. While this sounds technical, all you need to remember is that the pH of purified or distilled water is 7.0; thus any number lower than 7 is indicative of a more acid result while a number higher than 7 shows an environment more alkaline. Soils derived from eroded limestone are more alkaline, with pH numbers higher than that of water; basalt types of rock give soil an acidic nature. Alkaline soils also are saltier than acidic soils. Eastern woodland biome soils are generally more acidic, being derived from ancient basalt lava flows instead of the breakdown of alkaline calcium-carbonate-rich limestone from ancient seas as is found in the western portions of the Great Plains of North America. As expected, a huge number of biomes exhibit pH measurements either side of that of water. The optimal life zones for most plants depend upon the species: Heaths (Ericaceae family) prefer a pH of 4.0, where as the lilac (*Syringa* species) prefers a pH of 8.1.

Soil organism communities change based upon their tolerance for different pH because nutrient availability is often predicated on the electrical charge of the molecule a mineral food is in. Under specific conditions, a nutrient is available that even a slight alteration in the acid and alkaline balance would render unavailable. Hence with a more alkaline measurement, Ericaceae family plants such as blueberries or rhododendrons will fail to thrive and after a few seasons die — they are adapted to pH levels of 4.6 to 7. Species thriving in more alkaline soils include olive (*Olea europaea*), walnut (*Juglans* species), Asian persimmon (*Diospyros kaki*), figs (*Ficus* species), and Mediterranean herbs such as rosemary (*Rosmarinus* species), lavender (*Lavandula* species), and thyme (*Thymus*), along with common garden plants including daffodils (*Narcissus* species) and members of the cabbage family (*Brassica* species).

Very dry soils often have a very high pH. In those not extremely alkaline possible crops, pomegranate

(*Punica granatum*), date palms (*Areca* species), and *Acacia* species will grow well.

Soils and Salt Tolerance

Throughout history one of the greatest impediments to the survival of any human civilization has been the deterioration of agricultural land from a buildup of mineral salts leading to excessive alkalinity levels. This has been a direct result of evaporation of mineral-rich irrigation water from land in regions of insufficient rainfall. From ancient Sumer in the Mideast to the Anasazi culture in the American Southwest, farming cultures have faced collapse of their food infrastructure due to salinization of the soil.

Today the Colorado River has had its flow interrupted to power large hydroelectric dams, water vast desert farms in Southwest valleys, and supply drinking and industrial water to the cities of Arizona and California, reducing it to a trickle by the time it reaches the ocean. Recurring drought and high temperatures have resulted in an increase in evaporation; the resulting increase in salinization has effectively sealed the fate of some of the nation's most productive farmland unless changes creating a more water-wise agriculture occur.

Catastrophic Occurrences

Extreme disturbance events such as extended droughts, fires, and floods can destroy the capacity of the soil to sustain plant life for an extended period of time. Soil may be washed away by floodwaters, blown away by the wind, or have its microorganisms reduced by bad management practices. Periodic disturbance such as fire and flood have management solutions. Flood-tolerant plants break the flow of water and anchor soil to the streambank or shore. This reduces the velocity of the stream flow in a flood event, along with the deposition of suspended soil particles downstream.

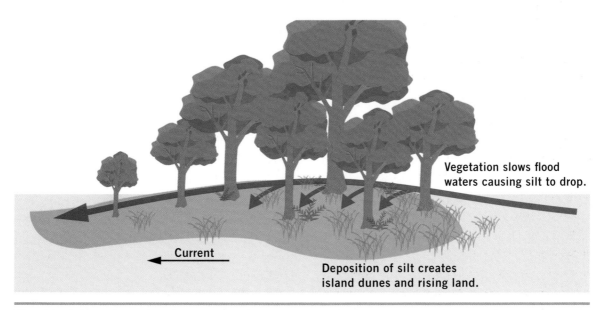

River island dunes. Illustration by Daniel Halsey and Kellen Kirchberg

Each flood occurrence builds a higher level of deposited soil in the area of the stabilizing plants such as willows (*Salix* species), and alders (*Alnus* species), causing deeper roots and more plant growth. Forested river islands do this naturally, creating eddy breaks that allow silt to be deposited. This builds up the island for more woody species to become anchored and collect more riverborne soil.

Wayne witnessed the devastation following Hurricane Sandy in New Jersey: Large rocks and even boulders were moved out of the river channel and piled along the new high-water mark. This catastrophic event realigned the river's banks and created a new high-water mark for future but hopefully less severe floods. His advice to residents was to immediately cover the rocky bank with plants to stabilize it with the fibrous roots, then add ground cover plants.

Floodplain species can tolerate periodic flooding and are often present in seasonal or diurnal ponds in eastern forest woodlands where the ash tree (*Fraxinus* species) is a major component. Spring flooding has less damaging effects than catastrophic events such as hurricanes and major summer floods because at that time plants either are still dormant or have adapted to water stresses. Proper species selection will ensure less damage from high-water or long-term inundation.

Deteriorated soil can be brought back to fertility through sound management practices. Incorporating organic matter by cover-cropping with plants grown to be plowed under at a lush stage of growth, incorporating carbon and nitrogen back into the soil, is one such technique. Hardpan compacted soil layers can be subsoil-plowed or keylined, and planting deep-rooted forage species or prairie native forages with the dual purpose of facilitating passage of nutrients up and down the soil profile and providing nutritious forage for livestock species that enrich the soil further with their wastes.

Agricultural Toxins

Another form of soil destruction occurs when industrial or agricultural toxins have been allowed to build up on a site whether from deliberate application, dumping, or natural disaster. Species that can bioaccumulate specific toxins — breaking them down into safer compounds — can remediate a site, allowing it to sustain soil life again. Plants that can gather up and hold certain toxic compounds in their tissues are engaging in a process known as phytoremediation, accumulation and subsequent storage for later disposal, transpiration to the air, or transformation via a metabolic breakdown of toxins to stable, safe, or less dangerous substances. A few examples are *Hydrangea* species, which accumulate aluminum; cabbage family (*Brassica* species) members and lead; willows (*Salix* species) and cadmium, zinc, and copper; and sunflowers (*Helianthus* species) with lead, arsenic, zinc, chromium, copper, and manganese. Mulberry (*Morus rubra*) roots exude a compound that supports a bacteria able to break down polychlorinated biphenyls — a serious industrial waste product of concern in many contaminated urban sites. The water hyacinth (*Eichhornia crassipes*), considered a noxious weed in many states because it spreads exponentially, can clean up polluted waterways by absorbing coliform bacteria and breaking them down as well as absorbing cadmium, lead, mercury, zinc, pesticides, and radioactive cesium and strontium isotopes.

Filtering water through a mass of roots is a natural process used in nature to remove excess nutrients and toxic chemicals from waterways. It is a process mimicked in human waste settlement ponds and provides about fifty billion dollars annually in ecosystem services, yet is without any cost to us. Constructed wetlands, catchments, and ponds are used by landowners, and cities, to collect and filter sediments and toxic runoff from urban and agricultural landscapes.

Fungi are also able to break down complex organochlorine pesticide compounds and hydrocarbons such as oil products. The common oyster mushroom is used to break down waste oil spills by mixing the contaminated soil with wood chips and then sowing mushroom spawn or sprouted spores into that mixture. The oil wastes are then broken down into simpler hydrogen and carbon compounds and readily returned to the environment in forms usable by soil microorganisms.

Planting of gardens on a raised soil bed of composted materials is a common practice in some urban areas where heavy metals such as lead and arsenic may have accumulated in the soil. The beds are wooden frames about a foot deep and often only thirty-two square feet in area, made of unpainted and untreated wood. The logic here is that using these frames filled with a foot of soil will allow a plant to take up nutrition from a root zone above the contaminant level. However, plant roots — especially those of root crops and tomatoes — are able to draw foods from soil at a depth of five and a half feet for tomatoes at maturity and seven feet deep for carrots. To be sure, the deepest roots are the moisture-drawing taproot portion, while shallower roots collect foods laterally at a depth of only a foot or more. Still, the intended purpose of bypassing any soil contaminants is not attained.

Get a soil test for several sites in an urban homesite where you may wish to locate a garden; if the soil shows dangerous levels of any toxin, then either phytoremediate with some of the plant species mentioned earlier or talk to an extension soil scientist about measures to alter the contaminants' ability to exist in plant-soluble forms by changing the soil pH. While there are plants that can reduce the amounts of toxic compounds in soil, depending on how heavy the contaminant load is it may be better to look for a different site for your integrated food forest.

Many urban areas have the soil removed and new soil, usually as compost, brought in. Sometimes the compost has not been mixed with other soil and is deficient in the clays needed for proper mineral absorption by plant roots as well as soil partners, the bacteria and mycorrhizae. When this happens, diseases may be more prevalent — for instance, the early and late blights that affect all Solanaceae species such as peppers, eggplant, potatoes, and tomatoes. Similarly planting a tree in pure compost might result in a specimen with a weaker root system unable to withstand winter-season winds and prone to toppling over.

Strategies that are used to protect areas prone to disturbance by fires include the planting of species that will deter the forward advancement of the flames and reduce the amount of available fuel. A rich understory of moist living plants burns more slowly, and a spongy soil layer with mycorrhizal moisture reserves resists fire. Ground cover plants with watery sap like rock cress (*Aubrieta deltoidea*) and bugleweed (*Ajuga* species) burn more slowly than those with a resinous pitch such as creeping juniper (*Juniperus horizontalis*).

Fire-resistant perennials include thyme, yarrow (*Achillea* species), columbines (*Aquilegia* species), and sedges (*Carex* species). Shrubs that are fire-resistant include *Cotoneaster* species, dogwoods (*Cornus* species), Russian sage (*Perovskia* species), and serviceberries (*Amelanchier* species). Surprisingly two coniferous trees, the ponderosa pine (*Pinus ponderosa*) and western larch (*Larix occidentalis*), are both resistant due to their thick barks and moist foliage. Many deciduous trees are fire-resistant. The oak species of the prairie oak savanna biome (*Quercus macrocarpa*) has thick, corky fire-resistant bark that allows the tree to survive fast fires that sweep across its habitat.

TABLE 3.2 FIRE-RESISTANT PLANTS

Ground Covers

Ajuga reptans—carpet bugleweed
Antennaria rosea—pink pussytoes
Arctostaphylos uva-ursi—kinnikinnick
Aubrieta deltoidea—rock cress
Ceanothus prostratus—mahala mat
Cerastium tomentosum—snow-in-summer
Delosperma cooperi—purple iceplant
Delosperma nubigenum—yellow iceplant
Dianthus—garden carnation, pinks
Fragaria species—wild strawberry
Lamium species—dead nettle
Pachysandra terminalis—Japanese pachysandra
Phlox subulata—creeping phlox
Sedum species—sedum, stonecrop
Sempervivum species—hens and chickens
Thymus praecox—creeping thyme
Veronica species—speedwell

Perennials

Achillea species—yarrow
Allium schoenoprasum—chives
Aquilegia species—columbine
Armeria maritima—sea thrift
Aurinia saxatilis—basket-of-gold
Bergenia cordifolia—heartleaf bergenia
Campsis radicans—trumpet vine
Carex species—sedges
Coreopsis species—coreopsis, tickseed
Delphinium varieties—delphinium
Echinacea purpurea—coneflower
Epilobium angustifolium—fireweed
Gaillardia varieties—blanket flower
Geranium cinereum—grayleaf cranesbill
Helianthemum nummularium—sun rose
Hemerocallis species—daylily
Heuchera sanguinea—coralbells
Hosta species—hosta lily
Iris hybrids—tall bearded iris
Kniphofia uvaria—torch lily, red-hot poker
Lavandula species—lavender
Linum perenne—blue flax
Lonicera species—honeysuckle
Lupinus varieties—lupine
Oenothera species—evening primrose
Papaver orientale—Oriental poppy
Penstemon species—penstemon, beardtongue
Ratibida columnifera—prairie coneflower
Salvia species—salvia, sage
Stachys byzantina—lamb's ear
Yucca species—yucca

Shrubs—Broad-Leaved Evergreen

Ceanothus gloriosus—Point Reyes ceanothus
Cistus purpureus—orchid rockrose
Cotoneaster apiculatus—cranberry cotoneaster
Daphne × burkwoodii—Carol Mackie daphne
Gaultheria shallon—salal
Mahonia aquifolium—Oregon grapeholly
Mahonia repens—creeping holly
Paxistima myrtifolia—Oregon boxwood
Rhododendron macrophyllum—Pacific rhododendron

SELECTING PLANTS FOR GUILD DESIGN

Shrubs—Deciduous

Acer circinatum—vine maple

Acer glabrum—Rocky Mountain maple

Amelanchier species—serviceberry

Caryopteris × *clandonensis*—blue-mist spirea

Cornus sericea—redosier dogwood

Euonymus alatus 'Compactus'—dwarf burning bush

Holodiscus discolor—oceanspray

Perovskia atriplicifolia—Russian sage

Philadelphus species—mockorange

Prunus besseyi—western sandcherry

Rhamnus frangula 'Asplenifolia'—fernleaf buckthorn

Rhamnus frangula 'Columnaris'—tallhedge

Rhododendron occidentale—western azalea

Rhus species—sumac

Ribes species—flowering currant

Rosa species—hardy shrub rose

Rosa woodsii—Woods' rose

Salix species—willow

Spiraea × *bumalda*—bumald spirea

Spiraea douglasii—western spirea

Symphoricarpos albus—snowberry

Syringa species—lilac

Viburnum trilobum—American cranberry

Trees—Coniferous

Larix occidentalis—western larch

Pinus ponderosa—ponderosa pine

Trees—Deciduous

Acer ginnala—Amur maple

Acer macrophyllum—bigleaf maple

Acer rubrum—red maple

Aesculus hippocastanum—horsechestnut

Alnus rubra—red alder

Alnus tenuifolia—mountain alder

Betula species—birch

Catalpa speciosa—western catalpa

Celtis occidentalis—common hackberry

Cercis canadensis—eastern redbud

Cornus florida—flowering dogwood

Crataegus species—hawthorn

Fagus sylvatica—European beech

Fraxinus americana—white ash

Fraxinus pennsylvanica—green ash

Gleditsia triacanthos—thornless honeylocust

Gymnocladus dioicus—Kentucky coffee tree

Juglans species—walnut

Liquidambar styraciflua—American sweetgum

Malus species—crab apple

Platanus racemosa—western or California sycamore

Populus tremuloides—quaking aspen

Prunus virginiana—chokecherry

Prunus virginiana 'Schubert'—Canada red chokecherry

Quercus garryana—Oregon white oak

Quercus palustris—pin oak

Quercus rubra—red oak

Robinia pseudoacacia—purple robe locust

Sorbus aucuparia—mountain ash

Pacific Northwest Extension, 2006. *Fire-Resistant Plants for Home Landscapes: Selecting Plants That May Reduce Your Risk from Wildfire*. PNW 590. Oregon State University, Washington State University, University of Idaho.

Plants that are fire-resistant will have the following characteristics: Their leaves are moist and supple. Plants have little deadwood and tend not to accumulate dry, dead material within the plant. The sap is water-like with no strong odor, and the sap resin materials are low.

When the Wind Doth Blow

When soil has been exposed by fire and flood because its vegetative cover is destroyed, it can suffer further destruction from wind erosion of the dried-out and bare upper soil layers. In the long term the deposition of windborne soil can enrich a biome; think of the way ice age glaciers carried soils and ground rock, which were swept by winds to give fertility to the vast steppe savannas of northern temperate grasslands globally. This loess, or wind-driven soil, built up to depths over a hundred feet and was the backbone of the American prairie farm until its deterioration by poor soil management practices in the early part of the last century and since.

During a drought and wind erosion event such as the 1930s Dust Bowl, many thousands of tons of topsoil are blown away, reducing the amount of fertility in a soil. You can lessen the effects of wind erosion by planting rows of trees to reduce the force of wind. These windbreaks slow the wind at ground level and increase the deposition rate of windborne soil at that level. This has the net effect of increasing soil fertility because of the added soil, but it may come at the expense of an upwind neighbor.

Terra Preta: The Dark Earth

Traditional societies around the world used a soil-building technique that incorporates partly combusted wood or charcoal—called biochar—into the soil. When it has been colonized by soil microorganisms, biochar has a crystalline matrix structure that allows the absorption of nutrient chemical compounds. The pores in the wood are colonized by fungi and bacteria that hold soil moisture, and continue to manufacture more nutrients. Using biochar in a disturbed soil environment will lead to a marked increase in fertility. Moreover, biochar's capacity to act as a storehouse of nutrition does not diminish greatly over time. Some sites in the Amazon River Valley have suffered no reduction in capacity for hundreds of years. The dark carbon-enriched soils there are known as terra preta, or dark earth. The form of carbon in the biochar is extremely stable and is always attracting new moisture and plant nutrients as they become available, holding them in the soil food web so that they are not leached from it.

Soil resources for plants contain huge loads of sequestered carbon from the atmosphere, originally created by decayed plant and animal sources. This carbon attaches to other nutrients and holds them tightly within the soil for later use. The deliberate planting of plant species that enhance carbon absorption—whether through the addition of biochar or by the natural uptake of atmospheric carbon and later decomposition with storage by soil fungi—needs to be an integral part of any long-term and comprehensive plan to reduce global atmospheric carbon levels.

Growing Zones

The US Department of Agriculture designates growing zones based on temperature extremes and associated plant tolerances. The USDA zone for a plant is tied to the plant's tolerance range and used as a guide for where it can grow. All plant tags

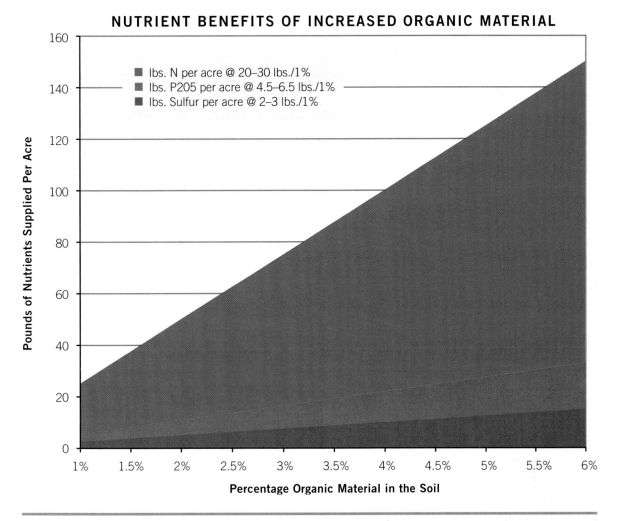

Nutrient benefits of increased organic material. Improved soil structure helps gas exchange, rooting, and water infiltration. Organic material (OM) can also hold up to 90 percent of its weight in moisture, increasing the available water capacity of the soil. It takes ten units of OM to decompose one unit of fresh material; without existing OM, it is a very slow process to build it. Perennial roots produce twice as much OM as aboveground foliage.

in stores and nurseries are marked with a growing zone indicator. And while zones change on the landscape, the plant tolerances are consistent and predictable. The zones only move on the landscape as thirty-year averages change. Every thirty years the USDA recalculates zone limits based on climate records for the years since the prior calculation. A region may be designated Zone 4b for some years, and then change to 5a. Plants that are suitable for the new growing zone designation for an area can then be chosen for inclusion in the guild, understanding that existing species may need protection or replacement while the new temperature norms reach an equilibrium and the plants adapt to the changes.

General trends at the time of writing this chapter are for temperatures to climb and zone numbers to increase, so plant selection needs to be based on a range of zones for the planting area. If the area to be planted is designated as 5a, choosing plants with a range from 4a to 7a would be advisable. The biggest concerns with the lower numbered zones are freezing temperatures, as they constitute the biggest cause of plant mortality. Drought periods and high heat can be mitigated with watering, but freezing temperatures are difficult to manage on a large scale.

Selecting Plants for Resilience

Selecting plants for resilience and a wide tolerance range will ensure their survival during periods of extreme climate change. When developing a plant list, start by designating planting areas by their USDA growing zone. Then look for areas of microclimate where warmth might be captured to limit freezing temperatures. Usually this is on the south side of a structure or in a small area surrounded by trees that face the winter sun.

Microclimate niches in the landscape may allow for plants that normally do not survive the climate to become established. But take care in any event to protect the plant from extremes in weather, understanding that freeze damage may occur without warning. You can enhance a microclimate with items that retain heat, such as large stones, fencing, and water features. Within the landscape, low areas that collect cold air are called frost pockets; avoid placing tender plants in these areas. However, others might benefit from the cool air. Fruiting shrubs in a frost pocket will have chillier nights and create sweeter fruit thanks to the cold air. North sides of hills freeze deeper and longer in the northern latitudes, causing fruit trees to bloom later in spring. This prevents a late frost damaging the blossoms. Selecting a late-blooming fruit cultivar for use in this situation can delay even further bud break and blooming in northern climates.

Within a biome you'll find many layers of variation. Rain falls just out of reach of one area as another is pummeled by hail. It is important to view the biome as a broadly patterned space for general and initial awareness. Understanding the site helps you define the plant palette, but there's always room for interpretation. Your space might feature swaths of dry creekbed dappled by patches of rich, moist soils, for instance. Define the site's niches and patches for opportunity, then retreat to books and plant database resources to build a plant guild or a forest garden that fits the local ecology as well as your intended purpose. Keep in mind the local intellectual resources others have created. Generations of people may have been doing the same thing you are and have great advice — maybe even perennials they'd be willing to share. Tour the local ecology in others' gardens — you'll learn more than you would by just relying on a zone map.

Understanding Sun Exposure

Most plants are categorized as: full sun, partial shade, or full shade. Full sun means the majority of the day, the plant receives direct sunlight. Light shade is when the plant is in shade two to three hours of the day. Dappled or partial shade is when the plant is in shade four or five hours of the day, or it's in an understory

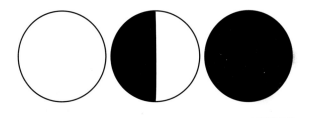

Sunlight symbols.

SELECTING PLANTS FOR GUILD DESIGN

Create a Plant Mind Map

We recommend that every designer or gardener create a mind map to identify the plant components to consider when building a guild polyculture. A mind map is a diagrammatic sketch showing the relationships among components in a guild. It's useful as a way to outline the framework of the guild design by condensing the material to an easily seen format. Ecological requirements (found on the upper right of the mind map on page 78) are the non-negotiable; these are the biome, niche, patch, climate, and environmental conditions the plant must have to emerge and thrive. Variation away from a plant's tolerance will cause it to die back or die altogether. Short bursts of extreme weather can cause a plant to lose leaves and branches as it responds to the changing conditions. Plants that have evolved with these variations will probably require them to fruit or set seed. Winter is the prime example. Many northern plants must have a cold period for a set number of days to complete their season or life cycle.

where shifting leaves and sun produce a patchy sun exposure. Full shade means no direct sunlight hits the plant, although some light may be reflected. Dense shade — such as you'll find under a pine or large oak tree — has no reflected or direct light.

When you're using a seven-layer forest structure for light partitioning, sunlight needs are a primary consideration. The duration of sunlight or season of exposure differs with some plants. Woodland ground covers, for example, can maintain health with just a few hours a day of dappled sun when an opening in the canopy allows a ray of light to spread across the forest floor. Bleeding heart plant (*Dicentra spectabilis*), ferns, and wintergreen (*Gaultheria procumbens*) are examples of woodland shade plants. Spring ephemerals such as wake robin (*Trillium* species) and dogtooth violets (*Erythronium americanum*) get all their light in early spring before the forest canopy closes with leaves. They are first to emerge with broad leaves; they bloom and then recede in the shadows until the next year.

Determine Your Soil Types

Few places on a property or landscape have a continuous soil type that is unchanged, and just as a biome can include many different variations, so it is with soil: Move a few feet to one side and you might see a change in type and quality. The combination of silt, sand, and clay in a niche defines the soil type. Plant roots are designed for a particular type of soil. Deep taproots, like those of dandelion and chicory, can pierce compacted and clay soils. But plants with shallow, fibrous roots might require sandy or loamy soils. For most plants there is a primary soil type for best growth, although many plants can actually grow in a variety of soils and soil conditions. Still, as a general rule, plants not suited to saturated soils will suffocate there for lack of gas exchange, and plants that do not mind "wet feet" in saturated soils will be less drought-tolerant if the soil should dry out. In other words, using the soil type as a parameter for plant selection is one of many factors that ensures you will choose the most suitable plants, but the ecological requirements of a plant are nuanced and numerous.

Soil pH or acidity is a product of the base material or surface decomposition. As bedrock breaks down into finer particles, its pH is dispersed into the resulting soil matrix. Limestone base material is high in pH, and is therefore alkaline. Basalt, raised from the earth's crust as a result of previous volcanic activity, has a low pH, and can create thin acidic soils or be a precursor for deep loamy forest soils. Soil organism communities change relative to the soil pH, and soil nutrient availability changes depending on pH. Again, plants have a tolerance range for acid–alkaline conditions. For example, blueberries (4.6–7.5 pH) thrive in acidic soils, as do pears ($\geq$ 5.0 pH) and apples (6.5–6.0 pH). This ability or inability to thrive in certain pH conditions is directly connected to the soil nutrients each plant needs and its soil organism associations. Some researchers believe that the organisms in the soil need the pH at a certain level; a plant just follows their lead. Bacterial and fungal mutualists that transfer nutrients to the plant root may limit the plant's resources based on their tolerance. Inoculating plant roots with mutualist organisms before planting may extend the plant's tolerance in site conditions and accelerate maturity. (Plant inoculants can be purchased from most nurseries.)

Understanding a Plant's Tolerances

Every organism has a zone of tolerance for each of its ecological needs and environmental conditions. If a condition such as temperature or soil moisture is present beyond the tolerance of the plant, the plant must respond by taking resources from its own rate of growth and propagation to compensate. During high temperatures a plant that does not fare well in heat will reduce respiration, which causes water loss. Extended periods of high temperatures will eventually kill the plant, as it cannot complete photosynthesis without respiration. Many plants in hot regions have adapted to this, however, by photosynthesizing during the day and using evapotranspiration to collect CO_2 at night when temperatures are cooler. These CAM plants, mostly found in the desert, have a greater tolerance for high-temperature and dry environments.

CAM — crassulacean acid metabolism — is a method used by arid-region plants to flip the normal respiration cycle on its head. Plants normally open their stomata, the breathing pores on the undersides of leaves, during the day and admit carbon dioxide at that time, using the nighttime hours to process the gas for photosynthesis. In CAM the reverse is true. To reduce evapotranspiration, the loss of water due to respiratory processes, the stomata are closed during the day and only open at night when the temperatures are cooler and winds have lessened. This is an adaptational strategy for dry land plants that greatly increases survival rates in time of drought. Agriculturally important CAM plants include corn (*Zea mays*), pineapple (Bromeliaceae family), and sugarcane (*Saccharum* species).

In addition, leaf structure has much to do with environmental adaptation. Succulent plants, those with thick waxy leaves, retain moisture in their leaf and stem tissue and tolerate dry climates. Some succulent plants die back to the ground using storage of root moisture to wait out dry conditions. Plants such as mulberry (*Morus rubra*) are widespread and adapted to hot, dry climates and arctic temperatures. If the guild design is for an area of drought or high temperatures, you will want a list of these types of plants on hand. Other examples are moss roses or purslane (*Portulaca* species), *Aloe* species, and even some plants such as the potato-like wapato (*Sagittaria* species).

Tolerance for Saline Minerals

One of the greatest problems for agriculture in civilizations from the Maya to the present-day Hopi is the destruction of soil by evaporation, which leaves minerals in high concentrations. All soil minerals are salts. Any time evaporation exceeds precipitation, these salts build on the surface. This is also true for land near oceans where salt water can contaminate the soil. Farmers and horticulturalists who use high tunnels and hoop houses struggle with mineralization: Irrigation leaves soluble salts in the soil as the water evaporates into the air, and over time the soil becomes unusable. Centuries ago it happened in South America to indigenous cultures that used canal irrigation in dry areas, and it happened hundreds of years ago in Casa Grande near Phoenix, Arizona, where local people used river water to irrigate their crops. It is irrigation without periodic precipitation that causes the minerals in the soil to collect at the surface.

Another source of salt in the soil is along roads, where snowplows throw salty road slush and snow onto property easements and ditches. Within five feet of any plowed roadway salt collects and builds in the soil, so salt-tolerant plants need to be used in this situation. But not to worry: Many plants are salt-tolerant. Hundreds of fruiting trees, shrubs, grasses, perennials, and annuals in the Natural Capital[1] Plant Database are listed as salt-tolerant (within limits).

Guild Design Basics

When you are beginning to design a plant guild, pay special attention to partitioning guild species based on available resources. As described earlier, plants respond to light in different ways (full sunlight, partial shade, or full shade), and germination rates are predicated on light availability, temperature, and placement. Once established, the response to these same processes changes based on each plant's growth rate, the nutrients available to it, and its ability to develop beneficial relationships with neighbors in the guild. A gardener or designer can plan as carefully as possible, but much of the understanding of these relationships among guild components comes through observation over an extended period of time as plants mature and enable new niche relationships to develop, ones of shade, moisture, and nutrient accumulation.

Indeed, as growth progresses, meticulously observe, and often, in order to make the proper adjustments in terms of spacing, access to fertility, and responses to changes in climate and precipitation. We cannot stress enough how important it is to be observant and thoroughly absorbed in the life of our plant guilds. As Permaculture founder Bill Mollison suggests, "protracted observation" is key to our sustenance and the well-being of the ecological matrix that surrounds us. Take notes (write it down!) of the changes that are taking place hour by hour and day by day. There is only so much that can be learned from a book, even this one. Having your hands in the full course of the life of plant guilds is synonymous with success.

First construct a guild design based on what you would like to grow. What's the point of growing something you won't enjoy? At the same time, be adventurous. Try something new, a new flavor or species of nut or fruit. Remember that the design and implementation of the guild are an ongoing project, carried forward in time, that will meet your future expectations and needs. Then consider the size of each plant, the niche that each plant will fill, all the needs (sunlight, nutrients, water, and so on) that the individual plant requires for a full life, the potential yields that will be harvested, and the many uses that each plant supplies in drawing together these functions for a rich and deep assembly of botanical variation. With diversity there will be a redundancy in the system. Should ten crops fail from a late frost, then the eleventh will succeed. If one plant needs nitrogen, another requires potassium, while a third needs a certain mycelial symbiosis, they are all going to share from their guild community palette of nutrients.

Types of Plants

The types of plants that can be chosen for an integrated polyculture guild will fall within these categories:

- Trees
- Shrubs
- Grasses
- Herbaceous species
- Ground covers
- Vines
- Mosses
- Lichens

Again, these are the components of a seven-layer food forest, an ecosystem where a yield for all creatures can be found at every height and depth. Plants for a Future (www.pfaf.org) is a database that lists the uses and growing requirements for thousands of species.

TREES

The eastern woodlands biome that stretches from the Atlantic to the Mississippi affords us ample opportunity to mimic the ecosystem services that a woodland setting demonstrates. Trees do not drop their leaves and till them into the ground. All the organic matter falls to the surface and is incorporated into the topsoil via weathering, and through the work of soil creatures, fungi, and microbes.

Deciduous trees lose their leaves every autumn, and conifers — for the most part — retain their leaves all year. Most conifers are evergreen, but a few species lose their leaves in autumn such as tamarack (*Larix* species) and cypress (Cupressaceae family). Typically the soils at the feet of deciduous trees contain a high percentage of organic matter, a dense mycelial web beneath the ground, and a springy duff (partially decomposed organic matter) that develops over time. The soil around the feet of conifers is usually devoid of herbaceous cover, and acquires a dense needle cover, acidic on the pH scale. There is immense opportunity for planting acid-loving plants in the understory of conifers. Blueberries (*Vaccinium* species), potatoes (*Solanum tuberosum*), and lingonberries (*Vaccinium vitis-idaea*) are a few possibilities.

The tree layer consists of two levels, the tall tree layer and a shorter one. Tall trees are those that will reach a dominant height on a site and — depending on the aims of the guild design — will be anywhere from twelve to a hundred feet in height. A dwarf fruit tree guild that includes a peach tree with pole beans, raspberries, comfrey, strawberries, and herbs could be considered a complete short-tree version of a plant guild. A beech tree with a lower story of maple, hazels, eleuthero, gooseberries, and woodland strawberries is an example of a guild utilizing tall-tree species as dominants. Often a guild needs to be based on existing large species already on a site. Adapting a design to do so can be a challenge, but by observing what grows in a natural setting you can draw up a design that mimics patterns found in nature.

UNDERSTORY TREES AND SHRUBS

Descending vertically from the overstory layer are found niches for lower-growing and shade-tolerant species of trees and shrubs. This layer is literally plugged in to take advantage of aboveground space. While you'd typically only consider the horizontal layout of a garden, there are boundless opportunities for filling the vertical niches allowing for available sunlight and proper spacing of plants so they don't outcompete one another for nutrients. Candidates for this layer of the guild include semi-dwarf and dwarf fruit and nut trees as well as tall shrub species such as highbush cranberry (*Viburnum trilobum*) and serviceberry (*Amelanchier* species). Many medicinal shrubs fall into this category also and can be coppiced to control their growth. Some of these are Siberian ginseng (*Eleutherococcus senticosus*), *Forsythia suspensa,* and white willow (*Salix alba*).

HERBACEOUS LAYER

Beneath smaller trees and shrubs an herbaceous layer of perennials, biennials, and annuals dominates the niches closer to the ground. In terms of natural succession, annuals colonize a "disturbed" site, followed by biennials and perennials. Construction site outlots, or damaged undeveloped parcels of land, are places where pioneer species whose seeds are lying dormant in the soil germinate. Annual species like amaranths (*Amaranthus* species) and mustards are among the very first plants to emerge. Biennial plants such as yellow and white sweet clovers (*Melilotus* species) anchor the bare soil in place and — through a symbiotic relationship with bacteria — collect atmospheric nitrogen that will meet the nutrient requirements of plants that follow in the succession process. Thistles, with their deep roots, draw up minerals from the subsoil. Tree and shrub species seeds such as mulberry (*Morus rubra*) and autumn olive (*Elaeagnus umbellata*) are spread by passing birds, then germinate and, as they grow, provide shade for less sun-tolerant species; after a score of years the bare original site is a thriving ecosystem of plants, or a guild designed by nature. If the focus is only on annuals, then the guild is tied to the very first steps in natural succession, and endless hours weeding and applying formulas for the suppression of undesired or weed species may be required.

TWINING AND VINING

Trees and shrubs provide a framework for species to climb toward sunlight, plants that do not have rigid enough stems to support themselves. These vining plants can grow along the ground but need to reach higher to gain the full benefit of available light for photosynthesis. Plants that vine along the ground include species such as bindweed (*Convolvulus* species) and some roses (*Rosa setigera, R. moyesii,* and others). But often a more sturdy support is needed. Human-made supports, fences, trellises, and poles can fill this function, too. In nature the riverbank grape (*Vitis riparia*) climbs up cottonwood trees (*Populus deltoides*) and willows (*Salix* species), growing high in their canopy layer. Mimicking that ability, a cultivated grape (*Vitis* species), hardy kiwi (*Actinidia* species), or pole bean can do the same on trees and shrubs in the integrated forest guild. This makes further use of the vertical layer, stacking and double-cropping species there.

CARPETS TO WALK ON

Grasses, ferns, mosses, and lichens make up a large part of the plant kingdom. Why plant grasses in a guild system when you've spent so much time removing them in order to grow neat and tidy rows of whichever species is in vogue? In an integrated forest garden grasses are considered to be neither a blessing nor a curse. Grass is an easy ground cover, blending well with short species such as ground ivy (*Glechoma hederacea*) and dwarf clovers (*Trifolium* species). That combination alone can aid a guild when planted to attract rabbits that would otherwise be feeding on more choice perennials and annuals. White Dutch clover (*Trifolium repens*) is a preferred food for rabbits, and a planting of it attracts them first.

Grass, like clover, regenerates readily after it's been cut or grazed upon. So how is grass incorporated into a guild system? On a large scale, of course, a farm pasture with rotational grazing comes to mind; but what about small scale, as in urban and suburban yards? The traditional urban and suburban monoculture of grass is a wasteland of monotony, but when interplanted with swaths of shrubs, trees, and ground covers, areas of grass lawn become places to play and rest, places to sink toes into cool grass on a hot day or after a spring rain and soak up life.

Rock gardens are sites where ground cover plants such as *Sedum* species, *Iris* species, and bulbs come into their own. Providing a path through the garden enables a place for reflection and meditation away from life's normal pressures. Here you can also add to the guild ferns. The rocks and moist spaces between them will allow micro niches for mosses and lichens, which paint vivid colors in the landscape year-round.

There are no cookie-cutter guilds, and these few examples are just some of many possibilities — groups of species that over time can become a community that is functionally vibrant and resiliently adaptive to environmental stresses. The more species are in a guild, the more adaptive niche possibilities it contains. A rock next to a fern adds moisture retention. The fern adds shade. A third species needing that extra moisture and shade then enters to fill that niche. That third plant has its own contributions to the system, which then enables another niche to open up, a place for a fourth, and maybe fifth and sixth species to enter and adapt, adding their contributions. And the guild grows some more.

PLANTS OUT OF PLACE

When a plant appears in a niche where it is undesired it is called a weed, which quite simply is a plant out of place, unintended and unwanted. This can be a dandelion, chickweed, or even a pawpaw tree. It can be a black walnut tree or a nettle. All of these have their uses but may not fit into the intended design or even the needs of the gardener. Once a polyculture has been established, and even earlier when the soil has been disturbed in readiness for putting in the guild, seeds of those species that may have been lying in the soil for years or decades begin to sprout as favorable conditions of light and moisture are in ready supply.

Gardeners all have nightmare stories of creeping charlie (*Glechoma hederacea*), chickweed (*Stellaria media*), or even buckthorn (*Rhamnus cathartica*) that has taken over sections of their plantings. While each of these plants and other weeds has uses as both food and medicine for animals and humans, the good news is that their seed emergence and early growth can be inhibited or even prevented by barrier plants that suppress weeds.

Using weed-suppressing species as a ground cover can give the early guild species time to get roots established and grow high enough to rise above any weed or other species shading them. Shading the ground where the weeds are growing is one way to do this. Plant a ground cover species with shallow roots that will not outcompete your guild plants. Canadian wild ginger (*Asarum canadense*) is a plant with shallow but very tough roots that can carpet an area of semi-shade, as do many ferns such as maidenhair fern (*Adiantum pedatum*). These ground covers will also cool the soil and reduce evaporation of moisture. Dandelions (*Taraxacum officinale*) emit ethylene gas that inhibits plant growth and can cause premature fruit maturation.

Annual grasses or field peas can be sown in the autumn to shade out and crowd out weed species that may germinate during the edge seasons of fall and spring; plow them under before you establish your main guild plantings. Buckwheat can be used during the warmer months of early summer, because it crowds out and shades weed species. It can also be turned under before flowering to add large amounts of organic matter to the soil for microorganisms to feed on.

Allelopathic plants exude natural chemical defenses to discourage competition from other plant species or even one another. White Dutch clover (*Trifolium repens*) has shallower roots than most perennial clovers and is mildly allelopathic to weeds. Just pull out a few plants and plug in a guild species in their place. Using plant chemicals such as juglone from the roots of the black walnut (*Juglans nigra*) and, to a lesser degree, from other *Juglans* species can discourage the growth of woody species such as buckthorn (*Rhamnus cathartica*) and herbaceous ones like garlic mustard (*Alliaria petiolata*), as well as many desirable species such as apples, roses, and many garden vegetables.

Using a weed-suppressing mulch, either of the chop-and-drop type or shredded leaves, is of great value also in forming a barrier to unwanted germination of out-of-their-intended-place plant pests. An added benefit of using leaves and barks as a mulch in the polyculture guild is that they often harbor some of the same microorganisms, both fungal and bacterial, that will make good neighbors to plant roots.

WALLS TO BLOCK, SHADE, AND PROTECT

Plants that function as a barrier need not be used only for weed suppression but can also block a view of a street, garbage cans, neighbor's hot tub parties, or any other undesired vantage. They can protect from sunlight and wind, absorb traffic noise from a highway, and absorb and block pesticide sprays in a farm field or next to a golf course.

Thorny plant: gooseberry.

These plants can be annuals such as *Sorghum* species; biennials; perennials like cup plant (*Silphium perfoliatum*) and sunchoke (*Helianthus tuberosus*); evergreen woody species such as white cedar (*Chamaecyparis thyoides*); or a mix of species collectively known as a hedgerow. The hedgerow has the added advantage of being a guild of its own at the edge of another guild, forming a barrier but at the same time hosting a diversity of life that can support all the other guilds by providing the pollinator and predatory species that enable a yield while controlling pest species.

A variation on the barrier plant is the trap or diversion plant. Species that are able to attract and divert pest species from the desired crop plant can be collected and disposed of — nasturtiums (*Tropaeolum* species) attract aphids, for instance, and radishes (*Raphanus* species) attract flea beetles. Planting a sacrifice crop of a variety of fruit to divert grazing animal pressure from the main planting is another way to save a crop; this is dealt with in more detail in chapter 5.

Ground cherry.

TLC SPECIES

Small transplants and slow-growing species often need to be protected and even nourished by another plant providing essentially some tender loving care. Examples include columnar cacti such as saguaro (*Carnegiea* species) paired with *Prosopis* species of nitrogen-fixing legumes in the American southwestern deserts. The legume shades the young cactus, provides nitrogen fertilizer to it in its early growth period, and cools the soil enough that available root moisture is not a problem. Similar pairings occur with columnar cacti in central Mexico and the legume *Mimosa luisana*. Thorny branches on the legumes can deter browsing animals such as rabbits from consuming the young tender cacti.

Other examples of nurse plants are fast-growing "weedy" trees and shrubs of the *Elaeagnus* or autumn olive family and *Morus* or mulberry species, both of which nourish the soil, provide shade, and draw birds to consume their fruit instead of the main guild species; both can also be shaded out by the canopy polyculture species later. The two nurse plants are also able to be coppiced to a shorter height periodically and are an animal fodder species, too. Small nut trees such as the Korean and Siberian stone pines (*Pinus koraiensis* and *P. sibirica*) benefit from growing alongside comfrey (*Symphytum officinale*), which provides a grazing barrier during all seasons but winter. They also share in the nourishment from minerals and other plant foods brought upward by comfrey's deep roots; these nutrients are concentrated in the leaves. Eventually the pine grows tall enough to push the comfrey aside and tower over it while the comfrey — which enjoys some shade tolerance — spreads to become a ground cover beneath the tree, continuing its nutrient support function.

SELECTING PLANTS FOR GUILD DESIGN

Korean stone pine and comfrey nurse plants.

George Washington Carver: The Plant Function Genius

From George Washington Carver's grave: HE COULD HAVE ADDED FORTUNE TO FAME, BUT CARING FOR NEITHER, HE FOUND HAPPINESS AND HONOR IN BEING HELPFUL TO THE WORLD.

There is not a plant in the plant kingdom that doesn't have thousands of functions and uses. This is the central reason that we always take a thorough and thoughtful inventory of a property in order to identify the species already present. At that point we begin rigorous research to identify the possibilities of maintaining these species and including them in our plant guild regimes. This is the Permaculture ideal of "making the least change for the greatest effect." Our observation, inventory, assessment, and documentation may not only mean having to plant less, but become an economic boon as well, saving money by narrowing our focus in the direction of already established species — plants that are, in essence, free for the taking.

George Washington Carver was the master of identifying the uses of plants. If we, as practitioners, follow his lead, we may find that the guilds we have in mind, the plants we would like to see on a property, take care of more needs than we envisioned initially. It is one thing to select apples for fruit. It is another thing to come to the realization, through rigorous research, that the apple tree may provide a hundred other uses and functions we could not have dreamed of. Carver himself was most noted for his work with the peanut — he uncovered 325 ways to use it! How many uses might the plants already present on our properties yield with a bit of observation, identification, study, and experimentation?

We present here the idea that every plant that we meet in our lives has infinite functions, whether it be for our use or for the ecosystemic services it provides. Even the pansy you bought from the local Walmart has a flower that makes an excellent addition to a salad.

As we carry out our initial assessment and inventory of a property, we identify and record what plants already exist, both within the perimeter and outside it, as far as we wish to explore. Even the most poisonous of plants has multiple functions. For example, poison hemlock (*Conium maculatum*) — which put Socrates into everlasting sleep — is used extensively in homeopathic medicine. Homeopathy is a medical practice founded on the belief that "like cures like." A patient's symptoms are matched to a remedy's symptomatology, described in the homeopathic *materia medica*. The similar symptom pictures of the patient and the selected remedy propel the body and mind into action, and the healing commences.

Corn and soybean researchers have turned these monocropped plants into multibillion-dollar industries. The chemicals and materials isolated from them appear in a never-ending list of products and services. And thanks to the depths of its research, the petroleum industry (that megalith) has synthesized countless products from the dregs of its own manufacturing processes. Now, this is not to condone the octopus-like arms of the corporate world; what we are getting at is the boundless possibility that lies hidden away in every plant. It would behoove us to take a page from this kind of deep research

SELECTING PLANTS FOR GUILD DESIGN

and get to know, intimately, the plants that surround us—whose yields, services, and functions are infinite.

A Short Biography of George Washington Carver

George Washington Carver was born in 1864 in Missouri on the farm of Moses Carver. He and his mother were kidnapped by Confederate night raiders and taken to Arkansas. Moses Carver found and reclaimed George after the war, but his mother had disappeared forever. Raised by the farmer, George found himself drawn to plants from a young age; in future years he would be known as "the plant doctor."

George moved to Newton County in southwest Missouri, where he worked as a farmhand and studied in a one-room schoolhouse. He went on to attend Minneapolis High School in Kansas. Entering college was a struggle because of racial barriers. At the age of thirty, Carver was accepted at Simpson College in Indianola, Iowa, where he was the first black student. Intent on a science career, he transferred to Iowa Agricultural College (now Iowa State University) in 1891, where he gained a bachelor of science degree in 1894 and a master of science degree in bacterial botany and agriculture in 1897. Carver became a member of the faculty of the Iowa State College of Agriculture and Mechanics (the first black faculty member for Iowa College), teaching classes on soil conservation and chemurgy.

In 1897 Booker T. Washington, founder of the Tuskegee Normal and Industrial Institute for Negroes, convinced Carver to come south and serve as the school's director of agriculture. Carver remained on the faculty until his death.

At Tuskegee, Carver developed his crop rotation method, which revolutionized southern agriculture. Decades of growing only cotton and tobacco had decimated southern soils. The economy of the farming South had also been devastated by years of Civil War and the fact that the cotton and tobacco plantations could no longer abuse slave labor. Carver, however, convinced southern farmers to alternate soil-depleting cotton crops with soil-enriching crops such as peanuts, peas, soybeans, sweet potatoes, and pecans. These suggestions allowed the region to recover.

Carver also worked at developing industrial applications from agricultural crops. During World War I, he found a way to replace the textile dyes formerly imported from Europe. He produced dyes of five hundred different shades and in 1927 invented a process for producing paints and stains from soybeans. For that he received three separate patents.

In his later years Carver devised hundreds of products from the lowly peanut, including ink, facial cream, shampoo, soap, milk, and cheese. Suddenly a new industry sprang up using surplus peanuts. Next, Carver looked at ways of utilizing the sweet potato; he was able to develop more than 115 products from it, including flour, starch, and synthetic rubber (the US Army used many of his products during World War I).

Carver did not stop there. From the inexpensive pecan he developed more than seventy-five products; he found dozens of uses for discarded cornstalks; and from common clays he created dyes and paints. As his fame grew, he was invited to speak before the US Congress and was consulted by titans of industry and invention. Henry Ford, head of Ford Motor Company, invited Carver to his plant in Dearborn, Michigan, where the two devised a way to use goldenrod, a weed, to create synthetic rubber. The great inventor Thomas Edison was so enthusiastic that he asked Carver to move to Orange Grove, New Jersey, to work at the Edison Laboratories at an annual salary of a hundred thousand dollars. Carver declined the generous offer, wanting to continue at Tuskegee.

George Washington Carver died in 1943.

Plugging in the Air Cleaners

Cleansing the air of dusts and odors is another ability of plants. Outdoors the hedgerow can absorb windblown dusts and pollutants, transforming some of them into forms that can be broken down or locked into molecules by bacteria and fungi. The moisture that condenses each night on the leaves of trees and other plants captures dust particles and anything carried in them; when that dew falls to the ground, the microbes and fungi go to work.

Indoors we can simulate this process by growing species adapted to a room's specific situation of light, temperature, and humidity. As a plant engages in transpiration, the rate of exchange of gases with the surrounding air is regulated by the area's humidity. Indoors this is usually rather low — in winter months, even more so. Increased transpiration or plant respiration means that the plant is filtering and exchanging more air, effectively increasing its function of cleansing the indoor atmosphere. Volatile organic compounds, or pollutants, are absorbed through the leaf pores (stomata) and sent to the root zone, where the actions of bacteria and fungi break them down or sequester them in molecular forms that pose less risk to the plant and other organisms.

Snake plants (*Sansevieria trifasciata*) absorb chemicals from the air. Photo from Creative Commons

SELECTING PLANTS FOR GUILD DESIGN

Leaves with a large surface area such as *Ficus* species and *Philodendron* species are able to collect a lot of dust on their leaves and so are very efficient at removal of airborne compounds.

Here are some of the toxic compounds most often found indoors along with the plant species that can be useful for their removal in an indoor guild:

- For **ammonia**, which is present in many household cleaning agents, try lilyturf (*Liriope spicata*), peace lily (*Spathiphyllum* species), flamingo-lily (*Anthurium andraeanum*), and broadleaf lady palm (*Rhapis excelsa*).
- **Formaldehyde** is present in many paints, plywood adhesives, glues, fire retardants, cleaners, flooring binders, carpets, and permanent press fabrics, among other things, and is ubiquitous in its presence in most homes. To remove it try plants of Boston fern (*Nephrolepis exaltata*), English ivy (*Hedera helix*), lilyturf, spider plant (*Chlorophytum comosum*), flamingo-lily, variegated snake plant or mother-in-law's tongue (*Sansevieria trifasciata*), heartleaf and selloum philodendrons (*Philodendron* species), dragon plant (*Dracaena marginata*), and weeping and edible figs (*Ficus* species)
- A class of chemicals similar in molecular structure to one another and in their deleterious effects on the respiratory tract and blood are the **benzene ring aromatic hydrocarbons**. Benzene, toluene, and xylene are the major sources of exposure to this group in any household and are present in detectable amounts in inks, paints, solvents, rubber, and plastics (where they are used to put a solid into a solution during manufacture). The outgassing of these chemicals can be controlled with English ivy, peace lily, bamboo palm (*Chamaedorea seifrizii*), snake plant, dragon plant, moth orchids (*Phalaenopsis* species), *Dendrobium* orchids, weeping fig (*Ficus benjamina*), and rubber plant (*Ficus elastica*).
- **Trichloroethylene** is a chemical used in the dry-cleaning industry; it outgasses from clothing cleaned with it. To remove it, try Barberton daisy (*Gerbera jamesonii*), florist's mum (*Chrysanthemum* x *morifolium*), rubber plant, and yet again mother-in-law's tongue. There is a pattern emerging here: Mother-in-law's tongue has many more uses than you might expect.

The best solution for indoor toxins, of course, is to reduce their presence by scaling down the use of products that contain them. Using natural fabrics and carpets instead of petroleum-based, seeking out low-VOC paints, using natural dyes and inks, and avoiding plastics whenever possible goes a long way to reduce the toxic load on the interior environment. You can also pair indoor air-cleaning species with culinary ones such as aromatic herbs, bay laurel (*Laurus nobilis*), dwarf lemon trees (*Citrus* x *meyeri*), lemon verbena (*Aloysia citriodora*), and many others that can thrive in the low indoor humidity.

The Myriad Other Uses for Plant Yields

Before the introduction of petroleum-based plastics and solvents, just about all products that were not either metallic or animal-based were derived from plants. Foodstuffs come immediately to mind and are a major focus of this book. Humans also use plants as building materials: wood as lumber, grasses as thatch, various plants as roofing shingles, insulation, plaster, pitch, water pipes, canoes, rafts, and boats.

Clothing is made from fibers of flax (*Linum usitatissimum*), cotton (*Gossypium* species), nettles (*Urtica dioica*), and other species. Wood can be used for darning needles, buttons, doorknobs, toilet seats, baskets, toys, and more. Extracts of oak wood are used for tanning leather. Inks and dyes are made from onion skins, beets, mallows, and so on. We can make an alcohol fuel from fermented wood; charcoal can be used as fuel to heat a home or fuel for an ore-smelting process in metal making or blacksmithing. Plants give us oils to use in the manufacture of candles as in bayberry or *Myrica* species; essential oils that can be useful in cosmetics, soap making, and medicines; and natural compounds such as salicylic acid for pain relief, found in meadowsweet (*Filipendula ulmaria*) and white willow (*Salix alba*) bark.

Scouring rush (*Equisetum arvense*) has silica-rich compounds that make it useful for scrubbing pots and pans as well as in soluble fertilizer preparations for agricultural compost. Broken twig ends of plants such as peelu (*Salvadora oleiodes*) or willow along with powdered eggplant can be used to clean teeth. Many musical instruments — especially stringed ones such as violins and guitars — are made of wood, as are broom handles. Straws can be fashioned from the hollow stems of edible plant flowers like daylily (*Hemerocallis* species). Just about every human need can be met by plants. Returning to a more comprehensive use of them in everyday life can reduce the environment's toxic load of synthetic chemicals and the disturbances that result from their manufacture.

Roots: Anchors and So Much More

Roots have three primary duties. First, they anchor or hold a plant onto a substrate — usually soil, but sometimes rock or another plant. Second, a root is the access point for water and dissolved minerals, whose uptake is enhanced by partnership with beneficial bacteria and fungi. Lastly, root structures are food storage reservoirs, especially for dormant biennial and perennial plants.

Just as succulent leaf structure can keep moisture in, a plant's root structure helps the plant seek moisture, either by sending out thin shallow roots to collect whatever moisture is available, or by using deep taproots that mine the soil depths for water. Prairie grasses, for example, have been known to send roots twenty to thirty feet into the soil, and mesquite roots have been recorded at eighty-foot depths, gaining moisture from the permanent water table. In other words, these plants' tolerance for surface drought has been increased by their root structure. Root structure is a plant characteristic used in partitioning soil resources when you select plants. Mixing root types can allow you to place plants closer together — they won't compete for resources below the soil, but they will play out their ecological benefits above the soil surface. Root tolerance to flooding, drought, and available water capacity in your soil is another element crucial for your site assessment and conditions for your plant list.

It is generally dicots — plants that, upon germination, feature two paired leaves — that create a more persistent rootlet, or radicule. This in turn develops into either a taproot or thinner branched roots. Examples of dicots include carrots (*Daucus carota*) with their taproot, and beans, which develop a branched root system. Monocots — plants that are single-leaved at germination — lose their radicule and develop a fibrous root system. Examples include corn (*Zea mays*) and grass (*Poa* species and others). The anchoring root system of a ryegrass plant can grow roots and branch roots up to four hundred miles long all within the space of twenty cubic feet of soil. The tiny root hairs vastly increase the surface area where plant–soil interactions take place. In Permaculture talk, that is a lot of edge for maximum opportunity.

Taprooted plants include many instances of multiple-rooted plants, not just those with the

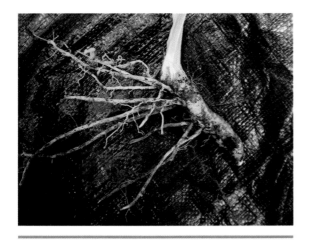
Iris rhizome.

Strawberry.

single root such as carrot, parsnip, or beet. The skirret (*Sium sisarum*) is an example of a branching taproot. Occasionally a carrot can grow multiple roots because of an excess of nitrogen or because it's growing in shallow stony soil. Some of the taprooted species grow very long roots, as anyone who has dug up a burdock can attest.

Monocots, such as corn and sorghum species, develop thick roots that serve to anchor the plant from lodging or toppling over in high winds. These roots are thicker than a pencil and extend deeply into the soil, enhancing the species' anchoring ability and, with their vast root hair systems, serve as dynamic nutrient foraging systems.

Roots with large starch reserves are found in plants such as the above, along with jicama (*Pachyrhizus erosus*), chayote (*Sechium edule*), four o'clocks (*Mirabilis jalapa*), and sweet potatoes (*Ipomoea* species). Each of these is a true root, not a modified leaf or stem tissue as is found in the potato, tulip, or iris.

In addition to fibrous roots and taproots, there are other adaptations that plants use for their survival. These include aerial roots, such as in epiphytic orchids, used for respiration in areas of waterlogged soils. Buttress roots are aerial root features near the base of some tropical trees such as mangrove (*Rhizophora* species) that help support the tree in shallow, nutrient-poor soils and reduce uprooting. Another adaptation is the haustorial root. In this instance a sticky disk from a parasitic plant, such as dodder (*Cuscuta* species), enables a sinker root to penetrate the host plant's tissue.

By now you are probably wondering about tulip and daffodil bulbs, potato tubers, and onion bulbs. Although they may be referred to as roots, these are not actually true roots but rather modified stems and leaves. Their modifications represent novel adaptive food and water storage strategies that plants have developed to adjust to environmental stresses.

Rhizomes are horizontal plant stems growing just beneath the surface. They have scalelike leaves and buds located at each node. True roots form beneath the rhizome. Many perennial grasses and ferns have thin rhizomes. An example of a fleshy rhizome is the iris; other rhizomatous plants include bamboo, ginger, turmeric, asparagus, and quaking aspen.

Stolons are also called runners. They differ from rhizomes in that stolons grow on top of the soil, not beneath. Plant buds appear along the runners and send roots into the soil. The most common example of a stoloniferous plant is the strawberry (*Fragaria virginiana*). Others are lily of the valley (*Convallaria majalis*) and field bindweed (*Convolvulus arvensis*).

Storage organs that form on stolons are called tubers, as seen in the potato (*Solanum tuberosum*). Each of a potato's eyes is actually a leaf bud; the tuber is merely a swollen stem storage organ with dormant leaf buds. The same is true of the common dahlia (*Dahlia* species).

Bulbs — whether tulips, daffodils, onions, or others — are leaf buds surrounded by folded, scalelike leaves. Think of peeling an onion until all that is left is the central core, the bud. Some bulbs — like *Lilium* species, or lilies — form specialized roots that pull the bulb into the soil each year until the bulb has reached the zone for optimal growth and survival. These roots are contractile; when an aerial bulblet along a lily's stem falls to the ground, it forms these roots and is pulled downward. Otherwise, the plant would reach a height that shallow roots could not support; it would fall from its own weight, have no protection from digging browsers, and be subject to the harsh extremes of the hottest and coldest seasons.

Nature loves variety, so there is a variation on bulbs known as the corm, a thick, hardened, compressed stem that is solid all the way through (unlike a bulb) and has no layers or scales. *Gladiolus* species, wapato (*Sagittaria* species), Jack in the pulpits (*Arisaema triphyllum*), water chestnut (*Eleocharis dulcis*), and bananas (*Musa* species) are the best examples. New corms form at the base of the old, and tiny cormels or mini corms form on them. Cormels are an adaptive strategy that ensures species survival in the face of foraging herbivores. If the corms are dug up and eaten, the cormels may be overlooked because of their small size and grow into new plants. As with bulbs, contractile roots pull cormels deeper into the soil.

The Fabulous Fungi

Mushrooms are the aboveground fruiting bodies of much larger organisms, fungi, some of which, like *Armillaria* species, can be acres in size though unseen belowground. Genetic research has found that fungi, formerly considered to be members of the plant kingdom, are neither plant nor animal but in a group all their own. Seeds of fungi are called spores, and they are carried on the winds. When a spore is deposited in a hospitable site, it can germinate and form a fungal body and mycelial or rootlike mat. Fungi are unable to photosynthesize their own food from atmospheric carbon; they must instead get it from the decomposition of other organisms or by partnership with plant roots. Fungi come in two basic forms: saprophytic and mycorrhizal.

Saprophytic Fungi

Everything gets decomposed sooner or later, and the saprophytes are the agents of that change. Secreting acids and enzymes able to break down woody tissues or lignins, they are able to reduce waste matter to the constituent building blocks for the growth of plants — and everything else. Some saprophytes are even adept at breaking the long-chain polymers of petroleum wastes. This decomposing fungal group is subdivided into three distinct groups, each of which functions best at a different time in the decay process.

Primary decomposers are some of the best-known mushrooms, including the oyster (*Pleurotis* species), shiitake (*Lentinula edodes*), and crimini (*Agaricus bisporus*). This group also includes many other species of interest for an integrated food forest. One that's circumpolar in its distribution is the turkey tail mushroom (*Trametes versicolor*). It forms on most dead tree species and forms a banded, stacked mushroom with both culinary and anti-cancer uses. True to their nature as the first fungi in the decomposer corner, primary saprophytes all need a fresh source of material on which to feed, such as newly downed logs and trees.

Secondary decomposer fungi go to work on decaying leaves and plant material already on the ground. These are the saprophytes that can be found in a compost heap spreading their white-colored mycelial root mats as the breakdown of woody brush and leaves begins. Common types are inky cap and parasol mushrooms. Intermediate between these and the primary decomposers is the garden king mushroom (*Stropharia rugosoannulata*). Feeding on decaying straw or leaves, its mycorrhizae form partnerships to exchange nutrients with common garden plants such as cabbage.

Last are the tertiary, or third level, of saprophytic decomposing fungi. They are nature's final act in the fungal repertoire. Rotting the detritus left behind from previous stages of decomposition, they are best represented by mushrooms of the fly agaric group (*Amanita* species), all of which are toxic if eaten by animals.

By the end of the tertiary fungal process, the original plant material has been reduced to forms from which it can be assimilated as nutrients for new plant growth in the guild.

The Ins and Outs of Mycorrhizae

Beneath the ground is a maelstrom of material and ecological processes. Plants must seek out resources; they cannot depend on the nutrients they need just floating by. Similarly, all organisms underground must be on the move to get their needs met. Plants are continually extending their reach underground with roots. Whether the roots weave down or horizontally through the soil structure, the goals are the same: water, nutrients, and air spaces for respiration. Along the way plants build interspecific relationships with other organisms. A symbiotic relationship between plants and mycorrhizal fungi is the most prevalent in nature. Mycelial mat–like roots of mycorrhizal fungi exchange nutrients with plant roots in exchange for carbohydrate the plant obtains through photosynthesis. Fungi are unable to make food from carbon dioxide and sunlight and are classified by how they engage in mutualistic relationships with plant roots.

Endomycorrhizal fungi form a sheath around a plant's root and exchange nutrients with the plant across that membrane. They are outside (*endo*) the host plant. Fungi that penetrate the root tissue and set up shop within the plant are known as ectomycorrhizae. The ecto group is also known as vesicular arbuscular mycorrhizae, quite a mouthful of words. What is important to know is that the ectomycorrhizae are the primary storage site for soil carbon, holding 30 percent of the total. Their hyphae — the collective term for fungal mycelia — are coated with a sticky gluelike material called glomalin. This adhesive seals gaps in the hyphae, preventing loss of water and the resultant loss of fungal and plant nutrients that would otherwise occur. Glomalin is also the primary site for soil carbon storage, keeping it in the ground for forty years or more. These two fungal groups are able to penetrate the soil at pore space levels a plant's roots cannot, granting greater access to a full array of nutrients while holding water in the mycelial sheath for use by the plant. Both groups of mycorrhizae afford a plant protection from pathogens and soil animal parasites such as nematodes. In return the plants offer to fungi soluble enzymes and carbon that they cannot make for themselves.

Additionally, the fungal mycelial "root" mat functions as a communication and distribution

Mycelial threads.

network among connected plant species. Sharing of nutrients occurs when one plant within the network is deficient in a nutrient or in moisture. In this way the functional health of the guild community is maintained.

Plants that have a healthy relationship with a fungal associate grow faster and healthier. Pine trees, for example, benefit greatly from fungal associations — to the point that commercial growers must inoculate the soil with mycorrhizal spores in order to produce healthy trees. Korean stone pine (*Pinus koraiensis*), a slow-growing tree with valuable nut crops, can reach maturity and produce nuts in half the time of an unassociated tree when it has been paired with the correct species of mycorrhizae. Many times pine trees are found growing in poor soil and on the ridges of mountains, and soil nutrients may be lacking. Having a fungal association in these cases makes it possible for the pine trees to prosper. Usually the mycorrhizae are found in forest soils where native trees are present. This soil and organic material from the forest floor can be transported and mixed with new soil when you plant a tree.

Most conifer and hardwood trees form an association with specific ectomycorrhizal fungi, for instance oaks (*Quercus* species) with the *Armillaria* fungi. Grasses, vegetables, and many shrubs, on the other hand, pair with endomycorrhizae. An example is the association of common garden vegetables such as cabbage family crops with the wine cap *Stropharia* mushroom. Pairing a plant with the correct fungal symbiont will result in a larger and healthier plant.

Because many of these fungal associates are naturally occurring, it is only when a non-native species is planted in a landscape that an inoculant is required. Pouches of inoculant are sold by nurseries as a soil amendment for use when planting a tree. Plants with fungal associations have higher drought tolerance: Mycelial fungal mats and their root sheaths hold soil moisture and release it to the host as needed. In a severe drought the fungi will give up the bulk of accumulated moisture and then go dormant until soil moisture levels increase to a level that revives them. Heavy mulch and high soil organic content will increase habitat for soil fungi. Looking at a freshly turned compost pile brings to view the mycelial thread mat and its hyphae. An integrated food forest, then, is a guild partnership among plants, fungi, and animals — and the whole is greater than the sum of its parts. Beneficial mutualisms enhance the health of the larger network.

Nitrogen-Fixing Plants

All plants need nitrogen (N) for growth, but it's often lacking in the soil because of agricultural practices that allow for it to be washed away rather than stored by binding to soil microorganisms. Nitrogen-fixing plants are a major component of any polyculture, plant guild, or alley cropping system, and all nitrogen-fixing plants have an association with bacteria in order to fix nitrogen in the soil. The plants may be trees, shrubs, annuals, or perennials. At any level of the integrated polyculture guild, nitrogen-fixing plants may occupy a niche and assist in the nutrient health of the anchor plants. In order for the nitrogen to become available to unassociated plants, the plant roots need to die back, which releases the nitrogen to the soil organisms and allows it to be taken up by the roots of other plants. Perennial nitrogen fixers do this seasonally as they die back to the ground and self-prune their roots. You must prune the branches of tree and shrub nitrogen fixers, however, for the roots below to die back and decompose, releasing the nitrogen to the soil. Or you can harvest the plants and add them to the mulch layer beneath your anchor plants, or to compost. Growing plants that dynamically accumulate nitrogen in their organic material is the most efficient and organic way to increase the soil fertility.

One nitrogen-fixing group is the legumes, which have an association with bacteria of the genus *Rhizobium*. These bacteria take up residence in the roots of legume plants and trade nitrogen resources to the plants in exchange for carbohydrates. Legumes are often used as a cover crop at the end of the season so that nitrogen can be scavenged from the atmosphere and tilled under for the new crop next season. The legumes can also occupy the spaces between crop and plantings. Tree and shrub species of legumes for North American biomes include mesquite (*Prosopis* species), honeylocust (*Gleditsia triacanthos* var. *inermis*), and black locust (*Robinia pseudoacacia*). A second group of bacteria that forms a partnership with plants to provide N-fixation tasks is *Frankia alni*. These soil microorganisms form this relationship with some important food and medicinal woody species, such as autumn olive (*Elaeagnus* species), buffaloberry (*Sheperdia* species), sea berry (*Hippophae rhamnoides*), and bayberry (*Myrica* species).

Rhizobial nodules develop on the outside of the root. Photo by Harry Rose

Incorporating N-fixing species into the guild provides the natural fertilizer that other species need for plant growth, too.

Interplanting with pine and fir trees can help with their growth in an agroforestry situation. For the backyard gardener or market farmer, interplanting annual vining legumes such as pole beans with corn puts back some of the nitrogen that corn draws down from soil reserves. The ratio of N-fixing species to non-N-fixers in the polyculture can be as high as a hundred to one. N fixers have only a limited effect on the current year's growth of the plants they are paired with. In order for the benefit to be added to the soil for uptake, the contributing species must release the accumulated N from its root nodules to the soil after it dies or has been pruned enough to shed some roots. Having these service plants in a design creates a support system for a guild's anchor and canopy species by improving soil with the fertilizer and carbon products necessary for the other species' relationships with soil microorganisms.

When the canopy species close the available direct sun, many N fixers will begin to die out from lack of light. The tree species mentioned earlier can fill some intermediate low-canopy niches, but there are also some lower species that are legumes and tolerant of the dappled light found on the forest floor.

The vine groundnut (*Pediomelum esculentum*) forms edible roots and is used to twine up shrubs and trees. The ground-hugging vine plant known as wood vetch (*Vicia sylvatica*) can grow in woodland shade. Having a diverse community of soil organisms including fungi and rhizobial bacteria — which can extract and convert nutrients from the atmosphere and decomposing plant materials — increases the availability of resources for associated plants.

Seasonal Considerations

During the progression of seasonal changes in temperature, available moisture, sunlight, and nutrient availability, the guild will draw upon different resources at the varying stages of a plant's life. Designing the forest garden to meet some of these needs will set into motion many of the resources. Again phenology comes to mind — the tolerances, growth rate, and other temporal characteristics of the individual species within its biome. Seasonal changes in the site will dictate which niche openings are available for plants at different moments of the year.

Early-emerging plants begin the nutrient cycle in spring. For example, daffodils (*Narcissus* species), are known for their early appearance and will emerge through snow. Their bulbs are best planted around fruiting trees; their roots exude a chemical that repels voles and mice, preventing damage to young trees. The early daffodil nectary resource is welcomed by awakening bees, and emergence signals warming soil conditions below. Tulips (*Tulipa* species) also appear in spring but have no rodent-deterrent ability. The two- to six-inch-tall *Crocus* species includes cultivars for spring, summer, and fall bloom, adding color to guilds at the beginning and end of the growing season and providing food or medicine, depending on the species. The use of plants with varied bloom times and functions ensures the guild has nectar and pollen services available to the ecosystem at all times.

Under an open spring canopy, small perennial plants get the light they need to flower and set seed. Their life cycle depends on a cool shaded area for summer with just a brief period of direct sun in spring. Later as ephemeral plants recede, other broad-leaved plants emerge and cover the soil. Summer shade plants thrive in the understory and require full shade for their growth. Within weeks, two plant types can shift dominance, with each inhabiting the same space but using different light niches.

Expanding Canopy Shade

When a forest garden is first planted, there is little shade, but as the canopy closes and the trees extend their branches opportunities for shade species expand as the understory conditions change. For this reason, consider a phased planting plan for your plant guild, with mostly direct-sun ground covers to start and partial-shade perennials under taller shrubs. Spring ephemerals, however, can be planted at any time; they will retreat to the soil each season before the sun and heat rise for summer. After the larger plants are established and consistent shade areas are created, start shade plants such as Solomon's seal (*Polygonatum biflorum*) and wild ginger (*Asarum caudatum*).

An initial installation usually leaves room for shade plants as the trees grow. Starter plants of annual flowers and vegetables fill in the sunny spots and take advantage of the light. In time these plants are moved to other areas and replaced with shade species. Choose plants that are sun- and shade-tolerant, and avoid the labor of moving plants by setting each in its prime conditions, which will ensure a harvest and the plant's long life. Most plants that are listed as sun-tolerant are also tolerant of partial

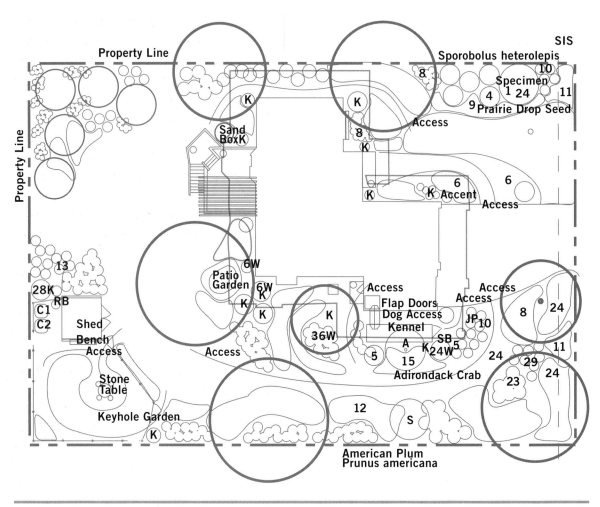

A small backyard corner is turned into an orchard space. With this initial structure of trees, shrubs, perennials, and ground cover masses, the selection of plant species is vast. The structure is developed to accept and support almost any plant suited to the site conditions.

sun or dappled woodland light. Apple trees, for instance, will do well both in full sun and in locations of only partial direct sun. When receiving direct sunlight for five hours a day and dappled sun from through a larger canopy tree the remainder of it, any fruit tree or shrub will still bear a crop while avoiding sun scald on tender fruits.

On a site plan the tree understory is drawn as relatively bare, showing the mature canopy. Viewing the canopy at installation size will show a vast expanse of available space for other plants. This is where bringing in daffodils and other perennials closer to the tree trunk will protect the soil and provide organic materials for nutrients. Make sure to space plants in close proximity, but allow for growth. It may be years before the tree canopy casts shade over wide areas; in the meantime you can use short-lived plants in the temporary gaps, including garden crops, herbs, and annual flowers. Low ground covers like short bulb root plants of camas

(*Camassia quamash*) and short rhizome plants such as creeping thyme (*Thymus serpyllum* 'Coccineus') are temporary placeholders and are easily divided and moved to new locations once there is more shade to take advantage of with other species.

Bloom Times

Keeping the garden's ecological services and functions high supplies needed resources to the insects and other animals that keep a garden healthy, in addition to building organic matter in the soil to feed thousands of beneficial organisms. When plants in a guild have bloom times that range over the entire growing season, even in drought and cold spells, a consistent supply of nectar to feed beneficial insects as long as they need it will be provided from a variety of sources. Parasitic wasps, for example, need alternative food sources when pest prey (aphids, et cetera) is scarce. It is one of the main goals of a guild to help build the beneficial insect population by having resources available as much of the year as possible.

The bloom calendar could include every flowering plant in the world. Better to check the plant list and make a simple bloom chart to schedule the weeks that nectar will be available. The type of flower is important, too, as there are short- and long-tongued pollinators. Long-tongued bats and hummingbirds frequent deep flowers, while small mason bees and wasps need shallow blooms. A deeper discussion of pollinators follows in chapter 5. Within a few years it is possible to observe polyculture guild plants in bloom during any given week. Two favorites of many gardeners are comfrey (*Symphytum officinale*), a late-spring flower that runs until fall frost, and chicory (*Cichorium intybus*), a deep taprooted plant with tiny blue flowers that blooms midsummer. Each has the added advantage of long roots that dredge mineral nutrients from deep in the lower soil layers. In the instance of comfrey, periodically cutting the large leaves and dropping them in place as a mulch adds potassium to the shallow root zones of other plants.

Timing of flowers and their bloom is totally controlled by weather. Growing degree-days, frost, precipitation, and stress may cause a plant to bloom sooner, later, or not at all. The life cycle follows the niche conditions so that the plant can do its best to set fruit and seed. A plant's main goal is seed; some may sacrifice much of the growth period to set seed for the next generation. Plants need nitrogen in the spring for good growth and then switch to potassium and carbon to produce fruit and seed, so fertilizing late or too much can cause a great amount of vegetative growth — but a relatively small harvest (of flowers and/or seed).

If climate conditions continue to change, we will see an annual average increase in temperatures of 1 to 3°F (0.5–2°C) over the next thirty years. The changes may be sudden or gradual, but phenological records of bloom times and associated temperatures reveal this steady, inexorable trend. That may not seem like a big deal, but when adding in the increased evaporation and transpiration, plants will need at least 20 percent more precipitation to offset the moisture loss. That is realistically not going to happen, so it's important to conserve water in the soil when it does rain. Building plant lists with resilient and drought-tolerant species will help protect the soil, and also feed the insects and other animals that will be affected.

Fruit Set

A plant's bloom time, flower duration, and reception of and distribution of pollen is a sequence of actions specifically scheduled for best pollination.

NATIVE FLOWERING PLANTS THAT ATTRACT BENEFICIAL INSECTS

COMMON NAME	SCIENTIFIC NAME	NATURAL ENEMIES	BEES
1. Wild strawberry	*Fragaria virginiana*	♦♦	♦
2. Golden Alexanders	*Zizia aurea*	♦♦♦	♦♦
3. Canada anemone	*Anemone canadensis*	♦♦♦	♦
4. Penstemon/hairy beardtongue	*Penstemon hirsutus*	♦♦	♦♦
5. Angelica	*Angelica atropurpurea*	♦♦♦	♦
6. Cow parsnip	*Heracleum maximum*	♦♦♦	♦
7. Sand coreopsis/lanceleaf tickseed	*Coreopsis lanceolata*	♦♦♦	♦
8. Shrubby cinquefoil	*Potentilla fruticosa*	♦♦♦	♦
9. Indian hemp	*Apocynum cannabinum*	♦♦♦	♦
10. Late figword	*Scrophularia marilandica*	♦♦	♦♦
11. Swamp milkweed	*Asclepias incarnata*	♦♦	♦♦
12. Culver's root	*Veronicastrum virginicum*	♦♦	♦♦
13. Yellow coneflower	*Ratibida pinnata*	♦♦♦	♦♦
14. Nodding wild onion	*Allium cernuum*	♦	♦♦
15. Meadowsweet	*Filipendula ulmaria*	♦♦♦	♦♦
16. Yellow giant hyssop	*Agastache neptoides*	♦♦	♦♦♦
17. Horsemint/spotted beebalm	*Monarda punctata*	♦♦♦	♦♦
18. Missouri ironweed	*Vernonia missurica*	♦♦	♦♦
19. Cup plant	*Silphium perfoliatum*	♦♦♦	♦♦♦
20. Pale Indian plantain	*Cacalia atriplicifolia*	♦♦	♦♦
21. Boneset	*Eupatorium perfoliatum*	♦♦♦	♦♦
22. Blue lobelia	*Lobelia siphilitica*	♦♦♦	♦♦♦
23. Pale-leaved sunflower	*Helianthus stumosus*	♦♦♦	♦♦
24. Riddell's goldenrod	*Solidago riddellii*	♦♦♦	♦♦♦
25. New England aster	*Aster novae-angliae*	♦♦♦	♦♦
26. Smooth aster	*Aster laevis*	♦♦	♦♦

Bloom calendar. Data from Michigan State University Extension

SELECTING PLANTS FOR GUILD DESIGN

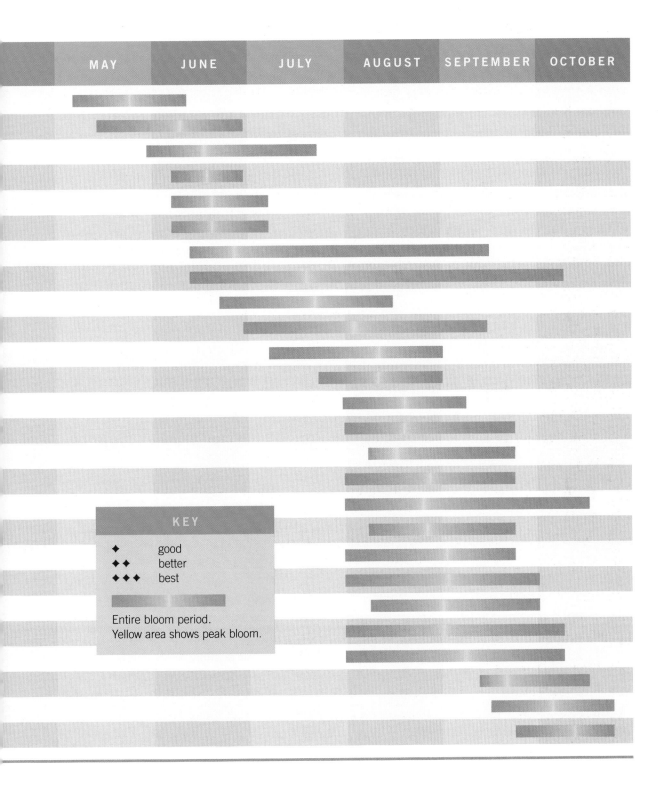

Fruit begins with the pollination of a healthy flower. Although many plants are self-pollinating (including beans, peas, peanuts, lettuce, eggplant, peppers, and tomatoes), others must be cross-pollinated with genetic materials from another member of the same species (pears, apples, and corn). Some plants are wind-pollinated, such as corn (*Zea mays*), wheat (*Triticum* species), rice (*Oryza sativa*), and pine trees (*Pinus* species). Wind-pollinated plants produce copious amounts of pollen to fill the breezes with opportunity, and may be cross- or self-fertile. The timing is specific to each species and influenced by evolution for insect availability, frost dates, or cross-pollination from companions. Insect pollination by bees is best known, but many organisms aid plants in the exchange of genetic material. Butterflies, hummingbirds, bats, and millions of flying and crawling insects deliver pollen to the receptive anthers in the flowers. See chapter 5 for more on insect roles in pollination.

Once the pollen has traveled down the anther to germinate the egg, the embryo begins its cell division into the components that make seeds and fruit. The fruit seen on the trees is an ovary for the seed inside. The purpose of fruit is to engage fruit-eating animals to ingest it for later distribution of the undigested seeds, usually with a good load of fertilizer. Fruit set happens on all plants in some way, typically in spring, although some have multiple harvests in a year. Cultivars of crop plants have been hybridized to set fruit in early summer, midsummer, late summer, or fall.

Currants (*Ribes* species) set fruit in early spring, flowering and fruiting before leaves appear. Even with this early onset of fruit, the berries can hang on the plant for months, increasing sugar content and size until they peak in sweetness after the first fall frost, assuming birds don't get them first. Elderberries (*Sambucus* species) bloom in late spring or early summer then set fruit, the berries slowly turning from green to black over several months. Again, the birds know when they are best eaten. Strawberries (*Fragaria* species) and raspberries (*Rubus* species) can set fruit multiple times through continuous flowering. They flower, set fruit, ripen quickly — and if left unpicked spoil even faster. Unlike currants, these fruits do not wait.

Stone fruits, such as pears, apricots, cherries, and peaches, are part of the *Prunus* genus. They have a fleshy skin and soft flesh underneath. These species are prone to damage at any time since they are soft and have early-blooming flowers. Cherries and pears are fairly consistent, but apricots and peaches are better suited to protected microclimates in northern climates. Unprotected, apricots and peaches may only fruit once every few years when spring frosts stop early, unless they're planted within ten feet of the south side of a house for warmth and frost protection.

All hanging edibles from a tree such as acorns and pine nuts, as well as grains and beans, are also fruits. Each has its special harvest and post-harvest processing techniques, which are beyond the scope of this book.

Patterns of Growth

The patterns by which plants actually grow are based on both internal and external stimuli. One of the most readily recognized ways that a plant elongates itself or grows is the way that a vine such as a morning glory, pole bean, or pea twines or spirals to a vertical support or along the ground, clambering over other vegetation or natural structures. This is called spiraling and is one of the forms most often replicated in Permaculture designs. It is one of the internal stimuli patterns for growth and occurs by an alternating distribution of plant growth hormones to opposing sides of the growing stem.

SELECTING PLANTS FOR GUILD DESIGN

Other internal patterns include nodding — a pattern legume seeds use to push their way out of the soil — and contraction, which is a mechanism that a plant uses to send young bulblets deeper into the soil toward a safe zone where they will have ample moisture and protection from temperature extremes. Nastic growth movements occur when there is an alternating distribution of growth hormones causing leaves to bend either downward or upward.

External stimuli, touched upon earlier, include thermoperiodism, or the range of optimal temperature for a given species. Dormancy is a state that occurs when a plant's environmental needs are not being met, and this can occur during any season of the year. During winter, of course, the plant is dormant, but in a severe drought situation during the summer it may also turn dormant — just holding in place until cooler temperatures and adequate moisture return. Quiescence is a state of seed dormancy wherein a seed will not germinate but conserve its stored resources until optimal conditions return.

In an external stimulus, a force or stimulus causes the release of a responding plant hormone leading to an asymmetric or uneven growth response on that side of the plant. A stimulus is then triggered, resulting in a corresponding growth response on the opposite side. And so it goes, alternating from one side to another as long as the stimulus recurs.

Photoperiodism is the response of a plant to the stimulus of light. A plant stem can be said to be positively phototropic, and the same can be said of the response of stem and root to the forces of gravity, or gravitropism. Another growth pattern used by plants in response to water shortage or high transpiration rates is leaf rollup or curling. In this instance the leaves roll up so as to present less surface area to intense sunlight and drying winds, thereby reducing evapotranspiration or water loss by as much as 95 percent.

Greens crops such as spinach (*Spinacia oleracea*), lettuce (*Lactuca sativa*), and kale (*Brassica oleracea* var. *acephala*) usually have their tastiest leaves when grown quickly under optimal conditions and with sufficient water, whereas many herbs and main-crop plants have evolved to deal with drought difficulties that occur with regularity every season. Stress can also have a beneficial effect on the nutrient composition of plants grown for protein and starch. Higher stresses result in a better nutrient load in the crop; if every need is met throughout the season, however, the plant may not reach its full nutrient potential. If growth is too lush as a result of too-plentiful moisture then an excess of green growth can result in a yield that, because internal plant resources were diverted to the leaves, is lower for the desired starch and protein components.

Which brings us to the old adage of *moderation in all things*. Forcing plants to grow too quickly can lead to them having fewer resources to deal with the tough times that inevitably occur each year. If you hold them back to achieve a very late-season crop extension, you risk losing all in a risky gamble should a cold spell come early. When you're trying to determine needs, be sure to think of the requirements of both the site and those species that you want to be a part of it.

While it's okay to tweak things a bit regarding yields and varieties, it's also best to stay within or near the hardiness zone parameters for the guild residents. It is all right to experiment a little, but never chance the yields on a single group of species that mature all at the same time. Do not gamble everything on an early or late harvest but instead ensure that a guild's yields are extended throughout all seasons so that every component of the design, every denizen of the larger guild, is supported, maintaining the intended resiliency of the site.

Growth rates vary among seasons and each species' cultivars. Rates of growth can be controlled by when and how a plant is set in the ground as well as by careful management of limiting environmental stressors. A season can be prolonged or delayed by withholding nutrients at critical growth times, resulting in a delayed harvest. An early or larger yield can be achieved by providing extra nutrients while a plant is young. However, a limited period of unmanaged stress can also help a species to set a root system that is deeper and extends feeding roots wider in search of food. This will lead to a guild of species more adaptable to floods or drought.

Optimum growth rates depend on a narrow range of conditions being met for any species. Germination and early growth differ among plants such as cool-soil-loving radishes (*Raphanus sativus*) and warmth-loving basil (*Ocimum* species). A radish may sprout and grow quickly in cool spring weather and mature as the days become warmer, but will go to seed in the heat of summer. Basil, though, will not germinate until soil temperatures have warmed and will add the most growth in warm summer weather.

All this sounds fairly simple except that there are many more variables at work governing plant growth than just soil and air temperatures. Day length controls flowering and fruit maturation for most crops. The production of plant hormones needed at specific times in growth is controlled by the factors of temperature, day length, quality of available sunlight, moisture, and the presence of beneficial mutualist bacterial and fungal species.

Populating the Guild

Stocking the Permaculture guild with plant material requires considerable expense if you're purchasing the material from a nursery. But you can avoid a large portion of this cost by growing the plants yourself. New plants can be acquired by growing them from seed as well as by division, rooting, layering, and tissue culture, which is not a technique used outside of a laboratory setting, but may be useful in those instances when you're propagating rare plants, especially orchids, which require a concise methodology. You can, of course, allow the species you want in your guild to spread on their own accord, culling those that occur in places or situations that are inconvenient. But very often the plants you'd most like to have more of won't germinate, spread, divide easily, or even grow true to the parent source material. By becoming a more active participant in the process of natural selection, humans have for millennia been able to propagate species of plant and animal that are helpful in the search for resilient food sources.

Simple selection for desirable characteristics and the planting of the saved seeds is the most common method farmers and gardeners use to change a plant. Over time an open-pollinated plant, one that shares genetic resources openly with other members of its species, can drift in its traits from the original. If you are willing to allow this wider expression of variation in the crop, then you can create a crop with a wide range of characteristics. If you don't further select these, a diverse cultivar with different ripening times and hardiness can lead to an adaptive and resilient new range of plants called a landrace.

Closed pollination, also called self-pollination, is the fertilization of a flower by pollen from the same plant or a clone of the same plant. This does not result in a new cultivar with different characteristics. Hybrid pollination is the mixing of genetic material from different cultivars or even species to select for traits, usually for disease resistance, fruit size, or hardiness. Further inbreeding of a hybrid line can lead to a stable version that can be considered self-fertile.

SELECTING PLANTS FOR GUILD DESIGN

Sexual Propagation

A seed is a dormant embryonic plant encased in a protective coating. When exposed to sufficient moisture in soil and temperature conditions favorable to that particular plant species, the stored energy in the form of starches is converted to sugars. This process is enabled by an enzymatic reaction created by a hormone, which allows the dormant plant embryo to awaken and burst out of the protective but softened seed coating to form roots, stem, and leaves.

All this seems simple enough, but each individual species has different requirements for the process and has set up various roadblocks to germination in order to protect the embryo from damage. If there is too much moisture the seed may rot; if there's too little, it may fail to germinate. Conditions of moisture, temperature, pH, light exposure, dormancy timing, and natural chemical inhibitors can all affect the process of germination.

Stratification

Careful observation of seed germination shows that some will not begin to grow until they have been exposed to cold for a period of several months in order to break the natural dormancy of the plant embryo within the seed. A seed may be dormant because favorable growing conditions are not present during a dry or cold season depending on the individual plant species' adaptation to its biome. Example of seeds requiring a cold stratification are wild rice (*Zizania palustris*) and roses (*Rosa* species).

Wild rice seed, though it ripens and drops from the seed stalk into water in late summer, will not grow until the following spring. It needs the winter season in order to break dormancy; it also exhibits the protective trait of multicycle germination, the staggering of growth over several seasons if optimal conditions are not met because of excessive cold or drought. Roses on the other hand have a very hard and durable seed coat that needs freeze-and-thaw cycles to crack open and allow the seed embryo to emerge. Very hard seeds like this can be nicked with a sharp knife or scarified with sandpaper; this will remove enough of the tough coating to allow moisture to penetrate and stimulate the enzyme reactions that will trigger seed embryo growth.

Other Factors Affecting Seed Growth

Some plants' seeds are natural multicycle germinators — they will stagger sprouting over several months, years, or even decades in order to ensure species survival. Queen Anne's lace (*Daucus carota*), the wild form of carrot, is an example: These seeds are viable for seven years or more and can germinate in any one of those years. Seeds may lie on the ground but be beneath a leaf cover, or might be on the surface of the soil; until either they are buried or exposed to light, they will not grow. Garlic mustard (*Alliaria petiolata*) is a woodland plant with the same ability to begin its growth over multiple years.

Some species require either darkness or light to sprout their seeds. Clovers, for instance, are best sown atop the soil, as they need light in order to germinate.

The fruit and pith surrounding some seeds contain chemical inhibitors that may need to be removed by decomposition, washing, or processing through the digestive system of an animal before the seed can grow. Excretion from an animal usually includes a coating of fertilizer with the seed, resulting in faster growth for the new plant — an example of mutual interdependencies between plant and animal. Small-seeded fruits such as mulberries, for instance, are readily consumed by birds. Other examples include the success of species such as European buckthorn (*Rhamnus cathartica*) and autumn olive (*Elaeagnus umbellata*), both consumed by birds and spread in the avian droppings at edge perching sites.

Some seeds have a very hard seed coat that must be exposed to fire in order to overcome the limitations to their germination. Jack pine, for example, can lie dormant in the soil for many years until exposed to heat intense enough to break apart the seed coat. Some prairie plants such as vine groundnut (*Pediomelum esculentum*) need fire exposure or scarification between two layers of sandpaper to be able to absorb water and begin the enzymatic transformation that leads to a successful plant.

To simulate fire for seeds that need it, you can plant them in a controllable area such as a large pot or small garden, covering them with a combustible material like pine needles or straw. Burning off that layer of straw or needles and leaving the ash on the seedbed, then adding water, allows the species to grow.

Still other plants are so inhibited in their germination that in order to speed things up we can soak the nicked seeds in a special solution of plant hormones and water. Every species has its own little quirks, and learning those needs can take a lifetime. Careful observation, replication of nature's processes, and deliberate note taking are essential to learning successful seed germination techniques. Many seed catalogs, especially those used in the nursery trade, give definitive requirements for the growing of plants from seeds along with the expected number of days until germination.

Asexual Plant Propagation

When there is only a single parent plant and thus no exchange of genetic information, identical copies of a plant result from division of root clumps, rooting of cuttings, and tissue culture propagation. The resulting plants are essentially clones of the parent. Because tissue culture involves complex laboratory procedures, we will not address it here; it isn't a commonly used procedure for the average farmer/gardener/nursery person.

Rooting and air layering use similar procedures in order to induce root formation at a leaf node. In rooting, a cutting is taken and placed in water or a moist medium such as sand at a depth that covers two to three leaf nodes. Roots will form at those locations while the above-medium leaf nodes will form leaves. You can speed up the process for reluctant species by dipping the end you'll be planting into a solution of a rooting hormone, like willow bark tea, that induces root formation. There are several hormones that can be used. Gibberellic acid 3 (GA3) is a natural hormone found in germinating seeds and is also used for rooting cuttings. Indole butyric acid is another rooting hormone though not used as often. Varying strengths of GA3 solution are used depending on the natural reluctance of woody plant material to root.

With air layering, an incision is made into a green and flexible branch of a woody plant. The wound is then treated with a rooting hormone and wrapped in damp peat moss or coir and covered with a plastic or other moisture-retardant wrap. With time, roots will form; when there are enough of them, you can remove the cutting from the parent plant and replant it as a separate organism. Many plants can be induced to form roots along a leaf node by simply bending a branch to the ground, pinning it there, and covering with soil. Annual crop examples of this are tomatoes and squashes, both of which will root at buried portions of their vines. Woody species that are shrubby, or vining, will form additional roots when portions of their stems are bent to the ground and pinned or tied there, then covered with soil or mulch. Currants, gooseberries, and grapes are all easily rooted species. Exact replicas of specific tasty or high-yielding cultivars are easily obtained at very low cost this way.

SELECTING PLANTS FOR GUILD DESIGN

Some plants are better propagated by division, whether of clumps, roots, or runners. Strawberries, for instance, are propagated by seed, runners, or clump division depending on the species. Division of root clumps can be as simple as separating dahlia tubers, or as difficult as splitting a cup plant (*Silphium perfoliatum*) crown clump.

It is worth noting that some plants are notoriously difficult to propagate successfully, and, as a result, they are rare in their natural habitats. For instance, lady's slipper orchids (Cypripedioideae family) are endangered plants best grown by stem tissue cuttings in a controlled lab environment. For this reason it is important to never collect orchid species in the wild; doing this will reduce the survival rates of a local population of that plant. Many orchid species are dependent upon specific associations with fungi to ensure their growth. Specific instructions on how to expand populations of plant species can be found in several sources. Some that are helpful include the *Prairie Propagation Handbook* by Harold Rock, published in 1971, and *Handbook on Propagation,* a 1957 book edited by Lewis Lipp.

COPPICING

Coppicing is a source of cuttings that can be used for a multitude of purposes. The wood repurposed from cutting down existing growth can be used to grow additional plants, for crafts, and to produce charcoal. Nearly any woody species can be coppiced or cut to the ground on a rotation cycle specific to that species' regeneration rate. Exceptions include many of the pine, spruce, and hemlock species, as well as others. Coppicing is achieved by cutting a plant six inches from the ground at an angle away from the center of the clump or trunk to direct water runoff away from the center (where it would induce rot and subsequent fungal and disease infections). The resulting stump is called a stool. It's allowed to regrow and form a new tree or shrub; it can be cut again at a later time.

Species suitable for coppicing include willow (*Salix* species), which can be cut at yearly intervals because of its rapid growth; hazel (*Corylus* species), cut at eight-year spans; and chestnut (*Castanea* species), which is harvested for cut poles every fifteen years.

GRAFTING

Many superior-tasting or otherwise valuable fruit and nut cultivars have a problem with disease susceptibility or are weak-rooted. Additionally, it is sometimes desirable from a grower's viewpoint to have a tree that will fit into a specific space or soil profile. Special varieties of rootstock trees for specific species are used to address these concerns.

To induce earlier fruiting and lower tree heights, scion wood (the fruiting portion of wood) is grafted or spliced onto another cultivar whose rootstocks produce a tree that varies from five to fifty feet in height with tolerances built in to wet, windy, cold, sandy to clay soil site characteristics, all based on your needs as the grower. Sometimes a second or even third cultivar needs to be grafted between the rootstock and scion wood. This is called an interstem graft and is used when the scion and rootstocks are incompatible. To make it even more complicated, some growers will graft several species of apples or pears onto the same tree. These are called fruit cocktail trees. The more grafts are on a tree, however, the more stress it will be under and thus the greater the possibility of a yield failure if something goes wrong.

Other fruits in addition to apples can be grafted. Pears (*Pyrus* species) are grafted onto mountain ash (*Sorbus* species), medlar (*Mespilus germanica*), hawthorn (*Crataegus* species), and quince (*Cydonia oblonga*). Plums and cherries (*Prunus* species) can share the same rootstocks, and peaches and

An apple branch is grafted to a scion from another tree.

nectarines (both are *Prunus persica*) and apricots (*Prunus armeniaca*) also have some commonalities. Many nut trees get grafted onto separate rootstocks that have superior tolerance to specific site conditions such as soil and root hardiness. The most common example is commercial cultivars of English and black walnuts (*Juglans* species).

The Natural Range of Plants

The range that any particular species, plant or animal, occurs within—how it is arranged in the larger landscape—is its species distribution. This pattern changes in response to resource availability and seasonal weather factors. Species distributions have been altered by humans as they spread favored species around the world through trade and agriculture.

There are three basic types of distribution. The first, clumped distribution, is the most common. Distance between members of a species is minimized. Examples include plants that sucker from the roots or stolons forming a clump such as the pawpaw (*Asimina triloba*). Solomon's seal (*Polygonatum* species) is another example, as are most of the bramble berries (*Rubus* species).

A second form of species arrangement is uniform distribution, where members of a species are evenly spaced in an area. Many allelopathic plants such as walnuts (*Juglans* species) discourage growth not just of other species nearby but also members

SELECTING PLANTS FOR GUILD DESIGN

LANDRACE/HEIRLOOM VARIETIES

Landrace plants are sexually propagated and open-pollinated by many plants within a limited climate space. High genetic variation causes variable traits, making some plants resilient to disease, herbivory, nutrient deficiencies, drought, and/or cold weather. Over time all plants are especially suited to the local ecology. From a production or commodity standpoint, this causes lesser yields of any one cultivar, less "appetite appeal," and inconsistent "quality," reducing predictable outcomes. On the other hand, this is the genetic pool for all other plants from the processes below. In seasonally unfavorable conditions, at least some plants are likely to yield a good crop. These numerous landrace plants are sources of genetic material for hybrid, cloned, and GMO plants.

HYBRID

- Sexual propagation as plants are cross-pollinated by hand from two known parent plants.
- Seeds are trialed for best characteristics. High-volume output yields consistent quality for commodity use.
- Postseason collected seed unlikely to perform and may contaminate heirloom varietal lines.
- Out-crossed seeds diminish parent traits. Inconsistent seed "quality."
- May be high-maintenance or require high inputs.

CLONE

- Asexual propagation by grafting scion or rooting plant material with no variation in plants and developing duplicate plants by default.
- Used for predictable, consistent quality, and commodity crops.
- Short-lived in some cases, with high labor and low volume.
- Used mostly in woody plants.
- No seed value.

GMO

- Asexual propagation by grafting genetic material for controlled traits.
- Each plant is a duplicate by design and mass-produced for high-volume output.
- Identical genetics yields consistent quality for commodity use.
- Collected seed is unlikely to perform and may be illegal to plant.
- GMOs may contaminate heirloom varieties, may have herbicidal tendencies, and may be harmful to bees and pollinators.

Landraces.

of the same plant family. This reduces competition for resources on a site by the same species. A similar process occurs when agriculturists plant a crop in uniform rows, and when orchard trees are set at even distances from one another.

Random distribution, or unpredictable spacing of members of a given species, occurs rarely in nature. When it does, it's only because environmental conditions and available resources are present at consistent levels. Examples include wind- and water-dispersed seeds. Animals are a factor in a species' seed dispersal, but the manner — while unpredictable — involves moving plant material by following a regular foraging pattern. Just because the pattern is not readily apparent does not mean that it has no value to either the animal or the plant.

On a larger scale, species useful to humans have evolved in regions across the world and still have wild relatives serving as a resource base for further crop diversity and resilience to disease and weather variance. Eurasia is a home for much of the fruit grown by humans, including apples, pears, plums, strawberries, and more. The ancestral homes of many fruit tree species are the steppes of Central Asia and China.

South Asia is the source of paddy rice, sugarcane, and other species, while Africa is a home for lesser-known grains such as teff as well as many well-known crops such as watermelon, wheat, sesame, and coffee. The Americas hold several areas of major plant origins. The eastern area of the United States was one of several points of origin for the common bean (*Phaseolus* species). Perhaps its greatest contribution was the domestication of the sunflower.

Central America, also called Mesoamerica, is the source for all maize (corn) genetic diversity and where an ancestor called teosinte (*Zea mays* ssp. *parviglumis, Z. mexicana, Z. diploperennis,* and *Z. luxurians*) still grows wild. Squashes and beans also originated there. In South America the Amazon forests were the original source of rubber trees; their dispersal has transformed forestry throughout tropical Asia. Recent diseases affecting rubber plantations worldwide have led to a renewed interest in the genetic diversity and resilience of some of the related species in the Amazon forests. Another game-changing plant species that originated in the American tropics was the common sweet potato (*Ipomoea batatas*). Its cultivation at higher elevations in China and the Philippines led to a rise in the quality and quantity of foodstuffs that was to a large degree responsible for the surge in population in East Asia over the last two hundred years.[2]

Peru is home to a wide variety of useful high-altitude plants that formed the basis of sophisticated cultures spanning thousands of years. Potatoes of course are the most significant example; dependence on a very narrow genetic resource pool was an integral contributor to the famine that beset the European island of Ireland in the early nineteenth century, setting off a wave of migrations. Other plant contributions from Peru include quinoa (*Chenopodium quinoa*), mashua (*Tropaeolum tuberosum*), and lesser-known root plants that may all have an important place in human and animal nutrition.

Landraces, as mentioned previously, are selections of commonly grown useful species that through continued natural and to a lesser degree human selection have stabilized genetically to enjoy a high degree of adaptation to a particular set of environmental conditions. They usually retain a greater degree of genetic variability, allowing the group to adapt to changes in growing conditions. They are unlike either common open-pollinated or hybrid varieties in that they lack a high degree of uniformity. Where their main value lies is in the ability to survive a wide range of conditions and still give a yield. As a reserve of genetic information, they are a valuable resource for resiliency that can be put to use to stave off diseases such as the potato blight.

Many of the landrace maize, squash, and bean collections originated with First Nations peoples of the Mesoamerican region. Preservation of these cultivars by local farmers has kept alive varieties that are locally adapted to drought and to low nutrient availability, and that exhibit high nutritional values. Conditions of drought in the western plains of the Americas over the last decade have led to renewed interest in the resiliency of these native cultivars.[2] In the Near East and Africa are landraces of ancient grains and fruits that can be of use when breeding nutrition and resilience to diseases and pests back into the overbred modern versions of many crops.

How does this all relate to the plant guild? You can enhance diversity and resiliency in guild relationships by adding locally adapted varieties of desired species. This is especially true for sites where environmental conditions can be difficult due to flooding, droughts, and pest predation. Maintaining a high level of diversity also creates more niche opportunities for inclusion of further species.[3] Most important, landrace varieties have the ability to survive environmental stress and still provide an intermediate yield when grown in a reduced-input agriculture.

The Importance of Diversity

Alpha diversity occurs within a guild's niche when a few species provide all the functions necessary for growth. The classic annual example of the Three

SELECTING PLANTS FOR GUILD DESIGN

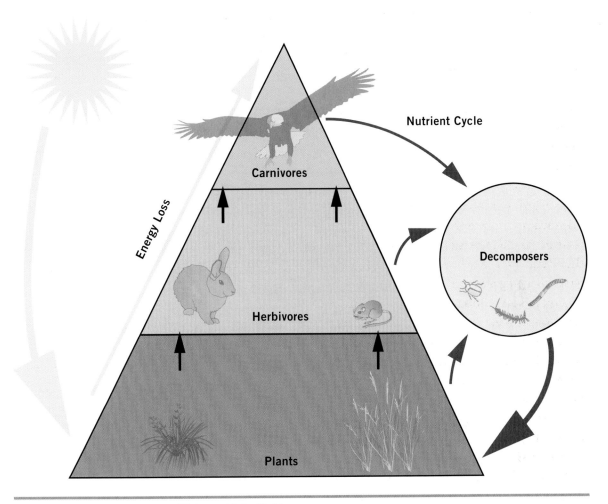

Energy transfer. Illustration by Daniel Halsey and Kellen Kirchberg

Sisters Guild — corn, beans, and squash — is an example illustrating this level of diversity. Corn provides the support for the beans to grow on, in return putting atmospheric nitrogen back into the soil. Squash shades the soil, preventing evaporation and thereby cooling the roots of the corn as well as suppressing weeds. A happy but small family group.

On the other hand, there needs to be more diversity among niches to provide services to other residents of a functional integrated garden. Beta diversity is filling of the spaces between the niches, and develops a strong and redundant environment with patches of resources and services for each ecosystem need. Returning to the example of the corn, beans, and squash guild, it can be improved by adding a few more species to enhance available pollen for insects. Catnip (*Nepeta cataria*) can also act as a deterrent to pest species by masking the odors of corn and squash. Crimson clover (*Trifolium* species) can be sown in the open spaces between the squash before they have grown large and will cover and shade the soil, add nitrogen and biomass to it, and attract even more pollinators. Rocky Mountain bee plant (*Cleome serrulata*) attracts pollinators and is an important food plant grown in dryland corn guilds. Sunflowers (*Helianthus* species), amaranth (*Amaranthus* species), and quinoa (*Chenopodium quinoa*) can all be grown on the edges of the corn guilds and add to its beta diversity by attracting bird species, butterflies, and bees. A perennial hedge of *Silphium* species will die to the ground each year but grow back to a seven-foot height of pollinator-attracting and goldfinch-feeding wonder by late summer every year.

The same holds true for any perennial mixed-species garden or orchard. Previously the role of comfrey (*Symphytum officinale*) in aiding the growth of a young tree was mentioned. The comfrey provided shade to the young pine while also bringing subsoil minerals into its root zone. Looks like a good relationship, but how can it be improved? Adding in early-blooming, even shorter ground covers such as wild ginger (*Asarum caudatum*) and violets attracts and sustains early-season pollinators and protects the soil from erosion. Fungi such as maitake (*Grifola frondosa*) can set up beneficial relationships in the soil with the young tree's roots. Other tree and shrub species can be used to reduce the wind velocity or direct windblown snow to another location. All of these changes enhance the functional niches among the members of a guild in the forest garden, the beta diversity.

When you step outdoors or look out the window at your integrated forest garden, polyculture guild, or whatever you choose to call it, the view is going to be different each time. It is alive with life, teeming and teamed with species plant, animal, and fungal all changing with every input of time and visits from out-of-guild energies: the actions of wind, water, birds, insects, and of course you. The light levels, seasons, and microclimates are always altering the niches available, offering new opportunities to include species you had not previously considered. Go for it. If you see a site where something new may fit, then add it in. The new species or feature — a rock, a log, what have you — will enrich the guild with even more niche sites.

The integrated forest garden is the goal, and beginning the planning of it is only a very small portion of the journey. After a few years the garden may have taken on entirely new characteristics that you did not foresee. Getting there you will realize that you have more questions than answers. When that happens you will be close to achieving the real goal: listening to and acting on the messages that the natural world provides. Those messages are that you are not alone — all of nature is there to help.

CHAPTER FOUR

Trees: The Essence of the Plant Guild

Much can they praise the trees so straight and high,
The sailing pine, the cedar proud and tall,
The vine-prop elm, the poplar never dry,
The builder oak, sole king of forests all,
The aspin good for staves, the cypress funeral,
The laurel, meed of mighty conquerors
And poets sage, the fir that weepiest still,
The yew obedient to the bender's will,
The birch for shafts, the sallow for the mill,
The myrrh sweet-bleeding in the bitter wound,
The warlike beech, the ash for nothing ill,
The fruitful olive, and the platane round,
The carver holm, the maple seldom inward sound.

— EDMUND SPENSER, THE FAERIE QUEENE

Trees perform countless functions in an integrated food forest, filling a major niche in plant guilds and polycultures. They are the centerpieces of our food plant cultures that all other plants in the polyculture revolve around. Trees are the oldest living, biological creatures on this planet, some anchored to the earth for millennia. Trees, in general, are diversified in time and space, dynamically stable, productive and food self-sufficient. They conserve and regenerate natural resources (water, soil, nutrients, germplasm), have stunning economic potential, and can be utilized for socially and culturally acceptable technology. Trees create soil cover at leaf fall. Nutrient cycling is ongoing, as is sediment capture with water harvest and conservation. They have productive diversity not only within the species itself, but with the functional relationships that are created in plant guilds, in food forests, and in the wild. They afford protection to other plants and crops.

Trees are a major player in the ecological order. If it's true that plant guilds are multifunctional, then it would also be true that each plant in a guild is itself multifunctional — almost a guild in itself. Trees are part and parcel of diverse polycultures, require no tillage, do their own selective weeding, chop and drop themselves, use their own residues and those around them from their neighbors and the soil fauna, food, and nutrients. There are many niches for production in trees and places for animals to nest and feed, vertically and horizontally. Trees create living and nonliving barriers. They are very much a part of regional biodiversity. Forest enrichment is all about trees.

Trees are naturally built into crop mosaics. They are the ultimate goal in the process of natural succession, and they create the mechanics of the ongoing give-and-take, in-and-out of human breathing cycles. Trees fix carbon, a huge issue in today's willy-nilly climate extremes, and they create the biomass that feeds and clothes us all, giving us shelter and pleasure beyond anything that we could cook up artistically.

There are many nut- and fruit-bearing trees that could be included in an integrated forest garden. In this chapter we focus on the oak, the pine, the maple, and Old Man Hickory. These four trees are widespread across North America and in many other countries as well. Not all of the four will be included in our plant guilds, but the delineation of these species is a good place to begin the process of understanding how trees work in all their multifarious guises. In this chapter the central focus is on the oak tree and its many uses. It is prevalent throughout the world, and is found in many biomes on all continents. The uses of the oak tree, whether for human utility or the services that it offers to an ecosystem, are innumerable, so the oak is a good place to begin in our research efforts. It can give us direction and insight into how we go about selecting plants for all of our polycultures.

J. Russell Smith, in his landmark 1929 book *Tree Crops: A Permanent Agriculture,* was the first writer in North America to recognize the economic and ecological potential of a perennial agriculture based on the tree. He spoke about the value of such tree crops as the honey locust, mesquite, maple, mulberry, persimmon, chestnut, oak, walnut, and hickory as food crops for humans and forage for animals. We may find that all of these species, and more, are already in existence on our property, doing their work in the ecosystem of the backyard and producing an immeasurable yield of food, medicine, and utility for our use; we do not have to start from scratch in order to begin the harvest!

Trees are the fundamental place to begin with when designing and installing plant guilds. Plan for them first before you plan for anything else in the guild, and plant them first, before you plant anything else. Work out from there, and you are well on your way....

Duir: Opening the Door on the Oak Tree

Wayne was three years old when this oak tree was planted in front of his house in suburban New Jersey. He was small and the oak seemed a mighty giant, its pointed leaves and unusually shaped fruits a delight to a child's passionate eye. It became a neighborhood challenge to see who could scale the oak and hold on for dear life as the wind blew and the top swayed. It's been fifty or so years now. He's gone back to visit that tree. It's still larger than life — the massive trunk many times the diameter of his torso, offering enough acorns to feed a thousand squirrels.

The oak tree is a powerful symbol in many cultures across the world, and for good reason. Its ability to grow and survive in just about every climate, and its strength and beauty, are unmatched in the world of trees. Oak (*Quercus* species) provides sustenance for many, something that human beings should take to heart. It is, like many other tree

TREES: THE ESSENCE OF THE PLANT GUILD

The majestic oak tree.

species, the centerpiece of any guild, and it shares its bounty with all that moves across the landscape.

In Permaculture systems oaks are planted as an overstory tree, the tallest plant in the guild arching over all other herbaceous, shrub, and tree species. In many neighborhoods across the United States there is already a proliferation of oaks that were planted when these neighborhoods were constructed. The oak-hickory forest that stretches from the Atlantic Ocean to the Mississippi River is the most predominant type of forest biome between these great bodies of water. It is up to us to insert a diversity of plants in the understory of the oak for our plant guilds, or to forage the neighborhood in autumn when the acorns come ripe.

The oak is found on every continent. There are fifty-eight tree and ten shrub species of oak in the United States, and they feature countless leaf shape variations and have a tendency to hybridize.

The Indo-European family of languages that ranges from Sanskrit in the east to Gaelic in the west has in common the root word for oak. In Sanskrit the name for oak is *duir;* that word is also used for trees in general. In Gaelic, too, the word is *duir.* The English word *door* comes from this ancient link as doors were traditionally made from the strongest available wood, oak.

There are over sixty references to oak in the Bible. Druid rituals revolved around the sacred tree, and Zeus's oracle was set in the midst of an oak grove at Dodona. Stability, longevity, strength, beauty.

Because acorns are so bitter to the taste, their tannic acid content must be leached out. This same tannic acid, as we shall see shortly, can be used as a medicinal. The white oak (*Quercus alba*) contains less tannic acid than the red oak (*Q. rubra*) and its acorn is much sweeter. Nonetheless, the white acorn must also be leached of its tannic acid for consumption.

Red and White Oaks

Oak is a long-lived tree. The red oak can survive to an age of three hundred years, and the white oak to six hundred. Most oak species fall within these two subsets, red and white. The red oaks are so named because the wood has a reddish color; that of the white oak is lighter. They can also be distinguished by the shape of their lobes: White oaks have rounded lobes and red oaks pointed, most of the time with small hairs projecting from the point of the lobe.

RED OAK

One of the major differences between the two groups is the length of time that it takes for the seeds to mature after pollination of the flowers. Red oak group acorns take two years, and although the acorns are higher in fats they are also higher in bitter tannins that need to be leached out or processed before we can use them as food. Deep-rooted and with strong lateral roots, the red oak's natural forest companions are sugar maple and hickory, particularly in the eastern woodlands biome of the United States. In some areas of the US this is changing. Maples, which grow readily in the shade of the understory, are overtaking oaks in some instances. The oak can grow as an understory tree since it is somewhat shade-tolerant, though not as much so as maples. Squirrels and chipmunks cache away the bitter acorns to eat later after the tannins have leached; indigenous peoples have buried the bitter acorns along sandy streambanks for the same purpose, and also crushed them, placed them in baskets or woven bags, and hung them from a branch submerged in the moving water.

Woolly hairs line the cup of the red oak's acorn. The bark is gray to reddish brown and vertically fissured. Red oaks grow up to ninety feet tall, with alternate, seven- to eleven-lobed leaves. They are

TREES: THE ESSENCE OF THE PLANT GUILD

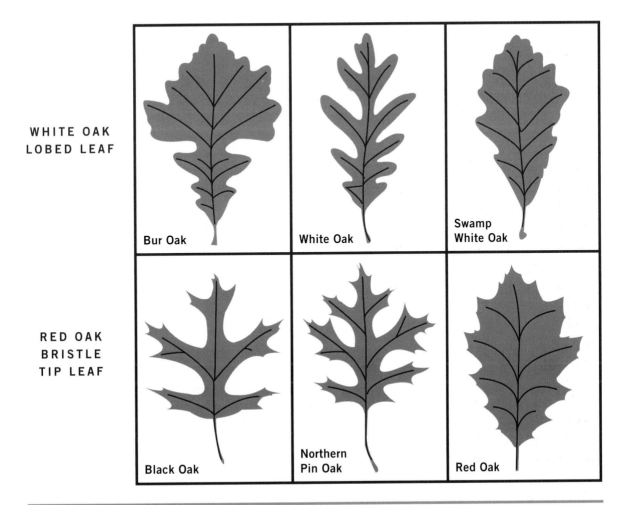

Oak leaf types: red and white.

the fastest growing of the oaks, adding two inches in diameter and twelve inches in height per year.

WHITE OAK

The white oak and the species in its group bear annual acorn crops. These sweet acorns contain less tannin and require less processing than those from red oaks. For most wildlife in the Northern Hemisphere, acorns from the white oak are a preferred food source. Most trees in this group begin to crop acorns by fifty years of age, but selected cultivars can begin to yield as young as age ten. The acorns are 6 percent protein and 65 percent carbohydrate. While the taste of acorn meal or flour is fairly bland, its nutritional content makes it a great carrier for other flavors such as those from nuts and berries.

Some of the trees in this low-tannin, tasty-acorn group are the chinkapin, bur, swamp white, chestnut, and cork oaks. Shade-intolerant, the white oak is a co-dominant species in eastern North American forests, sharing space with hickories, white ash, walnut, black cherry, and other oaks.

They grow upward of a hundred feet. The leaves sprout alternately with five to nine rounded lobes.

The acorn cup is shallow; it's smooth on the inside, has knobby scales on the outside, and covers 25 percent of the acorn, which matures in six months. The bark is whitish gray and scaly. White oak lumber is the preferred wood of cabinetmaking, finish work, and veneers among woodworkers.

Patterns of Growth

All oak species are veritable ecosystems of diversity with symbiotic insect and fungal relationships that form the basis of much of a forest or savanna biome. Their seeds are spread by birds and small mammals that harvest and store the acorns, or drop them elsewhere in the area. The delectable acorn is a mouthful for birds and mammals during the lean winter months. Jays, grosbeaks, wild turkeys, ducks, scarlet tanagers, quail, grouse, woodpeckers, titmice, pigeons, larks, nuthatches, grackles, varied thrushes, squirrels, chipmunks, raccoons, bears, and deer all feast on them. Gray squirrels will go to any acrobatic length to acquire them. The squirrels tend to eat more white oak acorns than red (apparently the higher tannic acid content of red acorns is a mild deterrent), burying any excess for later use. The forgotten acorns of the buried batch help to disperse and propagate new stands of oak. Bears, rabbits, squirrels, and other fur-bearing and game animals also eat the bark and wood of the oak. Hoofed browsers such as deer, elk, peccary, and sheep munch on twigs and foliage.

Oaks are excellent habitat indicators, mostly proliferating in hydric-mesic (wet and well-drained) areas, but some can be found in dry (xeric) areas. Oaks inhabit eastern oak-hickory forests, mixed Appalachian forests, northern pine-oak forests, alpine communities, Great Basin desert-mountain slopes, midwestern savannas, Californian canyons, and grasslands. They are the most widespread broad-leaved trees in America. Species include scarlet, willow, blackjack, bur, live, scrub, swamp, chestnut oaks, Virginia live, turkey, myrtle, post, pin, chinkapin, red, white, shumard, shingle, overcup oaks . . . the list goes on and on.

Functional Uses of the Oaks: The Department Store of the Forest

Other than habitat or centers of diversity for nonhuman species, there are many purposes to which human societies have put the oak and its products and services. Whereas the cattail is called the "supermarket of the swamp," the oak could be called the "department store of the forest." Different parts of the tree are used as wild or cultivated edibles, as medicinal and first-aid remedies, for crafting splint baskets, as a dye, and to build structures and make tools.

As a source of durable wood products, the white oaks are almost unsurpassed among tree species. White oak wood is rot resistant and is a premier carpentry wood; it was used for ships' masts for hundreds of years. Red oak is less durable and more prone to splitting, but it's used also in furniture making and plywood veneers.

Oak bark was traditionally a source of tannins for the leather industry. The tannic acid content of any part of the oak tree is used to process animal hides when the brain of the animal (traditionally used for tanning) is not available. After the hide is scraped of fat, meat, and hair, it is soaked and wrung out several times in a solution of brains and water or oak tree decoction. It is kneaded until dry. The natural brain tanning or plant tanning process ensures the softest and most beautiful of all leathers.

OAK FIRST AID AND MEDICINE

Indigenous Native Americans used the bark medicinally. As a first-aid and medicinal remedy oak bark has internal and external applications. Prepare a decoction of oak bark as a gargle for sore throat after a cold night in the brush or illness resulting from a

Tannic acid from the oak is used for tanning hides. Photo from Huarache blog

drafty window. As a skin wash for cuts, scrapes, and insect bites, it is especially soothing. After sloshing through a mosquito-dense swamp, chew a bunch of oak leaves, rub and poultice them on the bites, and enjoy quick relief. As an astringent for all kinds of ailments, oak bark is applied topically to fever blisters, herpes sores, ringworm ulcers, varicose veins, and various other skin disorders in the form of a fomentation or salve. Apply fomentations throughout the night to swollen glands, lymphatic swellings, goiter, and mumps. Oak bark decoction is an excellent douche for vaginal infections and a retention enema for hemorrhoids.

Internally, decoct and drink sparingly for diarrhea; bladder weakness; bleeding in the stomach; lungs, and rectum; hemorrhoids; prolapsed uterus; and varicose veins. A teaspoon of the decoction is snuffed up the nose to stop nosebleed. Oak bark is highly antiseptic. It will clean the stomach, clear mucous discharge, and help remove gall- and kidney stones.

OAK FOOD

Food, of course, is the best use for the oak as the harvest of acorns is a renewable resource. Around the world people have been eating acorns for thousands of years; some societies based their entire food culture on the acorn. Mediterranean cultures such as Greece and Rome used the acorn for food, as did much of Western Europe and elsewhere. Even today some Asian supermarkets carry acorn flour in the United States as well as in their native countries.

Leaching of the tannins can help to make the acorn meal palatable. This meal needs to be blended with another flour, nuts, fruits, vegetables, or herbs because of its bland flavor. Native Americans mixed it with cornmeal to make into bread. They blended acorns with dried or fresh berries, meat and fats, and maple sugar and fried it as cakes. John Muir said that the acorn bread he was taught to make by Native friends was a most compact and strength-giving food. To make acorn meal, simmer the acorn kernels in several changes of water until the water does not turn its characteristic brown color. The kernels can also be dried, mashed or ground, placed in a tightly woven, porous bag, and then simmered or left in a swift-running stream for a number of days to remove the bitterness and astringency. Once the tannic acid has been removed, the acorns are eaten whole or dried, ground into flour, and used in breads, muffins, soups, or ashcakes. There is no gluten content in acorn flour, so if you're making bread, acorn flour will not hold together as wheat does. It's better to use it as an additive than attempt to use it by itself in bread making.

Ashcakes are a delicacy not to be missed by backcountry travelers and collectors of acorns along the byways and front lawns of America. Mix acorn flour with water until a doughy consistency is obtained, knead into a flat, round pancake, fold over (you can put raisins, brown sugar, whatever pleases the palate inside before folding, as in a turnover), then place in hot ashes and coals on an open fire. Endless variations are possible. Native peoples would make ashcakes and take them along on excursions as a delicious and nutritious food that would last for many months.

Oak Tree Crafts and Social Gatherings

Because white oak (*Quercus alba*) wood is strong and close-grained, it makes excellent splints for basket weaving. It is split several times to suitable thickness, shaved down smooth, and allowed to dry. Soak when needed.

The bark of the black oak (*Q. velutina*) makes a brilliant yellow dye for basket elements or cloth. The black oak can be found in the Midwest, eastern Texas, Florida, and as far west as Nebraska. If you scratch a twig with a fingernail, a yellow or orange color will appear. Collect the bark from a fallen tree. Heat it with a mordant in water until it dissolves. Yellow color will be present. Place an article to be dyed into the bath until the desired depth of color is reached. Cool down and rinse.

For wood crafting, the oak is amazing. White oak is superb for creating a bow (remember the arrows!).

Probably the most recent North American cultures to use the food of the oak as a diet mainstay were nineteenth-century California indigenous peoples. Every tribe had one thing in common: the use of the oak acorn as a principal dietary food. The annual acorn harvest camp was an important social event in their culture and has its counterpart in the Upper Midwest in the wild rice camps of the Menominee, Ojibwa, and Winnebago peoples.

The galls of the oak tree are a source of sweet sap in the summer. The gall is not a fruit but an insect parasitization of oak leaves and stems. Lance a gall and suck out the sweet syrup. Edible and medicinal fungi that grow with or on oaks or their stumps include oyster, shiitake, turkey tail, nameko, chicken of the woods, white elm, cinnamon cap, lion's mane, maitake, and reishi mushrooms, all

TREES: THE ESSENCE OF THE PLANT GUILD

Making an oak bow on a shaving horse. Oak wood's close grain and resiliency allow for maximum tiller (bend). It is especially suited for the traditionally lengthier eastern woodlands Native-style bows. It's also excellent for fish and frog spears, pump drill weights (a Native drill), or any kind of project requiring a strong, rot-resistant, beautiful wood.

quite palatable and considered delicacies. The leaves of English white oak can be fermented to make a delicious wine. The tannins in the leaves of other oak species are used for their astringent medicinal qualities as an alternative to quinine.

Oak galls can be harvested for ink use. Red oak stems are used as a child's bubble wand. Wood is used for ships' masts and lumber, barrel staves and other cooperage, shingles and siding, fences, furniture and frames, home building, cabinetry, veneers, posts, and fuel. The logs and stumps of oak are even used as a timepiece in the science of dendrochronology, or telling time by counting tree rings. Acorns are also harvested indirectly by turning swine into pasture to feed on the fallen seeds. In much of Europe pigs were traditionally fattened on acorns and beechnuts.

Coppicing and Pollarding

The oak, like most broad-leaved tree species, can be coppiced and pollarded as a renewable source of fuel wood for charcoal production, poles, posts, and other wood projects. Oaks have a strong tendency to regrow from the base after fire or other disturbance. This tendency is most pronounced in the bur oak (*Quercus macrocarpa*) and the cork oak (*Q. suber*). The former has a corky, waxy layer that protects the tree from being killed by quick-burning prairie fires. The tree is able to regrow from its branches or base as long as the roots have survived. It produces a superior lumber and delicious acorns. Bur oak openings in the savanna regions of the Midwest provide an anchor that other species such as bramble berries (*Rubus* species) and hazelnuts (*Corylus* species) can attach to.

Similarly the cork oak has a thick layer of cambium or inner bark that serves the same function: to protect the tree from fire. This cork has been sustainably harvested for over two thousand years. The ancient Greeks and Romans used cork floats to aid soldiers in crossing streams. Non-military uses include bottle and jar stoppers, insulation, soda-pop bottle cap liners, flooring, corkboards, and, in prehistoric times, clothing. The cork oak savannas of Spain and Portugal are in danger of being completely destroyed as the substitution of plastic corks by the wine industry has reduced demand for the natural product. The mixed forest farms of cork oak, holm oak, grapes, annual grains, and rotational coppice grazing are giving way to modern high-yielding agricultural systems popularized in the United States and Canada. Curiously, the bur oak with its similar corky cambium could become a local source to begin a cork industry in the US Midwest.

Watershed Management

Another function that the oak — and for that matter most trees — serves in the landscape is watershed management. Trees help anchor soil by establishing symbiotic relationships with mycorrhizal fungi. These fungi act as a nutritional interface with the tree roots, providing nutrients and water; the tree provides carbohydrates, which fungi cannot make since they don't photosynthesize. The fungi also hold water in the soil through their network of mycelial hyphae, or rootlike strands, which literally glue the soil together.

The Montados and Dehesas: Restoring Oak Savannas Worldwide

Montado is a word for the wood pastures of Portugal and Spain. Its equivalent is the dehesa in surrounding Mediterranean countries. In a montado/dehesa the oak trees are grown in a naturally occurring savanna with grasslands, other oaks, and shrub species. Like the American bur oak savanna, it's one of the highest-yielding plant associations in existence. For all practical purposes the two systems of montado and dehesa are close enough in structure and function to be the same despite subtle differences.

The dehesa is managed as a silvopasture, a pasture where animals graze in rotation on tree leaves and branches. It also is home to several species of oaks that, because they are in the white oak family, bear tasty acorns. The cork oak additionally, like the American bur oak, has a thick cambial cork that insulates the tree against the fires that ravage the dry Mediterranean forests during the long, hot, droughty summers. This cambial layer, called cork, is stripped from the tree every nine years and readily grows back.

Trees are pruned on a seven-year cycle, with the prunings/coppice used to create a highly valued charcoal. In the past grapes were grown in the dehesas along with the oaks and understory brambles and herbs. Today wheat, chickpeas, lupines, and fava beans are more commonly grown in the oak openings. In autumn swine are turned into the dehesa to feed on the leftover acorns and fertilize the fields. Acorn-fed pork is a high-value commodity in Spain and Portugal. The earliest example of a montado/dehesa dates back to 4100 BC. Today this

TREES: THE ESSENCE OF THE PLANT GUILD

Woodland landscape.

ecosystem is an endangered form of agriculture; its yields are not up to modern profit standards. The forest savannas are being plowed under and the trees rooted out to make way for pulpwood forests of poplar and eucalyptus.

In North America restoration of oak savanna habitats can bring back diversity and productivity that has been almost destroyed by farming practices. The same practices of plowing down and rooting up bur oaks has reduced the original savanna to a few percentage points of what it once was. Restoring these areas for silviculture — along with edible acorns and potential use of bark for cork products — can open up new avenues of sustainable forestry or perhaps revive the old methods. Adding back bramble berries, hazels, black cherry, and understory species, along with edge species such as strawberry, comfrey, and medicinal and aromatic herbs such as dill, oregano, fennel, and others, could enable a diverse guild that would furnish habitat for predators that support the overall health of the forest by controlling or limiting pest species. Spring ephemeral plants and bulbs can be harvested for cut flowers in season and serve as reservoirs of nutrients before their leaves break down and they go dormant. An example is the trout lily (*Erythronium americanum*), which stores phosphorus in its leaves during the wettest season, spring, releasing it back to the soil when it goes dormant a month later. This keeps the phosphorous from washing away during heavy springtime precipitation events when the plant cover has not yet filled in.

Methodologies for restoration of montado/dehesa ecosystems are also useful in restoration of oak savannas in the American Midwest. Commonalities between species and functional niches between the two areas allow for similar species relationships to develop. The steppes of Central Asia and the veldt savannas of Africa, as well as the Argentine pampas, also feature similar tree, shrub, and grassland relationships.

The Oak Guild: An Example for the Integrated Forest Gardener

For a small site such as an urban or suburban yard, the question of what and how to plant around a larger preexisting tree can be tricky. If you're starting from scratch — that is, with an empty guild site — choose a species of oak that will yield within a time span that will work for you. In the white oak group that would include bur oak (*Quercus macrocarpa*) and chinkapin oak (*Q. muehlenbergii*), which bear earlier than a standard *Q. alba* or white oak. Both bear tasty acorns requiring little leaching of tannins and give habitat to other species. For a wetter locale the swamp white oak (*Q. bicolor*) is a good substitute.

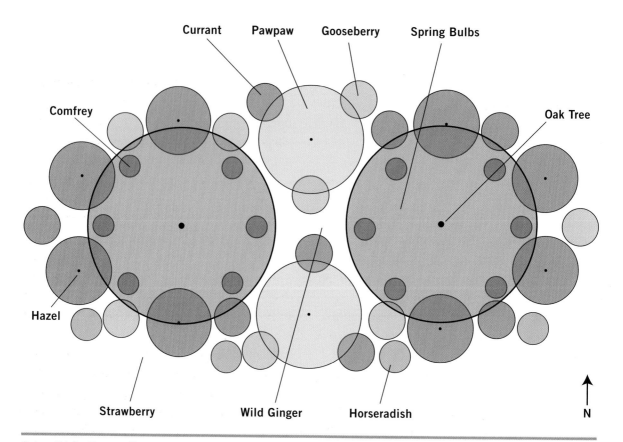

Oak guild for the smallholder.

TREES: THE ESSENCE OF THE PLANT GUILD

> ## The Great Swamp National Wildlife Refuge
>
> The Great Swamp National Wildlife Refuge in the Watchung Mountains of Basking Ridge, New Jersey, has always been a great source of study and enjoyment for me. Many years ago, in the dead of winter at the Great Swamp, while tracking a red fox through the snow on his daily mealtime hunt, I came around a bend and was stunned to see before me a massive trunk the likes of which only myths speak about. I would have fallen to my knees if it weren't for the enticement this old granddaddy set before me. Climb! Yes, come climb me! As though in a trance I did just that. As I climbed higher the cold wind slapped at my face, turning the few remaining brown leaves into ancient chimes, and the branches creaked like old floorboards. I climbed higher and higher into the crystal-blue winter sky. And just as I reached the top a gaggle of a thousand yawping geese flew just above in their characteristic V formation. Chills ran wildly up my spine, a combination of excitement, brisk cold wind, utter joy. The sacred oak stands at the center of its world in divine majesty.
>
> —WAYNE WEISEMAN

OAK GUILD LAYOUT AND POLYCULTURES

The original placing of the oak should be based on its spread at maturity and root system proximity to underground infrastructure such as water lines, power lines, sewer pipes, and the basement. Try to keep it away from the roof of the home unless you are planning to prune, coppice, and/or pollard the tree. Other species of trees that can work well with the oak will need to be planted as time goes on and the oak begins to gain height. Let it reach perhaps ten feet before you begin to install shade-tolerant trees such as persimmon (*Diospyros* species) and pawpaw (*Asimina triloba*), and tall shrubs such as highbush cranberry (*Viburnum trilobum*). A tall tree association with either sugar maple (*Acer saccharum*) or hickory (*Carya* species) will also work, as well as one with chestnut (*Castanea* species).

Tall shrubs for the edge with an open, airy leaf canopy include serviceberries (*Amelanchier* species), which flower early and ripen in June. Their heights can vary from a few feet to thirty, depending on species. Shrubs such as hazel (*Corylus* species) will grow to twelve feet tall but are usually shorter and have some shade tolerance. Currants (*Ribes* species), both native and European types, will yield berries in June and early July; gooseberries (also *Ribes* species) are a low thorny shrub yielding at the same time in the Upper Midwest.

Deep-rooted plants such as horseradish (*Armoracia rusticana*) and comfrey (*Symphytum officinale*) bring subsoil minerals up to the surface soil layers when their leaves decay in autumn. Comfrey can also be used as a chop-and-drop fertilizer in an oak guild. Spring bulbs provide visual appeal to the guild at a season when not much else is visibly happening. Strawberries and borage (*Borago officinalis*) grow well together with borage storing phosphorous and yielding edible flowers, with seeds that are a source of healthy nutrients for the nervous system. At the ground level the native dwarf iris (*Iris cristata*), a woodland plant, blooms in spring, along with wild ginger (*Asarum caudatum*), which is used like the common spice ginger.

If your oak tree is an older established tree and has reached a height above twenty feet, then just go ahead and begin to add some shade-tolerant species from the list above. Within ten years the ground level will have filled in and your oak food forest will be well on its way to fruition.

Most of the oak-associated mushrooms are saprophytes — that is, they feed on a dying oak tree. If you have these and your tree is old and you are comfortable with it being host to edible and medicinal mushrooms, then be sure to harvest when they are ready. These mycorrhizae work with the oak to mutually benefit each other. These organisms can be added either when you're planting the tree as a root dip, or as a liquid solution added to an existing tree's root zone.

The Precious Pine

Ah! The smell of fresh pine. Pinecones. Pine needles. Pine pitch. So familiar to all of us. There are nearly thirty-five species of pine, evergreen, ever beautiful, ever plentiful, throughout the United States, one hundred around the world. Eastern white, jack, shortleaf, longleaf, slash, loblolly, lodgepole, pinyon, ponderosa, red, scotch, sugar, western white, single leaf, bristlecone (oh ancient one!), limber, pitch, all pines!

Pines are classified into two groups, the softer white pines and the harder yellow pines, based on the density of the heartwood. White pines generally have more needles than yellow, and their cones have no bristles (yellow pinecones do). The male cones (catkins) emit pollen in the spring and fertilize female cones, which grow into larger green, then brown, seed-bearing fruits. The pinecone hangs down, as opposed to the fir tree whose cones stand upright. From wounds and around small openings oozes a sticky, gluey substance known as pitch. Pines are found at up to five thousand feet in elevation, sometimes forming timberline. From species to species, bark color and form, needle size and number, heartwood color and hardness, cone length, height, and trunk diameter vary, but pines, in general, are easily recognized and known by many.

The wood of the eastern white pine (*Pinus strobus*) is prized by carpenters, the jack pine (*P. banksia*) for furniture making and plywood manufacture ("knotty pine"), the lodgepole pine (*P. contorta*) for turpentine, resin, and pine oil, the pinyon pine for its delectable nuts, the ponderosa (*P. ponderosa*) for its "butterscotch" fragrance.

The seeds, needles, bark, and foliage of pines are a major source of food for many birds and mammals. Game birds, such as the spruce grouse, wild turkey, bobwhite quail, and mourning dove, and songbirds like the chickadee, jay, crossbill, warbler, nuthatch, woodpecker, and finch — to name only a few — feast on seeds and needles. Mammals including black bear, beaver, hare, porcupine, gray squirrel, chipmunk, mouse, rat, deer, elk, moose, and sheep eat seeds, bark, and foliage. Because of all this feeding frenzy in and around pines, trappers and hunters recognize pines as a sure place to study tracks and animal habits, and to hunt game (especially in winter when plant food is at a minimum). If the pine is so palatable for animals, would it not be a first-rate food source for humans?

Besides pinyon pine nuts there are other important food uses of pine. The soft inner bark is an excellent emergency food, especially during winter. It's peeled, dried, and ground into flour. The young male cones and new shoots can be eaten raw (they taste a bit like turpentine, but fill the stomach);

TREES: THE ESSENCE OF THE PLANT GUILD

Pine tree. Photo by Matt Lavin

they can also be boiled in stews to make them more digestible. The new shoots are candied by boiling them, then simmering them in sugar syrup. Even the pitch, if need be, can be ingested in small quantities, but only in extreme cases. Young pine needles are eaten, but it's better to extract their concentrated vitamin C content in a tea.

It is as a medicine that pine, because of its accessibility, is so important. For chest congestion (due to a cold), chew a very small piece of pitch. Expectoration will follow. A tea of boiled inner bark is also an exceptional expectorant for chest colds. An infusion of needles or twigs, either drunk or inhaled, is useful for bronchitis, sinusitis, and upper respiratory congestion, arthritis, and rheumatism. Added to the bath it calms restlessness and helps heal cuts and skin problems. Oil of pine is an outstanding external treatment for rheumatism, pneumonia, nephritis, and lung and bronchiole congestion (use a compress). Pine pitch placed over cuts and abrasions is antiseptic; it will help cleanse a dirty wound, and draw splinters and glass out of skin.

Pinecones make beautiful orange-yellow and brown dyes. Young, thin pine roots are split, softened, pulled apart, and used as weavers in basket

> ## Pines!
>
> Under a stand of pine trees, perhaps in an old plantation of pines in long rows, nothing grows. Acid soil. Nutrients gone. But pines perpetuate themselves in their own fallen biomass. They grow tall, majestically pushing pointed pinnacles above the tops of oaks and maples, cottonwoods, and tulip poplars, crowding out the white birches that begin to die, attacked by shelf fungi, turning different shades of brown and black, splintering and falling in the winter wind. A big old birch trunk leans on that big old branch of that big old pine, its final gesture before the acid soil beneath the pines claims it, burns it up, digests it. A pine forest is a unique ecosystem in itself. And around its edges, in the acid environment thus created, the opportunity to grow acid-loving species, such as blueberries or potatoes, affords us a significant yield.

making. Pinewood has vast application to any number of woodcraft and construction projects. For the campfire, pinewood is fast burning, though for longer-lasting fires, hardwoods are much preferred.

Pine woodlands and savannas are fire-adapted ecosystems. Mature trees have thick bark. Pine trees are well adapted to fast-moving fires that sweep through stands and burn grass, litter, pine needles, and low shrubs. Natural wildfires naturally recur every five to forty years, a frequency that significantly reduces the amount of fallen branches and pine needle litter on the forest floor. It takes a tremendous amount of heat for a pinecone to open and drop its seeds. After fire these seeds contribute to the regeneration of the pine forest.

Old Man Hickory

Of the sixteen species of hickory the shagbark (*Carya ovata*), with its "shagged" strips of bark peeling outward from the trunk, and the pecan (*Carya illinoinensis*), with its delectable nut (ah, pecan pie!), are the most familiar. Other varieties include the bitternut (*C. cordiformis*), mockernut (*C. tomentosa*), pignut (*C. glabra*), and shellbark (*C. laciniosa*) hickories. A close relative of walnut, the hickory has seven to nine leaflets — the final three larger than the rest — per shoot, with toothed compound leaves. The husk of the nut, mature in late August and autumn, is split into four to six sections (more commonly four). Hickories range in height from approximately fifty feet (mockernut and bitternut) to eighty feet or more (shagbark and pecan). They are found in a variety of habitats, including moist bottomlands with well-drained soil, streambanks, dry upland slopes, swamps, and sandy ridges. Hickories reach greatest abundance on deep, rich, moist soils.

Basically a Midwest to East Coast hardwood forest tree, hickories are almost exclusively North American, with very few species growing elsewhere. They may live for 250 years, some not bearing mature nuts for 80 years or more.

TREES: THE ESSENCE OF THE PLANT GUILD

Hickories flower in late March at the southwestern edge of their range, and as late as early June in the North and Northeast. Vegetative sections of hickory are susceptible to fire damage, after which they resprout from the root zone. In an oak-hickory forest, oaks survive fire more readily than do hickories, with only the larger hickories (such as shagbark) continuing to produce nuts. The hickory grows slowly and is partly shade-tolerant. Saplings can persist for many years beneath a forest canopy and respond rapidly when released. They are late-successional trees, and form a major part of the canopy in the oak-hickory biome.

The oak-hickory forest is the most widespread of all the eastern hardwood forests, rich in animal and plant species and varieties. The hickory has usurped the niche left after the demise of the American chestnut in the 1920s, which was caused by a virulent fungus, and possibly the result of the extinction of the passenger pigeon, the most prevalent bird species in the United States, which fed almost exclusively on the chestnut. Passenger pigeons supplied nutrients to trees through their feces after eating massive quantities of chestnuts. Researchers feel that the downfall of the passenger pigeon, and the chestnut, go hand in hand because of this.

Hickory nut. Photo by John Beetham

The shagbark hickory occurs as a principal dominant in drier parts of the Upper Midwest with oaks (*Quercus* species) and other hickories. It is a minor component of regions that support bur oak (*Q. macrocarpa*), chestnut oak (*Q. prinus*), white, black, and northern red oaks, pine (*Pinus* species), sweetgum (*Liquidambar styraciflua*), loblolly pine (*Pinus taeda*), swamp chestnut oak (*Quercus prinoides*), and cherrybark oak (*Q. falcata* var. *pagodifolia*). Many oaks, including white oak, northern red oak (*Q. rubra*), black oak (*Q. velutina*), northern pin oak (*Q. ellipsoidalis*), southern red oak (*Q. falcata*), chinkapin oak (*Q. muehlenbergii*), bur oak, and other hickories are prominent overstory associates. Red maple (*Acer rubrum*), sugar maple (*A. saccharum*), hophornbeam (*Ostrya virginiana*), shortleaf pine (*Pinus echinata*), American basswood (*Tilia americana*), redbud (*Cercis canadensis*), and sourgum (*Nyssa sylvatica*) also commonly occur with shagbark hickory.

Raspberries and blackberries (*Rubus* species), blueberries and huckleberries (*Vaccinium* species), rhododendron (*Rhododendron* species), serviceberry (*Amelanchier* species), gooseberries (*Ribes* species), hawthorn (*Crataegus* species), hazel (*Corylus cornuta*), muscadine grape (*Vitis rotundifolia*), common greenbrier (*Smilax rotundifolia*), western snowberry (*Symphoricarpos occidentalis*), common witchhazel (*Hamamelis virginiana*), wild ginger (*Asarum caudatum*), and nettles (*Urtica* species) are understory plants typically found beneath hickories. All these plants have significant edible and medicinal value for human use. Including them in our plant guild designs helps us mimic a hickory plant guild in the wild.

Many animals enjoy the nuts, bark, flowers, foliage, twigs, and leaves. Game birds such as the ring-necked pheasant, bobwhite quail, and wild turkey, along with waterfowl, munch on the nuts. Crows, grosbeaks, northern jays, nuthatches, and woodpeckers enjoy the nuts and flowers. Black bear, raccoon, white-footed mouse, wood rat, and white-tailed deer eat bark, nuts, foliage, and twigs. But the most familiar scene around the hickory tree in late summer is the commotion caused by prancing squirrels and prattling chipmunks. Who would not get pleasure from the playful and acrobatic bushytails tightroping to the ends of pencil-thin branches for treasures. Plunk! Plunk! Walking quietly at the edge of the forest. Plunk! Plunk! Plunk on the top of the head! Squirrels knocking hickory nuts to the forest floor (if it weren't for your head getting in the way). Chipmunks darting from underground tunnels to snare the ripened fruits. Some are eaten on the spot, some cached for the chipmunk's long underground winter slumber. It is always intriguing to come across "chipping beds" on top of rocks or stumps where squirrels and chipmunks have feasted. More fun than chipping arrowheads!

Hickory nuts are an outstanding autumn food and energy source for us humans, too. Like many other nut varieties they can be cracked open and eaten raw, roasted, dried and ground into flour, crushed, boiled and skinned for oil, even dipped in sugar syrup and eaten as candy. Similar to the maple tree, the sap is collected in the spring and used as a direct water source or boiled down into sweet syrup. Along with walnuts and acorns, hickory nuts are one of the most important foods to cache for the winter. Their fat and protein content make them a first-rate defense against the cold.

Virginia Indians cut knots out of hickory trees, shaped them like cones, placed them on rheumatic joints, and, igniting them, let them burn to the skin to create a sore that allowed the excess humor in the joint to run off.

Prized by woodworkers for its strength, hardness, toughness, and reddish-brown beauty, hickory is utilized extensively for tool handles, athletic equipment, furniture, and cabinetry.

> ### Hunting for Hickories
>
> "Squirrels love hickory nuts." Doc was an old hunter from West Virginia. "Squirrels and deer make for the best huntin'." Bald from the radiation treatments he was receiving for cancer, thin, weak, tired, he continued to hunt. The first deer I ever saw shot was with Doc. And the way he handled it. It wasn't that he taught me how to dress a deer. I simply watched. I was devastated. He took the meat and left the rest behind. No care. No understanding for a life. *And what about the squirrels?* I thought. *What about the squirrels in the hickory trees?*
>
> "I love to wait for fall when the squirrels are eatin' the ripe nuts and pick 'em off outta the trees with my .22." And then what? Leave them for dead? Unfortunately, my introduction to hickories was not altogether pleasant. But the appreciation for such a majestic tree has easily diminished the disheartening experience of my initial acquaintance, and brought new levels of awareness and love for the hickory's stately strength among the trees of the hardwood forest. I have since become an experienced hunter, and the deer, all of it, is a treasure to be delighted in, and the squirrel, no less. . . .
>
> — WAYNE WEISEMAN

For the survivalist and primitive-weapons enthusiast, hickory, along with other hardwoods such as black locust (*Robinia pseudoacacia*), white oak (*Quercus* species), white ash (*Fraxinus americana*), osage orange (*Maclura pomifera*), and elm (*Ulmus* species), makes one of the most reliable bow woods. It must be air-dried for a minimum of six months (up to five years) for the best results.

Yes, Doc. Squirrels do love hickory nuts. And we love squirrels. What would a city park be without those happy-go-lucky, playful denizens of the trees? The sight of several squirrels running haywire after each other, chattering, leaping to their heart's delight, is enough to take the edge off any kind of city madness. And lo and behold! Over by the baseball diamond, here in the 'burbs, a single, very tall hickory tree. It is early autumn. Plunk! Plunk! Plunk! Ouch! Squirrels at it again. "Hey, wake up down there. Look up. Look at us. Look at how much fun we are having in this big old hickory tree!" Dead in my tracks. Look up! Look up! Plunk . . . plunk . . .

The Maple: Sugar in the Gourd

The maple trees in Vermont are dying. The price of pure Vermont maple syrup has shot up considerably. It takes a lot of sap to brew a quart of maple syrup, and it only happens one time a year. Some of the larger trees are simply collapsing, falling over, even in New York towns bordering Vermont. Chestnut blight, Dutch elm disease. Kind of makes you wonder. Real Vermont maple syrup, but the

Sugar maple tree. Photo by Evelyn Fitzgerald

maple is a prolific tree of many species growing to massive proportions in a variety of habitats. "Yeah, but Canadian maple syrup isn't pure Vermont," you say? Hey, what's in a name? Anyway, syrup is not the maple's only claim to fame. Beauty and utility combine to make the maple an extraordinarily lofty being.

The three most prolific maples — the red (*Acer rubrum*), silver (*A. saccharinum*), and sugar (*A. saccharum*) — grow from the northeastern United States and southern Canada south to West Virginia and Kentucky, and west beyond the Great Lakes. Maples average between 70 and 120 feet in height, and thirty inches in diameter at the trunk. They can grow to large dimensions and spread branches, when not obstructed by other trees, to great widths. The five-lobed leaf is a familiar sight. The leaves grow in pairs, with whirligigs — paired seeds — emerging from leaf axils. These are called samaras. Depending on species, samaras appear in either the fall or spring. Other varieties of maple include the Norway (*A. platanoides*), striped (*A. pensylvanicum*), and bigleaf (*A. macrophyllum*), along with one species in the western United States, the Rocky Mountain maple (*A. glabrum*). Several Oriental varieties are mainstays of American gardens, especially the red-leaved maples that add such a deeply luscious maroon color to many garden plots. In autumn these maroon leaves turn a brilliant scarlet, contrasting exquisitely with the oranges and yellows of other maples.

Flowers precede leaves in spring, red clusters on red maples, pendant yellows on Rocky Mountain maples, greenish yellow on vine maples (*Acer circinatum*). The most unusual of all the maples is the boxelder or ashleaf maple (*A. negundo*), whose leaves are compound (similar to ash and various nut trees), irregularly lobed, and toothed. Boxelders are prolific along swamps and waterways; they are, in many cases, pioneer trees in the western United States.

Maple wood, used for woodworking, is close or straight-grained, quilted, wavy, or bird's-eye in texture, white to red, hard or soft, and it is found on everything from dance floors to butcher-block countertops.

Squirrels and chipmunks strip the seeds of wings and eat or cache them for winter store. Birds build nests from seed or leaf stalks. Many birds prefer the maple as a nesting site. Grosbeaks, chickadees, finches, flickers, pileated woodpeckers, screech owls, and nuthatches cruise the maple for seeds, buds, and flowers. The leaf flycatcher also relies on the sugar maple, picking insects from the tree's leaves. When populations of sugar maples decline in a region, the leaf flycatcher species population shows increased stress.

Porcupine, black bear, rabbit, and raccoon eat bark, twigs, seeds, and flowers. Small rodents prefer the seeds. Deer, elk, and moose browse twigs and foliage. Aphids eat maple leaves and excrete a sweet dew digestate that coats leaves with a varnish. Maple leaves, in general, have tough sheaths around them, and they contain tannins that tend to limit ingestion by insects to very few species, although boxelder rollers (larval insects) roll leaf edges around themselves for protection while feeding on boxelder leaves. Sugar maples provide winter browse for white-tailed deer, moose, and snowshoe hare. The leaves that are shed in autumn retain high levels of calcium, magnesium, and potassium, making them nutritionally desirable browse compared with those of other tree species.

Sugar maple roots create a highly mobile form of nitrate ion that is more likely to leach out into water bodies, with negative effects on aquatic environments. Sugar maple populations may be on the rise as environmental factors are causing a decline in the American beech (*Fagus grandifolia*), another tree species that grows in similar habitat conditions. An increase in sugar maples may lead to increased nitrogen pollution in streams and ponds.

Earthworms are more diverse and numerous beneath sugar maples than beneath other tree species. The large, nutrient-rich leaf litter that is released in the fall by mature sugar maples contributes to this earthworm bonanza. We can use these maple leaves for compost and as a potent mulch for integrated forest gardens. Sugar maple roots also interact with a variety of fungal species. The maple tree uses the fungi to absorb more nutrients and water volume from the soil to support the growth of its enormous crown, while the fungi benefit from the nutrients and water pulled from underground by the sugar maple's deep roots.

Maple bark, with less tannin than oak, is astringent and is used, in decoction, as an all-purpose wash for irritated skin and sore eyes. Boxelder inner bark is emetic, and the inner bark of other maples is used for coughs, diarrhea, kidney infections, colds, and bronchitis, or as a blood purifier. The pure, watery sap is a spring tonic, and leaf and twig tea can allay nausea.

Maple flower blossoms are eaten in the spring as a trailside nibble, but the best-known food extracted from maples is, of course, syrup. Although thirty

> ## Out and About on the Farm
>
> The line of boxelders, maybe eight altogether, grew from the pond up the embankment, past the garage and shop, and out beyond the barn. It appeared that this line of trees was growing on a water vein that kept their thirst at bay and gave them plenty of the nutrients they would need. Around the farm, red, striped, silver, and sugar maples thrived. One brisk, clear spring morning, three of us walked outside, responding to grunts and a peculiar assortment of barks, screams, and whines.
>
> As it turned out, the boxelder near the barn contained one very large raccoon. The staring match commenced. We were delighted. Whether he was or not is another story. All day long he sat in that boxelder and stared. I had used many branches from that tree to make spindles for bow drill fires. And the local opossum frequented that same tree. Popular spot. By nightfall the raccoon still sat and watched us watching him, not moving a muscle. We watched him, he watched us. We got tired, went into the farmhouse, went to sleep. The next morning he was gone.
>
> Two days later I was wandering, somewhat drunk on spring, among the red maples in the woods surrounding the farm when I heard a branch crack overhead. Looking up, I spotted a raccoon eating maple blossoms. When he saw me, he stopped. I sought out my two observant friends. The staring match began again. It lasted most of the day. Not a muscle moved, not a sound was uttered. I guess the moral of this story is: When you see a raccoon in your boxelder tree, do not expect much. And two days later, when the same raccoon shows up in your red maple tree, do not expect much, either. Just sit quietly, watch, listen, and be still.
>
> — WAYNE WEISEMAN

to forty gallons of sap are required to produce one quart of syrup, it is well worth the effort. From January to April spiles (tubes) are inserted into the tree, with buckets beneath, and the raw sap is collected. It is boiled to a thick consistency and readied for the breakfast table.

Maple sap is an important survival water source when stream, pond, or groundwater is not easily accessible. Other trees, such as birch and sycamore, can also be tapped for sap. Grapevines, when sliced a few feet above the ground, yield a flow of moisture for mouth or container.

CHAPTER FIVE

Designing for Optimal Species Integration

There is, for example, an interesting wasp which builds in the following way: when it finds a rather stiff leaf on some branch, it fetches small particles which it bites off from the bark of neighboring trees, or some similar substance; these it permeates with its saliva, and then proceeds to build a number of small stalks which it attaches to the leaf. When it has completed these attachments the wasp goes on working, mixing these substances with saliva and building on to these stalks something very similar to the single cell of the honeycomb.

— RUDOLF STEINER

For many of us, our first exposure to tales of human and animal interactions are the stories by Beatrix Potter, especially *The Tale of Peter Rabbit*. In the story a family of rabbits lives beneath an old fir tree growing in a sandbank. Peter is the youngest of the rabbits and has an affinity for the fine vegetables growing in Mr. McGregor's garden. This does not go over very well with the farmer or his wife. In fact, Peter's father had ended up in a meat pie for the same indiscretions. Avoiding the fate his father had encountered, Peter hides in a bucket and later behind a wheelbarrow before making his dash to the garden gate and the sanctuary of the forest. This story has all the elements of nest, shelters, refuge, food strategies, and relationships between species.

In his essay "Anthropocentrism," John Seed states that when human beings begin to examine their self-centered view as members of the dominant species, ". . . you start to get in touch with yourself as a mammal, a vertebrate, as a species only recently emerged from the forest. As the fog of amnesia lifts there is a transformation in your relationship to other species, and in your commitment to them."[1] To put it clearly, we are not nearly as special as we think we are, but are a part of a greater whole.

Beatrix Potter and John Seed are making similar points. They are telling us that humans are not the center of the universe or the pinnacle of all creation. Other creatures are here, too, and they are equally

Bumble bee on anise hyssop.

deserving of nature's bounty. This chapter will address the idea of sharing the yield with the creatures that react with the plant guilds we as designers and gardeners plan and enable. Recognizing the relationships among all elements in a guild helps us remember that connections are the glue responsible for resiliency in any plant–animal community.

An animal can be defined as any living thing that is not a plant or fungi. Both wild and domestic animals have impacts on our plant guilds — browsing on foliage, chewing plants to the ground, digging up roots, even moving plants to new locations by transplanting roots, tubers, and bulbs. Animals also store seeds, some of which get forgotten, then germinate and become new guild members. The human animal, too, affects the ecological balance of the plant–animal community.

A functional plant community or guild includes not only the plants that we have placed in it but also the species that animals bring in from their travels. Everything from garlic mustard seeds from a mouse's feet, to walnuts planted by a squirrel saving food for leaner times, to fruiting trees and shrubs whose seeds have passed through the digestive tract of a bird begins to show up in an integrated forest garden soon after it has been implemented.

DESIGNING FOR OPTIMAL SPECIES INTEGRATION

To quote the movie *Field of Dreams,* "If you build it they will come." *They* are the nonhuman creatures of the animal world, the Bambis, Peter Rabbits, Puddle Ducks, and all the rest. And they will want their share of the guild yields in the form of nesting sites, refuges, food, and water. As designers of a plant guild, we have to recognize that a failure on our part to provide surplus for nonhuman species will inevitably lead to those creatures establishing their own niches within the community regardless of self-perceived needs of humankind.

Leaving a few gaps in a design, either intentionally or by plant succession strategies, gives you a chance to observe and interact with animal partners also seeking to impact the formation of an ecologically adaptive integrated forest garden. All species — plants, animals, and humans — are active participants in the dance of life, and they all need to occupy a place on that ballroom floor. The designer's duty is to let it happen in a manner that allows everyone to participate harmoniously.

Beneficial Behaviors in the Permaculture Guild

Plant and animal communities are based on a system of mutualistic actions — patterns of behavior that benefit both groups. Animals have four basic needs for survival: food, water, shelter, and space. The latter need, space, is that territorial range that a species requires to forage for its nutrition; this can be much larger for some species than others. A hawk's territory may be several square miles, while for a white-footed mouse a few hundred square feet may suffice for food gathering. Animals' food foraging and nesting activities result in services for the plant community including pollination, seed dispersal, cultivation, thinning, and soil fertilization. In turn, plants provide the basic needs of sustenance and shelter for animals both domestic and wild.

Food Sources

Plants provide the sugars, vitamins, other nutrients, and carbon for all other living things. Offering food to wildlife in the landscape in the form of forage plants surrounding annual crops and flowers can create an alternative food source for would-be plant pests. This invites a large variety of herbivores as well as their predators — an abundance of niche pest control services. The integrated forest garden gives herbivore animal species the chance to forage on fruits, seeds, nuts, leaves, twigs, bark, roots, pollen, and nectar. For omnivores, the less fussy eaters, all of the above foods as well as insect and small animal species are gleaned from the guild. Predators are the animals feeding toward the top of the trophic feeding pyramid or food chain. They in turn are preyed on by larger predators. For instance, insect parasites such as flies, ticks, and mosquitoes are eaten by birds, which are in turn eaten by larger birds. Scavengers clean up the dinner tables and eat everything left. Everything is consumed and returned to the larger guild community — even bones and antlers, which rodents gnaw on for their calcium content, reducing them over a few years to a tiny fraction of the original deposit. This is based on personal observations of squirrels gnawing on moose antlers and rodents reducing deer bones at Wehr Nature Center in Hales Corners, Wisconsin.

Secondary food resources are those less desirable except in times of shortage due to drought, crop failure, overbrowsing, or late-winter depletion. A resilient and diverse plant community provides an assortment of feeding opportunities. If a single component of the guild fails to yield a crop, then another species of plant can fill that niche need. In times of severe drought, for example, many animal species change feeding patterns to include foods otherwise less palatable. Chipmunks will consume frogs when other foods are scarce, when otherwise the two exist side by side in a suburban yard or woodland, and garter snakes will eat toads despite the protective scent glands on either side of the toad's neck.

Fruits, seeds, and nuts that persist throughout the autumn months and into winter and early spring do so because they are not a primary food source for animals until other foods are unavailable. By late winter and early spring the preferred foods are eaten and will not be available again until months later, when the next crop matures. Overwintered foods from the less desirable species then become the only available energy sources for animal foragers. As food becomes scarce the populations of the omnivore species drop, and consequently so do the numbers of their predators. Populations of both groups cycle up and down based on the availability of seasonal food. Both groups also experience seasonal recovery in spring and early summer when forage and prey increases.

You can attract animals to particular sites by either placing foods there or planting crops that they can harvest on their own. Birds, squirrels, chipmunks, rabbits, deer, and raccoons can be

Silvopasture. Photo by Lynn Betts, USDA NRCS

encouraged with feeding trays of seeds and fruits at varying heights depending on which species you're inviting. Suet cakes of beef fats mixed with insect and seeds will draw woodpeckers and chickadees year-round, and inevitably hordes of grackles, blackbirds, and starlings in season. In many states natural resource departments actively encourage the planting of wildlife food strips along woodland edges and unpaved access roads to feed deer, turkeys, and grouse.

When you're designing an integrated forest garden, take care to provide as much benefit as possible for as many participants or species as you can. For instance, a planting of hybrid poplar trees may provide some deer browse and a source of biofuel feedstock, but it does not serve as a shelterbelt — a group of trees that provides shelter from the wind and species food variety — as would a grouping of boxelder (*Acer negundo*) trees, which provide nourishment to at least five animal species. Boxelders can also be tapped for a sugary sap to be processed into sweet syrup much like that of the sugar maple (*Acer saccharum*). Again, it's important to think of stacking functions of shelterbelt, food for many, and fast-growing browse and biofuel stock using a species such as the boxelder, a tree some consider a weed.

Animals live by foraging in most circumstances, but in a Permaculture designed system care is taken to ensure that despite multiple adverse factors — say, disturbance by drought, flood, or fire — a diversity of food and water sources is built in or allowed to evolve independently. Some more specific feeding patterns will be addressed in the pollination and plant–animal interaction sections of this chapter.

Plant Guilds as Refuge and Shelter

When you're developing a plant guild, remember that beneficial insects and animals will help regulate pests and contribute to plant health. Aside from a few flowers for nectar, designs need to include living, resting, and overwintering space. An area as small as a hollow root in the ground or the underside of a log or rock pile can be enough to serve as habitat. Snake pits offer protection for the snake as it controls populations of small rodents and insects in an orchard, and roosts allow perching sites from which hawks can scout for food.

Certain plant structures allow for egg laying, nesting, or water collection. A cup plant (*Silphium perfoliatum*) holds dew and rainwater for days for small birds and tree frogs. An evergreen tree adds wind and predator protection for small bird species and shade when temperatures are hot. Dense tufts of grasses are habitat for beetles that eat aphids and other plant pests. Small ponds add both a water source and habitat for birds, insects, amphibians, and mammals. Adding places to live brings new species into our production space, the guild. It's an open invitation to nature, extending the polyculture resources deeper outside the planting beds, bringing a return of ecological services that aid the entire system.

Trees provide cavities for nesting squirrels, raccoons, and birds. Beneath the bark are many burrowing insect species. Nests are made in the angles between branches by both birds and squirrels. Hollow logs serve as nesting sites and shelter from predators. Tall grasses are nest sites for many birds and deer, which also use them for shelter from predators. In the detritus layer of the soil many species of insects live, feed, breed, and are fed upon by their predators. Brush piles, bramble thickets, snags, evergreen trees, holes dug in the earth, and the abandoned dens of other species are all used by animals. Tip-up mounds are sites where a fallen tree has left a hole after the roots have been heaved out of the earth. Look here for the dens of coyotes, wolves, and bears.

Designed shelters include birdhouses, bat houses, pollinator blocks and hives, and livestock

No-till plant guilds keep soil biology and habitat intact.

shelters. Also included here would be living shelters intended to ameliorate the effects of severe weather such as leeward-side windbreaks. The leeward side of a windbreak is the sheltered side, as opposed to the windward side that faces into the wind. An interesting shelter design is the snake hibernation mound detailed in the Minnesota DNR publication *Landscaping for Wildlife*.[2] Snake hibernation mounds are ten-foot-long logs that are cut from downed trees and, along with the brush and trimmings, inserted into a pit dug down about eight feet deep. When you mix together the dirt, trimmings, leaves, and logs, the site becomes eight feet high, eight feet deep, and nearly ten feet across. The crevices and edges in such a mound provide a beneficial habitat for snakes to overwinter and can also attract mice and chipmunks looking for nesting locations. You can plant the aboveground portions of the mound with beneficial herbs, pollinator-attracting plants, and food crops. Similar in potential for attracting different species are the beetle bank and hedgerow.

In more conventional gardening practices, soil is dug up annually when the yields lessen and the autumn weather turns colder. But *not* tilling the soil (a central tenet of Permaculture and forest gardening) leaves the habitats of ground-nesting pollinating bees, ants, and wasps undisturbed. Ground-nesting bumblebees are one of the most easily recognized native pollinators. Not turning over their nest sites ensures a continuity of pollinating bees for the Permaculture plant community.

A no- or reduced-tillage practice will also reduce overwintering pest populations by allowing foraging birds and small rodents to gain access to weed seeds and the larvae of plant-damaging insects. Other animals aided by reduced tillage are snakes, toads, and chipmunks — all ground nesters.

Plant Pollinators

The most important impact of animals on plant species is the fertilization of flowers. Pollination is the act of carrying pollen from the anthers to the stigma of a flower. Without it there are no seeds or fruit. While some plants are self-pollinating and others are wind-pollinated, many species have outsourced that role to animals — insects (bees, flies, wasps, beetles, butterflies, and moths); birds; even mammals. To attract a particular pollinator, you must provide shelter and food for both the young and the adult stages of the species.

Pollinating birds and insects are drawn to a guild by plants that serve their needs. Flowering plants offer pollen and nectar to pollinators in return for services rendered. Plant mechanisms for attracting pollinators are highly specialized and include bloom shape, scent, color, time of day, and season, all of which determine the species of animal attracted. Scent attractants include some that humans also find pleasant, such as the odor of a rose, as well as smells we may find unattractive — mountain ash (*Sorbus aucuparia*), horseradish (*Armoracia rusticana*), or hawthorn (*Crataegus* species), for instance. The pawpaw (*Asimina triloba*) has a scent that — despite being almost indiscernible to humans — can attract carrion flies and beetles because it's reminiscent of rotting meat.

Cross-pollination within a plant species is accomplished when a bee, wasp, or fly visits multiple sites; pollen from one flower may stick to its legs and then be transferred to the stigma of the next plant's blooms. Some bird species accomplish the same services when they visit flowers for their nectar. Hummingbirds and orioles are both nectar-feeding species that can pollinate flowers.

An important necessity for any plant community is a nearby reservoir of wild pollinator species. Many economically important trees are also wind-pollinated. These include pines, spruces, firs, and many hardwood trees, including several species cultivated for nut production.

Fragmentation of natural habitats and increased pesticide usage have greatly reduced the density of pollinators worldwide. By implementing integrated forest gardening techniques in the landscape, however, we can provide a habitat for the revival and maintenance of vital pollination services. For it is an assortment of integrated plants that makes available a variety of nutrients for pollinators.

BEES

Bees are the insect species we most often think of when pollinators come to mind. European honeybees (*Apis mellifera*) were brought to the Western Hemisphere by Spanish, French, and English colonists in the seventeenth century to provide honey for the settlers and to pollinate crops. Competition between native bee species and *Apis* as well as the replacement of many species of plants with non-native ones has led to a decline in native pollinators. Recently diseases, such as colony collapse disorder (CCD) and chemical exposure, have contributed to a decline in native and non-native bee species important to pollination in North America and elsewhere. Scientists studying CCD believe a combination of factors could be making bees sick, including pesticide exposure, invasive parasitic mites, an inadequate food supply, and a new virus that targets bees' immune systems. A clear determination has yet to be found, however.

There are four thousand species of native bees in North America. Some specialize in pollination services for specific plant species. Others are generalists, including the introduced *Apis* bees.[3] One of the advantages that *Apis* bees have over many natives is a high degree of social organization in a year-round active colony. Messaging as to the location and status of pollen sources is done through a series of movements called a dance. The movements communicate the location, status, and blooming stage of the pollen source to other *Apis* bees. Some native bees, especially bumblebees (*Bombus* species), have similar levels of group sophistication and cooperation, but for shorter seasons of the year. *Apis* bees can forage over longer distances, too, and so are able to pollinate large-scale agricultural fields in a monocropped system. All of these advantages serve *Apis* bees well in farm and orchard systems, as well as during late winter and early spring when fewer plants are flowering.

Solitary bees are not as social as *Bombus* and *Apis* species, so they do not serve as important pollinators of large field crops. This is an advantage for the gardener or designer in the context of specific niche adaptations for pollination. The perennial plantings in a guild system do not require disturbance of the soil as often as an annual vegetable garden or monocropped farm field. Ground-nesting species of pollinators and those that nest in hollow spaces in trees, logs, and biennial plant stems can readily adapt to the integrated forest garden, providing valuable species-specific pollination services.

Species of native bee pollinators include the alkali bees (*Nomia melanderi*) that pollinate alfalfa, clover, and mints. These native bee pollinators are located everywhere these plants are found. Solitary ground nesters, they are slightly smaller than a honeybee, with a bluish-, yellowish-, or greenish-and-black stripped abdomen. They live in vast communities, and take their name from their preference for nesting in ground that features a salty crust covering a damp, saline soil beneath.

A blue orchard mason bee on an *Allium* flower.

Carpenter bees (*Xylocopa* species) are adapted to pollinate passionflowers (*Passiflora* species) as well as eggplant and peppers. The blueberry bee of course services the blueberry. The digger bee does the same for cotton and fruit trees. Blue orchard bees pollinate almonds, apples, and sweet cherries. Squashes, pumpkins, and gourds are pollinated by, well, squash bees (*Peponapsis* species) and gourd bees (*Xenoglossa* species). Sunflowers are served by the sunflower bee (*Melissodes* species). The mason bee (*Osmia* species) is perhaps the best known of the solitary bee species. There are more than thirty-five hundred species of solitary bees that pollinate both wild and crop plants in North America. The generalist and more social

DESIGNING FOR OPTIMAL SPECIES INTEGRATION

Mason bee nest. Photo by Jack Dykinga, USDA

species of *Apis* and *Bombus* are able to feed on a wider variety of pollens.

Even though *Apis* bees were brought to the New World as a source for honey production, there was already a native species for that purpose. Mexican stingless bees (*Hymenoptera* species) were raised in logs in Mayan agricultural fields to provide both pollination services and honey. Due to severe habitat fragmentation and modern intensive agricultural practices, the habitat available to the stingless bees has decreased, leading to a reduction in their numbers by 93 percent in the last thirty-seven years.

Solitary bees do not form the large hive social colonies of the honeybee. When the solitary queen builds her nest and is done depositing her eggs, she flies off to start another nest. The new bees emerge in spring and feed and procreate, forming new nests. They are not prone to the same pests and diseases as *Apis* bees but do have some of their own. Females like to nest in tubular structures and crevices in bark, stems, and abandoned beetle holes.

Mason bees are an important native pollinator active from early spring until late summer. They are hardy enough to overwinter outdoors, even in Canada. North America has 130 species of mason bees. They are beneficial and benign insects that do not sting unless stepped on or startled. An integrated forest garden or guild can support mason bees by providing nests made of wood blocks with bored holes, stacked bamboo, or paper tubes, or by growing species of woody plants such as elderberry that have hollow woody stems.[4]

WASPS

Many digger wasp species are predaceous, but in their search for food they sometimes engage in pollination. Spider wasps (Pompilidae family) search out milkweed (*Asclepias* species) nectar and are a primary pollinator for that plant. Although milkweed is a native species, it may be considered invasive; actually, however, it's beneficial in terms of pollination for wasps and monarch butterflies. It spreads through an underground root system and by seed. As long as we keep an eye on it and create a diverse plant polyculture, we can keep it under control. In the North American Southwest, many cacti and other plants are pollinated by predatory Masarinae wasps, which collect and feed the pollen to their broods. Fig species are nearly all serviced by wasp pollinators.

FLIES

The insect order Diptera, more commonly known as flies, is the second most important group of pollinators. In North America there are 17,400 known fly species that visit and pollinate plants. This makes you think twice before reaching for the fly swatter. Flies can be trapped in a flower as it closes for the night. When the warmth of the morning sunlight causes the flower to open the fly is showered with pollen, which it carries with it on its search for the next bloom's nectar.

Plant species served by Diptera species are primarily those with small flowers that bloom in seasonally moist and shady locations. Many forest flowers and edge plants such as trilliums, irises, and saxifrages are pollinated by flies. One of the most commercially important forest crops served by flies is cacao, the source of cocoa and chocolate. Beargrass (*Nolina* species), a plant native to the US Northwest, is important to the florist trade and is also pollinated this way. Some flowers with distinctive odors are attractive to fly visitors. The stinking gladwyn (*Iris foetidissima*) was named for this trait. Others similarly attractive to fly pollinators include pawpaws (*Asimina triloba*), Dutchman's pipe (*Aristolochia* species), Jack in the pulpits (*Arisaema triphyllum*), and *Gladiolus* species. The common skunk cabbage (*Lysichiton americanus*) attracts flies with its fetid blooms in late winter and early spring.

BUTTERFLIES AND MOTHS

The order of insects called Lepidopterae is well known to enthusiasts of butterfly and moth observation. These insects are able to travel long distances between flower visits. Because of their wide-ranging foraging, they can ensure a mixing of genetic information within the overall gene pool of a distinct plant species. A diurnal or time-of-day specialization allows butterflies and moths to visit flowers at different times. Both groups have specialized feeding tubes for reaching the nectar of long, tubular-shaped flowers. Plants such as *Fuchsia* species, beardtongue (*Penstemon* species), four o'clocks (*Mirabilis* species), lilies (*Lilium* species), and *Verbena* species are some of their preferred food sources.

BEETLES

The beetles comprise the largest group of all insects, with approximately 350,000 species worldwide. There are fifty-two plant species pollinated by beetles in the United States and Canada, but none of them is a crop plant of economic importance. In Central American forests beetle pollination of custard apples (*Annona reticulata*) is common. Some beetle-serviced crops include pawpaw (*Asimina triloba*) in the tropics, Carolina allspice (*Calycanthus floridus*), mariposa lilies (*Calochortus* species), *Ipomopsis* species, and *Magnolia* species. As larger insects alight on flowers, they are able to pollinate as a by-product of pre-flight grooming. Discarded pollen lands on flower stigmas, which the beetle then stands on to launch. Thrips are also herbivorous and carriers of many viral and bacterial plant diseases.

BIRDS

Many of the species of birds that breed in the Northern Hemisphere — functioning as pollinators spring through autumn — overwinter in the tropics of Central America, where they are active pollinators as well. Wrens, sparrows, orioles, vireos, finches, blackbirds, and others work in force in Mexican forests. In North America, hummingbirds are specialized tubular flower feeders, while orioles dine on orchard flower nectar and, in so doing, pollinate many species of sweet-scented flowering trees such as pear (*Pyrus* species).

MAMMALS

There are several species of pollinating mammals in North America. One is the opossum. Two species serve plants in Mexican forests; in the United States and Canada the northern opossum is not an active pollinator. Others include kinkajous and bats. Bats are attracted to large-flowered, highly scented, night-blooming species, and are major pollinators in the North American dry lands, where they service cacti and other desert species.

Pollinator Temporal Relationships

One of the ways that plants provide for species continuity is through seed formation. Seeds are enabled by flower pollination through various means, both biological and physical. Many plants are insect-pollinated but some self-pollinate through physical mechanisms such as wind and water. Others pollinate by multiple vectors, such as maize or corn (*Zea* species), which, while primarily wind-pollinated, can also have its flowers fertilized by insects.

Many plants have evolved so that pollination can occur even when insect and other animal pollinators are not present. Pollen may be dispersed by wind or may simply drop onto nearby flowers, as in *Solanaceae* species (tomato, pepper, tobacco, potato) and beans. Pathways or vectors of pollination, such as animal or wind, are controlled by seasonal limiting factors such as temperature, day length, and moisture levels. Self-pollination results in seeds that show little genetic diversity. Additional limiting factors result when pollinating species of animals are unavailable. This occurs due to a loss of habitat or by a species being subjected to chemical exposures that weaken and kill it. Examples of this include the current scarcity of honeybee and native bee species due to mite predation and neonicotinoid insecticides.

Colony collapse disorder appears to have multiple causes including the removal of buffer strip plantings (beetle banks and hedges) meant to encourage habitat for native pollinators. These buffer strips, if restored, have the potential to bring back native populations of agriculturally important pollinators. Much of the CCD problem may have its origin in not just habitat destruction but the use of *Apis* species for hire, meaning the transfer of beehives across states to pollinate vast farms of single crop species, such as almonds. Stresses caused by moving commercial hives from state to state and relying on a single pollen source at crucial stages of insect development have led to a weakened pollinator base: When bee nutritional needs are not being met, it results in a compromised immune system.

Solitary nesting bees are exposed to the same limiting factors as *Apis* bees and have also suffered from population loss. Seasonal and daily variations in activity levels require different niches used by both plants and animals to maximize food acquisition, pollen dispersal, and cross-fertilization within species. Bumblebees are adapted to forage for available food earlier in the season than are other bees, due to their larger body mass and subsequent heat retention. Other bees engage in less activity at cooler temperatures early and late in the year, but are adapted to high activity levels during the hotter days of summer.

As temperatures rise in the midmorning hours, plant oils vaporize, and the scent attracts insect nectar feeders. Birds lack the keen sense of smell insects have and rely instead on sight, so they are drawn to the bright colors of the flowers from which they draw nectar and disperse pollen. Other plants are evening-scented, attracting species such as the night-flying sphinx moth. Beetles, bats, and small mammals are also attracted to fragrant night-scented plants. Night-flowering and highly scented plant species are usually richly fragrant to make up for their drab colors (the vivid hues of daytime are not necessary to attract night pollinators).

Movement of air as the daytime sun increases warms the flowers and causes the dispersal of the pollen of gymnosperms and related plants. These breezes also disperse the pollen of maize and oaks, all wind-pollinated.

The most critical factor in the partitioning of pollen resources is the staggering of pollinator forage across seasonal and diurnal (daytime) availability zones. By ensuring a steady supply of preferred foods for a variety of species and for different dispersal vectors such as wind and water, we can ensure a diverse and abundant yield from the plant guild.

Using Plants in Functional Pest Strategies

Animals also affect the plant community in manners that we can find annoying. They browse on foliage as well as dig up, chew up, even relocate plants to different locations. Occasionally these behaviors may result in the death of the affected plant. As designers and gardeners, we need to create strategies to divert an animal's attention along another route, away from a crop in which we have an interest, such as those in an orchard. Planting a preferred browse — which can be any of many species based on the food preferences of different fauna — can divert animal feeding patterns to a different location. In short, we can fill their bellies elsewhere and still get our yields of a preferred crop. This could also be achieved with fencing, but by planting species such as amaranth (*Amaranthus* species), corn (*Zea* species), or buckwheat (*Eriogonum* species), browsers such as deer will be sated by the time they reach crops of concern for the gardener. Deer require, as do most browsers, a diversity of foods to fill their dietary needs. So by providing some variety for them we ensure their health, a source of healthful meat we harvest during the autumn and winter seasons, and the survival of a larger portion of agricultural crops.

Trap plants are also useful for control of insect pests. Aphids, for example, are drawn to certain plants (such as goldenrod, *Solidago* species), which you can then gather and dispose of. Or birds will be attracted to that site and consume the aphids. Japanese beetles, in addition, are attracted to evening primroses (*Oenothera* species) and grapes, and can defoliate entire plants. To prevent this, gather beetles by knocking them into a jar, and then use them as food for fish or farm fowl. Other strategies include planting soybeans near corn to lure away cinch bugs. Also, planting radishes near cabbage will divert the attentions of harlequin bugs. On a larger scale, using trap plants in mixed-species buffer plantings along crop fields both attracts the pests damaging the main crop and provides a ready foraging site for species feeding on the plant pest.[5]

Beetle Banks

The function of the beetle banks is to increase supportive habitat for conservation biocontrol, reduce stress on predators of crop pests, and reduce the travel time from overwintering habitat to field center. Overwintering refuges (beetle banks) are placed midfield and allowed to regenerate foliage from seed and natural sources. Colonization from existing habitat by local plant, animal, and insect genera may take time, depending on the soil, distance and quality of field margins, and insecticide used in field areas. Generally, banks are seeded and left to natural movement.

Recent research on beetle banks has focused on barren strips that were seeded without the trophic system existing in the unplowed habitat of field edges. Comparing these with field margins as the banks mature gives a good baseline for the progress of the monitored insects. In practice it may be better to transplant dormant grasses and soils from the margins into the banks to inoculate and accelerate the banks' succession.[6]

Stinging nettles (*Urtica dioica*) can be a deterrent to raccoons and human pilferers of especially valuable crops such as ginseng (*Panax* species). Plantings of hairy-leaved squashes and edible thistles around corn can slow down and even reroute animals such as raccoons that have sensitive noses. Strong-smelling herbs — especially lavender, rosemary, thyme, and sage — can confuse the sense of smell for insects and deer when it comes to finding foraged foods. This is especially important as a pest diversion strategy for orchards.

Field margins and hedgerows planted with diverse species of herbs, grasses, and wildflowers attract pest species predators. These are called beetle banks, and are reservoirs of biological diversity. To accommodate large farm machinery and suburban lifestyles, many of these features of sustainable farming were removed during the last half century, but before that they served an additional function as a demarcation line between gardens or farm fields and orchards. Plants in a predator refuge such as a beetle bank or hedge can include grasses, forbs, shrubs, and trees. Flowering species that can provide the food larval nymph stages of predator species need include many taprooted species, such as Queen Anne's lace (*Daucus carota*), parsley (*Petroselinum* species), and carrot (also *Daucus carota*). Rock cover, found along a fence line, is often included in the refuge. From there foxes, snakes, toads, and predatory insects stage feeding raids to adjacent gardens, orchards, and fields to feast on crop pest species.

A berm or raised bed about ten feet wide serves as a deterrent to herbicide and pesticide drift from adjacent fields or lawns of neighbors using chemical sprays. These are also known as variegated buffer strips in conservation farming and have proven very useful in intercepting surface and airborne runoff of herbicides, pesticides, fertilizer residues, and excreted livestock antibiotics. These buffers do double duty by slowing and breaking down the runoff contaminants and providing beneficial habitat for plant and animal species, such as those found in the above beetle bank and hedgerow guilds. Normal fencing would not be of much use for the same function but would be a temporary adjunct until the buffering berm plantings are stable.

Agroforestry Techniques

Defined as an intensive land management system optimizing the advantages of biological interactions that occur when shrubs and trees are grown alongside crops and/or domestic animals, applied agroforestry can be summarized within five sub-categories: alley cropping; windbreaks; riparian buffer strips; silvopasture; and forest farming. The benefits practices include biodiversity and landscape enhancement; protection for topsoil, livestock, crops, and wildlife; reduction of energy and chemical costs; increased water use efficiency; improved water quality; and diversification of the local economy.

Alley Cropping

With alley cropping, trees or shrubs, or combinations of the two, are planted at wide spacing

Hazels in pasture. Photo by Lori Patotzka

with openings or alleys between the rows. Woody species used in an alley cropping can be chosen for lumber or veneers, nut crops, fruit, animal forage, or even pulpwood production. An agricultural crop is then sown in the alleys to provide a harvest until the woody plants mature. You can also use forage to be harvested or grazed by livestock, or annual or short-term biennial or perennial crops: grains, annual vegetables, grass forage such as alfalfa (*Medicago* species) or timothy (*Phleum* species), biomass crops such as switchgrass (*Panicum virgatum*), fast-growing trees, and specialty crops for fruit or holiday tree cutting.

Seasonal production of cut flowers such as daffodils and tulips, medicinal herbs and roots, and annual vegetables can all provide an income yield from the alley-cropped plant guild. Trees and shrubs will be browsed by livestock, but you can provide protection from overbrowsing for the newer plantings of trees and shrubs by using fencing or wire cages until the plants have reached a height where domestic and wild animals can no longer injure them to the point where they can't grow back. If you use nut or fruit trees in the system, you can turn hogs and/or chickens into the guild in the autumn to glean any leftovers.[7] Alley cropping can also be useful as a way to attract wildlife. Many states and landowners plant forage species on logging roads to attract grouse, turkeys, and deer as part of wildlife conservation programs. These forage mixtures often include bee forage species as well.

Windbreaks

In an integrated forest garden or guild where animals will be a part of the site, windbreaks can provide shade and shelter. Windbreaks are plantings of trees and/or tall shrubs that can block the prevailing winds, thereby protecting livestock or crops from either hot drying winds or cold ones.

A windbreak needs to be planted at right angles to the prevailing wind direction; in a farming system it's usually two or three rows wide. With time and multispecies cropping, any windbreak can serve as an edible and medicinal hedge and insectary. Your choice of species within the windbreak will determine its use for forage and habitat for animal species.

An interesting variation on the common windbreak is the outdoor living barn, a shelter made from living trees and shrubs. When oriented perpendicular to prevailing winter winds, the outdoor living barn provides a shelter for winter-pastured livestock protecting them from cold rain, snow, and windchill. The windbreak can be composed of trees and shrubs chosen for their ability to block winter winds and snow along with serving as a snow fence; in some spots the accumulation of excess moisture can not only protect the animals from harm but also allow for storage of moisture until the spring melt, at which time the water will infiltrate into the soil and return to the local aquifer.

Outdoor living barns can be located in open grasslands, livestock watering sites, or pasture sites. During hot and dry summer months they can be a refuge from excessive heat for livestock and a site where watering facilities are sheltered from evaporation. They pay for themselves by lowering feed and water costs and reducing stock losses during severe weather.

The objective of an outdoor living barn is to create a continuous vegetative barrier. Gaps in the planting will funnel wind and snow through the barrier and into the protected pocket. Replant gaps created by loss of plants as soon as possible. There are many acres of open grasslands and pasture that could offer excellent winter grazing for livestock, if adequate protection from adverse weather is provided. Investing in a long-term living structure that increases the survival of newborns, reduces winter and summer stress, decreases feeding costs, and at the same time provides wildlife habitat is a wise investment. An outdoor living barn may be the answer.[8]

Riparian Buffer Strips

A riparian buffer strip is a planting of perennial vegetation along the sides of wetlands, lakes, ponds, and streams, designed to protect the shoreline and water quality from agricultural runoff, both animal and chemical. It can serve as a windbreak, soil stabilizer, fodder and browse source, producer of fruit and nut crops, animal habitat, and more. Many waterfowl species, small birds, and insects nest in the tall grasses within these buffers. A buffer strip along a waterway can trap nutrients, using them for its own growth. Woody plants suited for a wet site such as the buffer include poplars (*Populus* species) and willows (*Salix* species), which can act as biological sponges absorbing the nutrients from the fields on higher ground draining toward the local watershed. In the case of a buffer strip guild, the issue of adjacent agricultural runoff is a problem that becomes a solution, enabling a multifunctional guild.

Wider vegetation strips are more effective than narrow ones at reducing flood impact and slowing streambank erosion. You can harvest willow and poplar coppice every few years for use as either fodder or biomass fuel. The stumps will sucker and within another few years be ready to harvest again. Additional species suitable for a riparian buffer strip coppice include dogwoods (*Cornus* species), *Viburnum* species, and many others. You can choose plants that provide browse for wildlife such as beaver, deer, and waterfowl, and that will yield a chop-and-drop forage for domestic livestock in the adjacent alley-cropped fields.

Silviculture

According to agriculture professionals at the USDA, *silvopasture* is "a form of agroforestry defined by the intentional integration of a system of symbiotic, planned interactions that produces wood, forage and livestock products while simultaneously creating sustainable environmental benefits and ecosystem services."[9] It is an integration of trees and livestock forage and grazing systems with all sorts of possibilities.

These are similar to the definition of Permaculture design offered by David Holmgren: "Consciously designed landscapes which mimic the patterns and relationships found in nature, while yielding an abundance of food, fibre, and energy for provision of local needs."[10]

Riparian buffer strips. Photo from USDA

DESIGNING FOR OPTIMAL SPECIES INTEGRATION

The silviculture system is perhaps the form of agroforestry most similar to that of an oak savanna, nature's most prolific biome. In each, grass, tree, shrub, forb, ground covers, and fungal layers dwell together in a stacked design accessing different soil layers for their nutrients. As part of a guild or integrated forest garden, a silviculture enables stacked resource acquisition zones. For instance, a grass layer primarily gathers its needs from the top eighteen inches of soil while a walnut or oak tree gathers its nutrients from a level much deeper — three to eight feet or more.

Overall grass, shrub, and tree design should be planned so that all inputs from each component are used. Every plant and feature serves multiple functions; waste is minimized as everything is consumed at different trophic levels of consumption. Animals graze the grass-and-forb ground story, while the shrubs and trees can provide browse for animals as well as wildlife forage, cover, shelter for the livestock, human food, and timber crops. All the plant layers serve as a deep carbon sequestration sink and as a destination for animal nutrients that would otherwise run off to adjacent watersheds.[11]

You will need to observe carefully: Sometimes you'll need to intervene to control overgrazing by the livestock you've chosen for the design implementation. Use of a paddock-style system of rotation grazing (turning the livestock into different fenced enclosures within a silviculture) allows a pasture to recover and gives the animals access to successive premium-grade food sources. At the same time rotational silviculture grazing discourages the buildup of plant and animal pathogens by

Animals grazing under trees. Photo by Steve Bevan

reducing the accumulation of disease organisms. Cycling grazers through different paddocks allows natural biological systems to break down pathogens.

Perhaps the best example of successful silviculture systems is that of the Spanish dehesa, where cork oaks (*Quercus suber*) and chestnuts (*Castanea* spp.) combined with understory shrubs and forbs coexist with pastures and row crops. The dehesa is grazed by cattle, and its acorns and chestnuts are harvested by the Iberian pig — a high-value niche pork species. This system has been in use for at least four thousand years and produces lumber, cork bark, chestnuts, herbs, row crops, honeys, and acorn- and chestnut-fed meats. It is probably the most successful example of a human-designed savanna system.[12] In North America we could learn a lot from the Mediterranean example (described more fully in chapter 4) and perhaps even find ways to harvest our native bur oak (*Quercus macrocarpa*) in a similar manner.

Forest Farming

In forest farming and forest gardening, on a smaller scale, the emphasis is less on livestock and more on wildlife habitat and timber, nut, and fruit production. Other yields of such a system include woodland specialty crops such as goldenseal (*Hydrastis canadensis*), ginseng (*Panax* species), and Siberian ginseng (*Eleutherococcus* species) — all medicinal herbs — as well as honey. Other possibilities include the production of high-value food and medicinal mushrooms such as oyster (*Pleurotis* species), shiitake (*Lentinula edodes*), maitake (*Grifola frondosa*), and reishi (*Ganoderma lucidum*),

Mixed crops.

which fetch high prices in the markets. Additionally charcoal and coppice wood can provide a sustainable income.

As always, careful planning is necessary regarding plant species. Animals such as deer, squirrels, and others will come in on their own and can be culled when necessary to ensure that their numbers do not exceed the capacity of the designed site to support both them and sustainable yields for the human community.

Comprehensive Integrated Agroforestry

Recently a fusion of all forms of agroforestry has been trialed in Iowa. Initiated by the Practical Farmers of Iowa, it combines practices including alley cropping, shelterbelts, riparian buffers, forest farming, silvopasture, and others using a mosaic system of tenth-of-an-acre blocks of fruit- and nut-bearing trees and shrubs planted with a single overstory tree species per block. This allows sufficient plant concentration to ensure the cross-pollination needed to set nuts and fruit while allowing for easy management of each block. Ten different species of trees are planted per acre — a practice that discourages the easy pest access common in a monocrop orchard system. Beneath the trees are planted a mix of perennial herbs and shrubs with high medicinal value as niche crops. A variety of wildlife use the guilds as food, fodder, and shelter. Livestock can be grazed in the open areas between the blocks.[13]

The advantages of designing your guild using a comprehensive integrated agroforestry plan include: less soil erosion; lower fertilizer and chemical costs; less equipment needed for harvest; lower runoff with high water infiltration back to the aquifer; a more efficient use of solar inputs; higher biodiversity; ecosystem stability; higher profit potential compared with costs; greater carbon sequestration; and the production of more energy than is consumed.

Prairies

Windswept grasslands obviously do not fit into a definition of an agroforestry site but can nevertheless be functional landscapes used for grazing and wildlife habitat as well as other benefits. Traditionally the great prairies were home to tall- and shortgrass from horizon to horizon, punctuated by shrubs and cottonwood trees at rivers and creeks. They were neither savanna nor desert yet were home to many species of animals. Later the grasslands were severely depleted through overgrazing by beef cattle and overuse of the plow. However, methods of restoring the biodiversity and productivity of native grasslands are being pioneered by visionary farmers who realize that only bringing back that which was lost can restore the land to its full potential.[14]

In Australia the Communities in Landscapes Project Benchmark Study of Innovators[15] studied five farm properties, pairing each with a neighbor. Each of the five was a site where innovative restoration practices based on restoration of native grasses, shrubs, and forbs were being used to bring back balance to the nutrient and water cycles. The adjacent sites tested for comparison were each on the other side of the fence from the innovators but were using standard high-input agricultural methods. Results showed that the innovator sites had a higher degree of soil and plant diversity with higher grazing yields for both sheep and cattle on carefully managed grazing systems. Wildlife species were also much more evident on the innovator sites. Other benefits included increased soil moisture levels, lower costs due to decreased pesticide and herbicide use, lower veterinary and feed costs for livestock, and a greater overall biodiversity.

Bringing back the levels of diversity that were present in agroforestry and prairie systems before European settlement would go far toward restoring resilient natural systems to counter climate change, and species extinction. Adaptive guilds and integrated forest garden plans using techniques developed by innovative landholders around the world will be a design challenge based on site-specific issues in the very near future, if they aren't already.

Specific Plant and Animal Interactions in the Plant Guild

A newly established guild is an open invitation to animal participants to add their own inputs into the community. They can move some plants around by digging and replanting, such as when a rodent transplants a bulb to hide it from other animals in the vicinity, or serves as an unwitting carrier of seeds that adhere to its fur then drop off at a favorable location. Bird droppings along fence lines and below trees and shrubs hold seeds encapsulated in fertilizer, which germinate quickly to grow into a hedge. Seeds and nuts stored in the ground by rodents go forgotten and become the basis for an entirely different design from the one originally anticipated by the gardener/designer — but one that includes the requirements for life of myriad nonhuman species.

Grazers

Grazers are animals such as sheep, cattle, horses, and deer that are able to process cellulose from woody plant species for food. They have an extra stomach called the rumen, where bacteria break down plant fibers or lignins, converting them to a form that can then be digested by an herbivore grazer's primary stomach.

Alley cropping, as described previously, is the practice of growing an annual or temporary perennial crop between rows of taller woody species. An example would be growing tomatoes, corn, squash, beans, strawberries, or raspberries between rows of nut or fruit trees. When the trees grow tall enough to begin to shade out the lower crops, you can introduce shade-tolerant species. Cattle, sheep, and swine can be used to harvest some of the crops from both parts of the alley-cropped guilds.

You may also want to implement chop-and-drop feeding systems for grazers and foragers. Cutting back excessive growth of the woody species allows for constant regrowth of fresh green fodder, which is more digestible than older growth woody branches and leaves. The three feeding strategies of alley cropping, silvopasturing, and chop and drop allow for a more complete diet nutritionally than does pasture alone.

Controlled cropping of vegetation by domestic stock animals is also managed through rotational grazing. This is the practice of moving the grazing species — whether cattle, sheep, goats, or chickens — to a different field or paddock with lusher growth than the previous one. Overgrazing in a field can lead to problems with erosion, overfertilization, and nutrient runoff from the site. A result of this can be pollution of downstream waterways. Rotating animals in a pasture leads to a managed guild system with fast regrowth for forage species. Portable fencing and shelter systems are used to rotate chickens and quail through the pasture, allowing for protection from predators. Such a design is called the chicken tractor.

In grazing and browsing guilds, animal wastes are dropped in place and processed by decomposer species, or detritovores, to become food for the

DESIGNING FOR OPTIMAL SPECIES INTEGRATION

Rotational grazing. Photo from USDA

Alley cropping with hazels. Photo by Amy Lykosh

plants that will in turn feed the entire system. As in any well-designed Permaculture or integrated forest garden design, all wastes are utilized.

Omnivores

The omnivore species — sometimes referred to as non-fussy eaters or opportunistic feeders — include bears, squirrels, chipmunks, raccoons, and ducks. They will consume insects, fruit, seeds and other grains, other animals, and (in the case of opossums) even one another in the form of road-kill. Omnivores play an even larger role than the grazers and browsers.

Squirrels and chipmunks, for instance, will eat birds' eggs, fruit, and many garden vegetables, as well as insects and grubs. Bears will eat most fruits,

Omnivore. Photo by Bruce Ruston

grubs, and even carrion left by predator species. In the north they are a major pest of bird feeders and difficult to discourage. Raccoons, with their almost-opposable thumbs, are able to climb, grasp, and improvise tools to gather food such as fruits, vegetables, small birds, eggs, and rodents. They are very dexterous and can gain access to shelter in places we humans really would prefer they didn't. Many homeowners have encountered issues with raccoons setting up shop in attics, basement crawl spaces, and garages. In urban areas they have even been found to have adapted to sewer living.

Ducks gather larvae, grubs, small fish, seeds, and tubers of water-growing plants. And they eat corn. Domestic swine, like bears, dig up the ground looking for worms, grubs, and roots to eat. A pig will eat almost anything. In the right season you can use them to clean and plow a garden guild or area where you want to establish a planting with minimal expenditure of human time and energy. The chicken, another omnivore, is of course every homesteader's dream animal. Eating grains, weed seeds, insects and their larvae, and even grit, the chicken is the ultimate living machine. Their system yields include meat, eggs, feathers, insect control, and garden cleanup.

Wild birds also have a big role in maintaining a thriving plant guild. Not only do they bring in seeds of other plants and drop those from roosting sites onto the soil, but they also perform system services in other ways. Woodpeckers and sapsuckers feed on overwintering bugs such as scale insects, a major pest of orchards. Chipping sparrows and white-crowned sparrows are adept at turning over dead vegetation searching for weed seeds and insects during autumn and spring months. They leave literally no leaf unturned. Chickadees, those cute little birds that visit bird feeders, also eat insects overwintering on trees as well as gnats and spiders. Hummingbirds will also fly into a cloud of gnats and mosquitoes and just chow down.

In the plant guild omnivores are probably the biggest challenge, as they can always find something to consume. Your goal is using that trait to advantage. By building habitats for them and by allowing their inputs to make the guild even more resilient, you can work with them toward that goal.

Placing bird feeders and nesting boxes on a site brings in bird species such as chickadees, sparrows, and nuthatches that will not only feed on suet and seeds provided to them but also forage on their own. While gleaning from the site they are sorting through the layers of duff on the ground, and of bark on trees and shrubs, consuming insects and seeds. Mammals such as chipmunks and raccoons will sort through everything and even turn the upper layers of the compost heap in their search for food. Skunks will dig up soil grubs at night, reducing the numbers of root pests in the guild.

Carnivores or Predatory Species

Carnivores are the top of the food chain. They are primarily consumers of other animals, and as such their numbers are fewer. They include the farm dog and cat, the coyote, the wolf, and the cougar. Owls, hawks, and eagles are three of the bird species in this group, although eagles are also carrion eaters. Smaller carnivores include caddis flies, dragonflies, wasps, and mosquitoes as well as some soil nematodes and other species.

Coyotes have moved back into the margins of urban areas, traveling along wildlife corridors such as rivers and creeks. They are a major predator of small rodents such as mice and rabbits. Red foxes also have been adapting to cities although they are fed upon by coyotes; the two species are not good neighbors. Both are major pests of any plant guild that includes domestic bird species.

Recently wolves and big cats such as bobcats, lynx, and pumas have been expanding their ranges to include suburban and city areas, and have even been observed in midwestern metropolitan areas of Milwaukee and Chicago. Their hunting ranges are very large, so sightings of them — while uncommon — are not unexpected. Wolves and large cats

Dragonfly. Photo by Nicholl Williams

are the major predators of deer, another species that has extended its range right into the cities and suburbs as vegetable garden pest.

Rodents (Small Omnivores)

Rodents are largely seed- and grub-eating smaller animal species. They include ground squirrels, mice, voles, gophers, and shrews. Many of them will also eat grubs and small insects. Like the chipmunk they are great harvesters of seeds for the winter and will spend much of the year amassing huge amounts of food. The tunneling of small rodents helps improve the texture of heavy soils as materials are moved from one soil zone to another.

Detritovores and Other Creatures of the Soil

The cleanup crews of the animal world are the detritovores — so named because they feed on the detritus layers in the soil where nutrient recycling begins. Whatever is remaining when carnivores and omnivores are done gets eaten by the least among us: the beetles, nematodes, bacteria, and a host of other players in the game of decomposition. The actions of omnivores in moving and turning leaf litter give decomposers better access.

Looking into a compost heap after a few weeks of decomposition, you will observe the character

DESIGNING FOR OPTIMAL SPECIES INTEGRATION

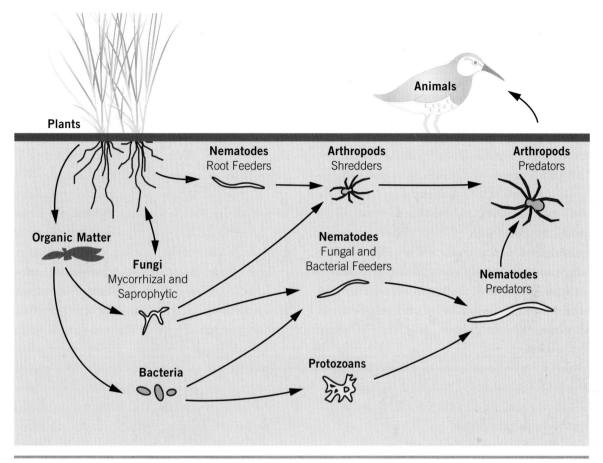

Soil food web. Illustration by Kellen Kirchberg

actors in the detritovore play. Pill bugs, centipedes, worms, and beetles all aid the decomposition of the guild's compost and leaf litter layers. Squirrels, chipmunks, birds, and even opossums all play a part in tilling the compost and aerating it in their search for sustenance.

The soil environment includes mammal species as well as insect, plant, and fungal ones. Larger ground-burrowing mammals such as ground squirrels, badgers, woodchucks, coyotes, foxes, and rabbits all dig dens in the earth. Their tunnels can be a home for other species when the nests are empty. Snakes, moles, voles, mice, and shrews will all move into abandoned dens and call them home.

Chipmunks and other rodent species will store seeds and nuts belowground and have separate burrows for each bodily need. You can sometimes see extensive colonies of ground-nesting mammals such as the prairie dogs of the Great Plains area of North America.

There are many creatures living in the soil smaller than mammals and reptiles. Each has a separate niche-feeding pattern. They can be divided into three major groups, the macrofauna (large); the mesofauna (medium); and the microfauna (small) of the soil. Macrofauna include earthworms, ants, termites, and spiders, as well as centipedes, millipedes, and pillbugs.

Earthworms feed on the composting waste layer in the upper few inches of the soil, turning over the leaf litter and transporting it down to the root zone. In the North American areas that were previously covered with ice sheets, earthworms were scraped from the surface along with all the soil layers; they remained absent from the landscape until Europeans reintroduced them. The forest soils of the eastern forest biomes of the Upper Great Lakes and Northeast evolved soil systems where leaf matter accumulated annually, building up a deep duff layer of half-rotted leaves that were a perfect nursery medium for tree seedlings.

With the introduction and subsequent expansion of earthworm populations, forest seedlings are unable to access the nutrients needed for growth, as those fertilizers are being taken deeper into the soil zones by earthworms. This is not to say that the worms are not fulfilling a necessary biological function — merely that the northern forests are not adapted for that nutrient loss and are suffering a decline because of it.

Ants harvest leaves, dead insects, their own comrades, and other detritus, as well as farming *Aphis* or aphid species, sap-sucking insects that secrete the sugary exudate some ant species use as a food. Their underground activities support water and nutrient penetration of the upper soil layers, increasing aeration.

Very social, ants can build nests that in human terms would be as large as a small city. Periodically they will engage in social behaviors of information

Termites. Photo by Filipe Fortes

sharing wherein they congregate in groups of four or more to communicate only they know what. Often in the summer we have seen them in groups of thousands, subdivided into groupings of four to six, early in the day before the sun heats up the bricks. We think of these ant associations as conventions; they certainly have all the trappings of one, right down to the parade.

Termites are another social species that lives in rotting wood like some ant species. Unlike the ants, termites can excavate good wood and so are considered a major pest of residential buildings with wood foundations. Like worms and ants, termites are a major nutrient recycler, breaking down more complex molecules to return them to the trophic systems for other organisms.

The actions of macrofauna mix and redistribute organic matter, neutralize soil pH, enhance microbial populations via moving materials, mediate physical soil properties, and reduce harmful species such as predatory nematodes. All of these functions serve to build and maintain a viable soil medium for guild fertility.

Nematodes are the soil mesofauna. Think of them as free-ranging omnivores and predators on a soil equivalent of the Great Plains. Tiny roundworms with unsegmented bodies, nematodes are found in all soil habitats and are extraordinarily diverse. There are ten to twenty million of them per square meter, and while some are visible to the unassisted eye, most are not. Some are free living and others parasitic. Some are major plant crop pests in the guild; others have evolved to prey upon fellow nematodes.

The most numerous soil microfauna are protozoans. They can be agents of specific plant diseases, but many have beneficial functions. Protozoans are important to the processes by which soil-soluble minerals are immobilized so they can be taken up by plants. They prey on bacteria and other microbes and can suppress the activity of fungi and bacterial pathogens.

Several other species groupings inhabit the soil layer but are not, strictly speaking, defined as either plant or animal. Bacteria are single-celled organisms related to actinomycetes. Related to them both are the fungi. Each group has huge roles to play in the ecology of the biosphere. Some of those functions include capturing and processing nutrients and symbiotic relationships with plant roots.

Everything and Everyone Is Lunch

One of the most interesting and enjoyable aspects of guild observation is the opportunities it gives us to learn from the masters of life, all of nature. While watching, friendships can form between species — or at least we can think of them that way. Mutualistic dependencies of wild creatures turning to humans for food and water extras, and humans turning to them for pest control and other ecosystem functions, enable recognition of common interests between species. We're all in it together. A web of interactions connects all species from the tiniest bacterium to the largest redwood tree. A plant guild design takes into account the truth that everything becomes food for something else, that everything is related across time and space, that not only are we what we eat, but we will be what ate us.

CHAPTER SIX

Guild Project Management

We abuse land because we see it as a commodity belonging to us. When we see land as a community to which we belong, we may begin to use it with love and respect.

— ALDO LEOPOLD[1]

The art of guild project management entails budgets, implementation, management, and maintenance. Basically, all of life is maintenance and management. Why would the care of a plant guild be any different? Management brings stability to the system. This obviates the fact that plant guild maintenance is about ongoing interaction with human beings. We might say that there is no invasiveness, simply lack of design and management. If the plan is for maximum diversity, a plant would very readily take care of itself: This is the ideal. But the reality is that as we establish our planting regime, it may take several years to reach the point in our work where most of the labor is gathering the fruit — the harvest.

Of course, the key here is the design, which is the basis, driver, and plan for installation. Based on the time line you create for establishing a guild, you can institute a concurrent, phased, or protracted implementation.

Another important factor is budgeting. The price of trees from most nurseries is within the range of ten to thirty dollars each. By doing your own grafting and plant propagation, though, you can eliminate this expense. This is where "divide and conquer" takes on a whole new meaning. The nature of the plant world is to continually add to its essential level of biomass. As we designers key into the biological intelligence and plant physiology that surround us, we find that plants are very forgiving — in some ways encouraging — of the fact that our process can disturb, cut, divide, and replant them in any number of ways.

Permaculture is design in the future tense. In other words, the plant world is a dynamic world, not a stationary, unchanging community. Based on our level of soil and water management, environmental modification, insect pest management, disease and weeds, agronomic details, and physical, chemical, and biological processes, our management and budgeting techniques play tenfold into the health of our guilds and landscape.

In this chapter we delineate the options in a step-by-step process of budgeting and accounting, and post-design implementation and management. For homeowners this is an intimate and detailed process, accounting for long-term consequences and benefits. Initial plans can be changed, adapted, and redirected with future learning. On the other hand, the professional designer must take all this into account in the short term and anticipate the natural progression of the design. Working with clients and their property holds a special responsibility, much broader and lasting longer than for conventional landscapers. A certain amount of education, training, and guidance for reasonable expectations is involved to prepare the owner for the transition. The foresight to visualize the cause-and-effect progression of plants growing in the landscape and to foresee in the mind's eye what the landscape will look like in the future is essential to a seasoned designer. This talent is highly valued. The reward is an abundant yield of nutrient-dense foods, sweet-as-sugar fruits, health in all its diverse guises.

The nature of Permaculture does not end in design work. Without implementation there is just an idea — a pretty picture, but one that misses the next step that will eventually bring a tangible harvest. The design is simply a conceptual representation of where we wish to go. The reality is that it cannot get there unless we take it there. But if we do this right, all maintenance is harvest. There is no waste in a systematically designed plant guild or a forest garden. All the energy flows and nutrient cycles are accounted for as plants are positioned to partition the resources, and share their ecological functions.

Add to that all the organic material that will be produced and stockpiled from pruning and trimming. These are also a harvest to be returned to the local ecosystem. This reduces imported resources since nature is creating the organic materials to build natural capital and cycling all of the nutrients back into the system.

Implementation Time Line

The success of a Permaculture project rests on how it is staged during implementation. The tasks can be divided between broad and coordinated site preparation and sequential tasks that allow for modifications as the design appears. For example, there is a supporting structure in the soil for the plants, thus soil resources that need to be present before the plants arrive. Even when plants are chosen for the soil type and climate conditions, organic matter and associated organisms for healthy soil may be lacking.

After an entire patch is planted, for example, the drip irrigation would be installed, then mulch applied to perennials, and seed spread for ground cover. Acting out of sequence can stall or disrupt the emergence or establishment of the plants.

What Are the Broad Site Preparations?

Access is key to all guilds. Being able to move into and out of the guild area without disturbing the plants and crushing the soil requires giving some thought to paths, entries, and exits. Build the paths and use them to define the spaces and edges within the planting area. Paths can also be depressions where water collects during rainy days and spring thaw. Swales make great access paths when dry.

Prior to making any progress in developing the plant guild, forest garden, or polyculture, the soil needs to be assessed and most likely amended to create conditions suitable for good plant growth. Except for existing woodlands, most sites lack a perfectly balanced and natural soil. Site preparation can begin at any time; the more time the planting bed has to collect and build the soil organism community, the more resources will be available to the plants. Soil preparation can be started on the entire planting space and, with few variations, involves similar materials and treatments. Soil in compacted sites should be broken through for drainage and root growth. This can be done with a broad fork, keyline plow, or subsoiler.

Add organic material and the soil organism communities create a rich biotic medium for the plants. In site prep and newly restored soils, this is very important. Large compost or leaf piles can be temporarily staged on top of a planting area to encourage organism growth and soil moisture.

Trees always come to mind. "Get the trees in first" is a common call. As in the seven-layer forest garden hierarchy, bring in the trees, auger or dig the holes, and put them in as a group. This helps define the resulting spaces and gives a sense of scale to the design space. When we work on relatively flat spaces, it's the trees that orient our perspective to the vertical proportions.

What Is the Sequence of Implementation?

It helps to work off a template when you begin the design and implementation process, not only for selected polycultures and plant guilds, but for an entire property. An integrated forest garden is what it says: *integrated*. This presumes that it is integrated with the entire land base in all we design and put on the ground. Here is an eighteen-point course of action to get you started:

1. Earthworks
2. Main irrigation lines
3. Access
4. Infrastructure
5. Hardscapes
6. Structures
7. Stockpile and staging areas
8. Site preparation

A Wildlife Corridor

During the initial year of our small orchard space, we discovered that the previously large hill of turf was a wildlife corridor. We had positioned tasty deer treats across their seasonal path. It was not until the animals had re-created their access and consumed a few tender trees that we became aware of the issue. We immediately turned our attention to fencing affected areas. A tall "trellis" now surrounds much of the orchard.

9. Soil conditioning
10. Flagging of plant positions
11. Procuring of plant materials
12. Overstory installation
13. Lower trees and shrubs
14. Perennials
15. Irrigation system emitters
16. Mulch
17. Ground cover
18. Tree tubes, staking, and plant protection

What Steps "Complete" the Design Implementation?

Frequent observation for the first few weeks is important. Transplant shock and the stresses on individual plants require immediate attention. Fresh plant stock may attract herbivores. The disturbed soil will germinate portions of the seed bank and cause weeds to emerge. This initial time, as new ecological interactions appear, is the best time to catch issues, such as ecological deficiencies and weeds in early stages of growth.

Add more plants as spaces appear. Even the most comprehensive design may leave opportunities for other plants. More nitrogen fixers, mulch makers, and insect habitat will only increase the diversity and ecological services available to the guild.

Think of the plant guild as a living ecological document, ever changing, shifting, and spreading in all directions. Completeness is a matter of needs being met. If each plant is getting its daily requirements, the system is complete. An "ornamental completeness" focusing only on aesthetics frozen in time will guarantee many days maintaining that socially determined "tidy" look.

Seasonal pruning, usually in winter, is the time to shape the plants for health and harvest. Summer and fall field and ground sanitation is the time for removing dead material and diseased fruit. Seasonal harvests of organic materials like this will reduce plant stressors and increase airflow while giving the guild a healthy look. Focus on species-specific practices (you can learn these from the literature) for the plants' best advantage and your best harvest

What Is Needed for Long-Term, or Protracted, Implementation?

A long-term vision is best aligned with a long-term harvest. Once an initial guild or forest garden is implemented, it will supply many of the resources needed to extend the growing area and propagate the plants. Small rootstock fruit trees are planted and grafted from the original stock of trees. Time is now on the side of a long-term strategy.

Although you may have purchased the first wave of plants from a nursery, many will grow to the point of dividing or supplying rooting stock for further plantings. Perennials especially are suppliers of new plants as they go to seed and multiply in years two and three. You are now the proud owner of a nursery, too; be ready to pot plants that can be pulled for later, and create a protected area to start harvested seeds, building your own proven line of plant material. The foundation planting area is a great place to keep an eye on new plants as they emerge. One comfrey (*Symphytum officinale*) plant equals forty sprouts by season end. Black currant (*Ribes* species) roots from multiple buried branches, allowing you to transplant for future yields. Cuttings from many plants can be rooted in small pots using wet sand and IBA rooting powder or willow bark tea as a rooting hormone. This is the best reason to start small and go slow. Learn the plants and develop seasonal guild expansion at low or no cost.

While all this is going on, keep in mind that the soil is the basis for all growth. Composting, worm bins, and piles of rotting wood chips act as the stocks you'll use to colonize and rejuvenate depleted soil resources. Mulch and chop-and-drop dynamic nutrient accumulator plants all around the guild area to feed the soil organisms. Healthy mulching and soil will buffer weather extremes and ensure plant health.

Budgeting the Financing

With the design process comes a detailed budget for the long haul of your plant guilds: rootstocks, trees, utilizing what is available, division, propagation, and so on.

All of this is predicated on the required level of soil and water, as well as insect pest, disease, and weed management. Then there are the agronomic, environmental adaptation, and management particulars; the physical, biological, economic, and labor resource base that can be supplied over time. It is a system needing stewardship.

Time Lines

The project management decision tree is a step-by-step process that guides problem solving with chronological precision; you make basic decisions first that support later ones. Use this in guild design and in many ways it is like the Permaculture principle, "Design from patterns to details." As mentioned in chapter 2, Yeomans's scale of permanence is also a guide to the process of decision making. It can involve stepping back to see what needs to be considered before the detail at hand. Logistics can help you keep the chaos to a minimum and focus on the order of tasks, rather than getting overwhelmed by the magnitude of the entire project. Break it down.

Design Decisions and Checklist

When you're planning a trip, party, or rocket launch, it's important to make sure all the *t*'s are crossed and *i*'s dotted. Missing any crucial point can cause a catastrophic failure — or at the very least, scrambling for resources. Within ecological design, we are working with nature and at the same time trying to guide the progression of the ecology to a certain yield. The final assessment of our polyculture guild design can get quite complex since it's being implemented not only in three dimensions, but also in a complex integration of relationships over time. If the resource supply is not prearranged, there may be gaps in the functional relationships; this can cause stress to the plant systems, affecting yields and plant health. Use the checklist below along with your design to confirm that all the ecological functions are present, and the plant requirements are supplied. A standard twenty-four-by-thirty-six-inch paper has space for you to make notes in the margins. Using a red pencil can accentuate the importance of changes to the design. Once you've completed the checklist, attach it to the design or keep it in a design/implementation binder.

1. Polyculture and Guild Resource Structure
    a. Are all needed ecological functions represented in the guild design?
        i. How many nitrogen fixers are supplied to each tree?

INTEGRATED FOREST GARDENING

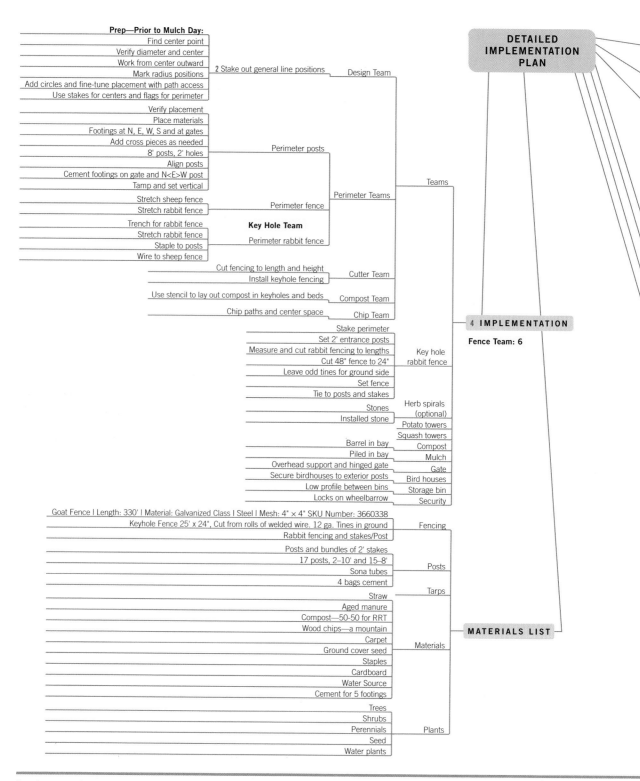

Implementation Planner.

GUILD PROJECT MANAGEMENT

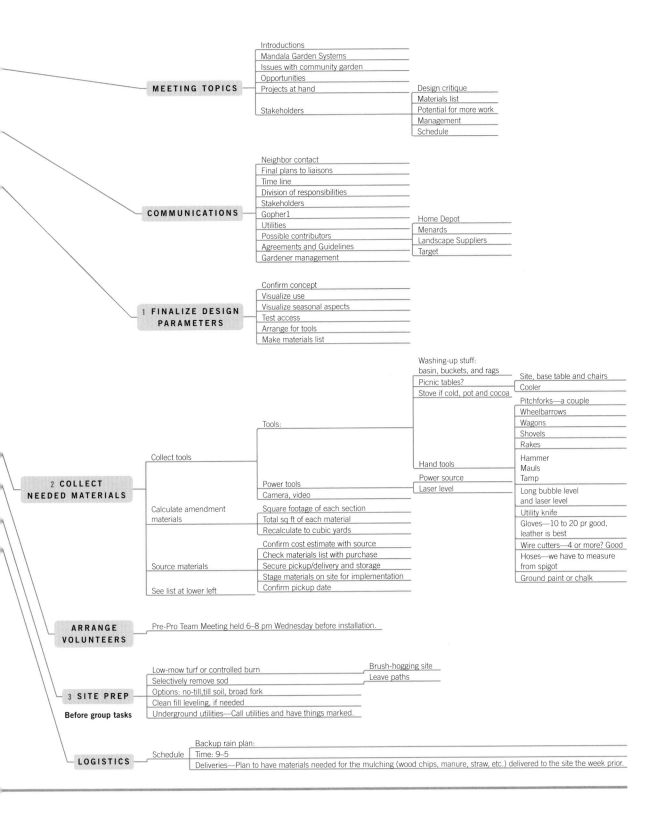

1. How many are perennial, shrubs, or trees?
   2. Do they cover an adequate area?
    ii. Are dynamic accumulators and mulch makers adequate for seasonal coverage?
    iii. Have taprooted and fibrous plants been interplanted to partition soil resources?
    iv. Is sunlight partitioned to the plants by positions relative to their height?
    v. Is there a seven-layer forest canopy or similar structure?
    vi. Does the plant list account for all plants and their numbers?
  b. Are berms and swales adequate for landscape access and water collection?
    i. How are they configured?
    ii. Placement, capacity,[2] proximity, distance, and slope?
    iii. Have all water sources been integrated into the design?
      1. Roofs, gutters, and hardscape runoff.
    iv. Is swale and pond capacity adequate for extreme weather events?
      1. Do spillways extend excess water to other areas for infiltration?
    v. Are swales usable for access by machinery and mowers?
      1. Is the width wider than the largest tractor, trailer, or mower deck?
      2. Can the area be entered easily without damage to plants?
    vi. What areas are more suited to diverse plantings?
      1. Do berm sides and top have different plants for sunlight and water?
  c. Are rain gardens or ponds adequate?
    i. Pond placement, access, size, and mechanics.
      1. Water source, overflow, circulation, filtration.
    ii. Determine pond use.
      1. Infiltration only, sand base.
      2. Beneficial plants and habitat potential.
        a. What are the local beneficiaries?
      3. Aquaculture pond.
        a. Aquatic plant propagation.
        b. Exotic fish sales.
        c. Fish for human use.
    iii. Circulation and aeration pumps.
      1. Power, solar?
      2. Plumbing.
  d. Do production spaces need to be enlarged for adequate access?
  e. Do spaces need to be recontoured for runoff containment?
    i. Can areas be inundated for short periods to retain moisture?
  f. What irrigation will be used to help establish the plantings?
  g. Determine location of materials.
    i. Storage while installing design components.
      1. Potted plants.
      2. Compost area.
      3. Topsoil.
      4. Water cans.
      5. Irrigation materials.
      6. Tools and supplies.
  h. Think twice, move once.
  i. Consider a stockpile location for long-term use.
2. Materials ordering timeline
  a. Irrigation materials.
  b. Soil materials.
  c. Soil restoration.
    i. Compost.
    ii. Peat moss.
    iii. Compost tea and amendments.
      1. Feather meal.

2. Humic acids.
   3. Powdered seaweed.
   4. KIS compost.
   5. Fish hydrolysate.
   6. Liquid plant dynamic accumulator ferments
  d. Hardwood or straw mulch.
     i. Rent chipper/shredder.
     ii. Purchase and stockpile.
       1. Delivery?
       2. Truck access and deposition?
  e. Topsoil.
     i. Preparation needed.
       1. Tilled to a good depth?
       2. Broad-forked for water infiltration?
       3. Subsoil plowed to break up compaction?
     ii. Purchased.
       1. Delivery?
       2. Truck access?
       3. Deposition (dump where)?
  f. Plants.
     i. Delivery or pickup? Which nurseries?
     ii. Truck access and deposition?
     iii. Storage space, protection.
3. Delivery
  a. Arrange delivery of all materials prior to install day.
4. Materials staging area
  a. Prep area for plants, pots, watering.
  b. Irrigation parts assembly.
5. Gathering area for workers
  a. Labor.
     i. Schedule equipment operators.
     ii. Manual labor: How many people? Hours, days?
6. Tools
  a. Shovels, spades, scoops.
  b. Rakes, garden- and landscape-style.
  c. Augers.
  d. Spade forks.
  e. Pick-ax.
  f. Gloves.
  g. Trowels.
  h. Watering cans.
  i. Knives.
  j. Hoses.
  k. Wheel barrows.
  l. Cooler, drinks, etc.

The Budget

The budget is an integral facet of the master plan for a property. It is one thing to dream (which we always encourage); it is another to get real about what we have in terms of financial resources to move a project forward. Figured into this are recycled goods, plants gifted to us, organic materials collected, and an assortment of bartered, scavenged, and traded goods.

The budget is under constant review. As long as the project is in process, it's a major player in the development of the site. Not reviewing the budget frequently brings needless risk into the project. Reviewing and adhering to the budget throughout will highlight any unforeseen issues. You can then make any necessary adjustments, keeping in mind that the bottom line needs to be constant and often money is moved around rather than added. The site budget should be a compilation of smaller project budgets and investments; some are annual expenses and others, long-term investments.

The infrastructure and supporting systems are initially your most costly investments. Having the resources to implement the projects and then the infrastructure that will ensure its success is a priority.

Unplanned resource time, hours and hours of unforeseen work and maintenance, colors the way

> ## Dan's Ninety-Minute Tasks
>
> Living on a moderately sized homestead, many of my projects are made of ninety-minute tasks that I can start and finish in that time — small, attainable advances toward the final goal. It also helps keep projects from creeping into larger unmanageable mega-messes. Some projects will take solid days to complete and need professional craftsmen, but taking the time to slowly progress helps the vision stay clear, lets me make small adjustments early on, and keeps the goal in my mind.
>
> — DANIEL HALSEY

we feel about our work and can make or break our willingness to proceed further. In all of our years of putting Permaculture and related systems on the ground, we have seen many projects go to the wayside, victims of a client's lack of time or the feeling of being overwhelmed. Time and labor, although these may seem too "businesslike," figure into anything that we do as human beings.

But how is this quantified? How do we "budget" for our time or those "volunteers" or staff that we bring into the mix? The natural world is so unpredictable that it seems almost impossible to calculate or enumerate work in numbers or finances. Most of us who garden do not think of labor as an expense. The work in itself is so satisfying and brings with it such a feeling of well-being that we would be hard-pressed to attach dollar signs to it.

How to reckon and estimate labor into the mix of a design is a personal choice. Some will not care in the least, and some will make a gallant effort to measure the time it takes to manage a plant guild, food forest, or gardening project. It is important for those who need hard facts and figures to also measure the time it will take to do so. As you might imagine, there is a huge gradient between the hobbyist, the CSA[3] grower, and the subsistence farmer. In any case the inputs are kept to a minimum so the harvested calories support the grower. If you live on a large piece of property with variable projects going on all at once, it's imperative that you block out a more or less sequential schedule to take care of the chores. You can plan this during the design process by prioritizing zones of work and dividing up tasks. In essence, the reality of the "work" involved as we proceed helps us to sincerely plan for implementation, management, and the risks that may ensue.

So when necessary, review the budget. If you are supplying the labor, it's important that clear channels of communication stay open and that you stick to the budget as closely as possible. The project resources are ultimately where most of your project budget is spent. Pay close attention to the resources and skills you've assigned to a project to ensure that they are being fully utilized and that the right balance of skill and resources is available.

Time, materials, and tools: As a budget is created, keep these three categories close in your planning efforts. Possible risks and existing issues have a way of making their way into projects, causing problems. It is of utmost importance to keep a tight eye on project risks and reduce the level at which they can affect your project budget. It is unrealistic to think you can stay clear of risks and issues, of course, but you can have a plan of action ready when and if they do arise.

Let's attempt to put the *eco* back into budget practice. It is imperative to take care of the "house" and the greater house of the ecosystem that we live in. Strive toward a zero-waste environment where all resources are turned back into the biology at the site. Nature prevails and takes care of business for us in many ways. Be cognizant of all that this entails. These elements — time, labor, implementation, finances, eco-nomy, eco-logy — all figure into the course of action. If you persevere to relinquish power (as stated by Bill Mollison), and to fuse your efforts with the biological intelligence of your site, it is quite possible that you can offset financial outlays and build real economy into your living space where the web of all life is intact and thriving. It then supports you.

Continuing on, the process of design homes in on designated areas and begins to detail out in terms of biomass and constructs. Each of the plants, paths, hardscaping elements, soil considerations, and so on will have a price tag placed next to it. Materials and plants obtained free through trade or gift will also figure into the final tally.

We suggest to folks who do not have the means to put everything on the ground at once that they stage their implementation over the long haul. Begin by planting woody overstory perennials; work out from there over time as funds or plants and materials become available. A budget chart specifically focused on developing a plant guild follows. The plant list can also be used for accounting for prices and quantity.

In this list you'll find the ID for locating and counting plants on the design, the base price (retail), and the total cost of plant material. Lines can also be added for mulch, compost, and other materials. Many of these plants were ultimately transplanted from other areas on the site. As you

How Climate Change Affects Our Plants

In recent years the seasons have been disrupted all over the world by climate change. The planting times for trees and shrubs have been affected by the stress of summer temperatures at night. During hot days plants shut down transpiration and wait for cool night hours to uptake soil moisture. Unfortunately, in some areas the nighttime temperatures have not been cooling to a point that all plants can return to transpiration. This is especially hard on plants stressed by transplant shock or another disturbance. Sugar maples in the Upper Midwest of the United States are especially susceptible, dying en masse across urban areas.[4] Because of this shift it is recommended that large potted trees and shrubs not be planted in spring, but rather in cooler months when soil water is available and diurnal temperatures allow for a period of water uptake.

TABLE 6.1 PLANT LIST WITH PRICES

Subtotal	Price	Quant.	ID	Name	Scientific Name	Ecological Function	Human Use/Crop
$180.00	$60.00	3	1	Evans Bali	*Prunus cerasus* 'Mesabi'		
$60.00	$60.00	1	2	Honeycrisp Apple	*Malus pumila*		Food
		6	3	Comfrey	*Symphytum officinale*	Chemical Barrier, Domestic Animal Forage, Dynamic Accumulator, Insecticide, Insectary, Mulch Maker, Water Purifier	Biomass, Compost, Food, Medicine
$30.00	$5.00	6	4	White Wild Indigo	*Baptisia alba*	Dynamic Accumulator, Erosion Control, Insectary, Mulch Maker, Nitrogen Fixer, Soil Builder	
$28.00	$14.00	2	5	Leadplant	*Amorpha canescens*	Dynamic Accumulator, Erosion Control, Insecticide, Insectary, Nitrogen Fixer, Nurse, Soil Builder, Wildlife Food, Windbreak	Insect Repellent, Medicine
		7	6	Alpine Strawberry	*Fragaria vesca*	Moderate 6.1–7.8 pH	Mulch Maker
$40.00	$5.00	8	7	Purple Coneflower	*Echinacea purpurea*		
$45.00	$15.00	3	8	Black Currant	*Ribes nigrum*		Container Garden, Dye, Essential Oil, Food
$45.00	$5.00	9	9	French Sorrel	*Rumex acetosa*	Insectary	Dye, Essential Oil, Food, Medicine
$30.00	$5.00	6	10	Chicory	*Cichorium intybus*	Dynamic Accumulator, Insectary, Mulch Maker	Biomass, Compost, Dye, Food, Medicine
$10.00	$5.00	2	11	Saskatoon Serviceberry	*Amelanchier alnifolia*	Erosion Control, Wildlife Food	Food
$60.00	$60.00	1	12	Summercrisp Pear	*Pyrus communis*		
$15.00	$5.00	3	7a	Sky Blue Aster	*Aster azureus*		
$15.00	$5.00	3	7A	New England Aster	*Aster novae-angliae*	Insectary, Soil Builder, Wildlife Food	Cut Flower, Medicine
$-	$5.00	0	Exist	Rugosa Rose	*Rosa rugosa*	Nurse, Wildlife Food, Windbreak	Aromatics/Fragrance, Food, Medicine
$15.00	$5.00	3	GC	Anise Hyssop	*Agastache foeniculum*	Aromatic Pest Confuser, Insectary, Water Purifier	Aromatics/Fragrance, Food, Medicine

Subtotal	Price	Quant.	ID	Name	Scientific Name	Ecological Function	Human Use/Crop
$15.00	$5.00	3	GC	Lupine	*Lupinus* spp.	Domestic Animal Forage, Erosion Control, Nitrogen Fixer	Fiber, Food, Soap
$48.00	$1.00	48	GC	Daffodil	*Narcissus* spp.	Insecticide	Cut Flower, Dye
$96.00	$8.00	12	GC	Common Yarrow	*Achillea millefolium*	Aromatic Pest Confuser, Dynamic Accumulator, Erosion Control, Insectary, Mulch Maker, Nurse	Aromatics/Fragrance, Compost, Cut Flower, Dried Flower, Dye, Essential Oil, Food, Insect Repellent, Medicine
$736.00							

can see, propagating your own plants will save a lot of money and is worth the time. Buying plants in the off-season and potting rootstock for growth gives you time to work new plants into the system.

Implementation, Management, and Maintenance

Implementation is key to taking the next step. When we begin the progression of placing elements into the landscape, it never quite looks the same as we planned on our drawings. We are now working in 3-D, but even when we have a computer-generated or hand-drawn 3-D picture in hand, it never quite lays out with reality. We cannot be locked into this "conceptual map" of guilds and the property as a whole. Things will change, as always. Regardless, proceed with abandon and take notes.

And now, here is where the initial work comes in, sometimes pleasant and easy and sometimes back-breaking. Depending on the details in the design, the devil may or may not be in them. Because the plan is dynamic, there is room for revision as work progresses. The plant world is very forgiving and will fill the niches unseen or unfilled in the design. Placing plants involves a tweak here and there; moving this plant to the south and the tall one to the back of the line. How about sheet mulching this area or digging a swale before anything else? Didn't think of this previously? Why not implement it anyway for the balance and integrity of the guild? You can always go back to the plan later and make the proper changes.

How do you plant the trees? Dig a large hole? Add compost? Simply slot the tree in and see whether or not it survives and thrives? There is no clear-cut way to do this. It's a matter of personal preference. Maybe you had success planting trees in a certain way at some time in the past. Maybe you wish to try something you researched recently. Or maybe you simply want to get the show on the road and not worry about the particulars. Stick it in the ground and wait.

Paths, mulch, signage, and all other elements will now be integrated into the whole of the guild. How does this plant guild figure into the entire property? As you implement the design, does it look like the functional relationships you need so much are functioning? Now that you're in "real" mode, does it all look and feel right to your senses?

Ask yourself: *What am I missing here? I know that the design is about what the guild will look like when it comes to maturity, but it all seems so empty at the get-go. Maybe I can plug in some other plants for the meantime as the trees age? Maybe I can intervene in the process just as people forever have been intervening in the process of implementing and managing the land?*

With plan in hand make your best attempt to stick to it. Whether you realize it or not, the map is not the place. Are you flexible and forgiving enough to make this plant guild happen? Can you make the appropriate adjustments? Upon completion you are now submerged in it for the long haul. Maintenance and management are next. The first few years will be more rigorous in terms of observation, hands-on care, and attention.

There is a plant that seems to be taking over. How do you handle it? Chop and drop it? Pull it out? Replant it somewhere else? Maybe you didn't include enough diversity in this guild to stop any kind of invasiveness. What is that dandelion doing over there? How can you make use of it instead of freaking out about dandelions? What else is showing up? What level of observation and attention are you willing to take on? How well do you know your "place" in all of this? If all of life, so to speak, is management and maintenance, then taking care of your plant matrix is no different than all life. Listen to what your guilds and property are telling you, the feedback they are providing. If a precious plant is being choked out by another precious plant, make a spot decision and move on it. You are, in essence, like Sherlock Holmes, searching for clues, constantly. What appears to be the problem may actually be something happening over there, or over here. Yes, you want the fruits of your actions to show up sooner rather than later, but the reality is that this will happen in its own time. Now is the moment to nurture the landscape and prepare the ground for what may come. Do you micromanage the site? We think not, but we also think that astute and rigorous attention and research are key. "The problem is the solution," says Mr. Mollison. True. Very true. There is no failure, there is only feedback from the organism that you have helped to create, this organism called a plant guild.

It's wise to develop a routine, a maintenance schedule that helps you organize your thoughts and tasks. This can be done by observing the movement of the seasons and applying intervention where needed, becoming more and more insightful into the whys and wherefores of your property, and knowing when and when not to intercede. You have become part of the plant polyculture that you have put into place. There is no separation. The human is now one of the organisms and it is you!

Typically, in annual gardening, it's easy to plan maintenance schedules at particular times of the year: seed selection, garden planning for rotations and placement during the winter months, planting in spring, harvest throughout the growing season, and so on. Much of the time you can watch the weather as a guide to stewardship. As you plant your guilds, however, the hard edges of maintenance scheduling become fuzzy. Until you learn about each plant that you incorporate into the guild in detail, document its characteristics and functions and where it is placed, you have little idea as to care, harvest, fertilizing, and pruning (if you do prune) — the list goes on. In general, as the guild is designed, you plan the work that is to be done across the year to the best of your ability (and knowledge). Think of guild stewardship as an intuitive exercise as much as a planned exercise. This starts and continues on seemingly forever, of course, with observation, from beginning to end.

CHAPTER SEVEN

Case Studies: Fifteen Plant Guilds

No occupation is so delightful to me as the culture of the earth, and no culture comparable to that of a garden.

— THOMAS JEFFERSON

Over the next pages fifteen plant guilds will be delineated in some degree of depth. Although there are no cookie-cutter plant guilds (we can't say it enough!), these polycultures are offered as a framework to work from, an infrastructure of sorts. Still, we cannot imagine that there is a single guild that applies everywhere — or anywhere. Guilds, as we've stated countless times throughout this book, are predicated on climate, landform, water wealth, microclimate, and countless other factors unique to a site. The fifteen case studies presented here each include a plant chart, a list of functions plant by plant, and delineation of uses and services offered by each plant in the polyculture.

So take these combinations of plants with a grain of soil and run with it. Find what works best for you and the place upon which you walk and observe every day. The guilds in this chapter may not make sense for your particular property, but the combinations of plants can be adapted to your site based on site conditions, with a few changes. Just as each human being is matchless and without equal, so are the guilds that find their way onto the land. Indeed, the plant guild is an unparalleled assemblage of green living matter configured for our overall sustenance and health. Make it happen! Explore to the ends of the earth!

These fifteen plant guilds, all of which may exist on any sized holding, are a beginning, a harbinger of sorts, that can, with a small push, implore the neighborhood or farm to action. A stone is dropped here and in other places, and as concentric rings form and expand, they touch and converge with other expanding rings from other neighborhoods and food forests of dynamic guilds come to life again, where sustenance, balance, health, and an agriculture of bounty reboots on what was once forest, then farm, town, and city. You can turn the historical succession back a few notches and re-embed what was once the matrix of all life in the region.

By way of example, a delineation of thinking about how to create guilds is presented here based on the Southern Illinois biome where Wayne and his family have made their home over the past seventeen years. The landscape is predominantly eastern woodlands, a once vast forested realm that stretched from the Atlantic to the Mississippi, a mix of varied terrains interspersed with wetlands, rivers, prairies, savanna, tidal flats, and mixed oak-hickory-basswood-maple mesophytic forests, all located in USDA Planting Zones 7a–7b.

Near the convergence of the Mississippi and Ohio Rivers, the Shawnee Hills, also known as the Illinois Ozarks, are primarily a sandstone/limestone escarpment that arises near Mount Vernon, Illinois, and falls off gradually toward the confluence of the two waterways. A meeting ground of several ecosystems, including the eastern woodlands, the Ozark Plateau, and the northernmost boundary of the Gulf Coast, the Shawnee Hills contain some of the most spectacularly diverse plant and animal landscapes in the United States, including 125 species of trees and over 100 species of mammals. Situated on the Mississippi Flyway migration route, an abundance of water and vegetation attracts three hundred species of birds annually.

The small city of Carbondale in Jackson County, Illinois — with a population of about forty-five thousand (when the university is in session) — is rather "forested" in its essential makeup. Interspersed among alleyways, empty lots, old homes, and a university are thousands of majestic, mature woody species that tower over streets, parks, river- and streambanks. What we have here, inherently, is an established overstory, not quite closed, with much opportunity to plug in a high-yielding and diverse understory. It is as though we are working with a backward succession from canopy to taproot.

In Wayne's yard there already exists an enormous female ginkgo, a towering white pine, a sweetgum, a cypress, and an Ohio buckeye; the neighbors' mature woody species include red oak, pin oak, dogwood, elm, maple, walnut, mulberry, stands of mature bamboo, white basswood, and many more. The surrounding blocks are lush with photosynthetically active tree crowns in abundance, with homes interspersed, and very few fences separating one yard from the next. There is something miraculous in all this: In spite of ourselves we have created a rather mature forest in and around us. In essence, all it will take is to supply plants for the many-layered niches below this ubiquitous umbrella of species.

The perimeter of a property is no limit. You can get to know every mulberry tree along the byways, and when they come ripe, they come ripe for a month and a half of continual harvest. You can literally farm the neighborhood and make use of all the functions of all the plants growing everywhere. The neighbors' yards surrounding this plot are teeming with nuts, fruits, and berries, edible perennials of myriad kinds, food, medicine, utility, for the taking. These are wild and semi-wild centers of the suburban wilderness. The possible yields that come from study and harvest in our suburbs, cities, and rural areas are unlimited if we take the time to seek them out. Remember that the pansy flowers bought at the local discount store and planted in your flower boxes make a colorful and tasty addition to any salad.

Water collection systems, chickens, stockpiling of organic materials, composting, sheet mulching — it goes on and on. No lawns, though! Yes, you may plant annuals in the backyard, intermixed with a

CASE STUDIES: FIFTEEN PLANT GUILDS

thriving perennial culture. You may simply broadcast annual crop seeds on the ground and see what arises, what thrives and what does not, and through the years select for the hardy crops. You're now branching out into the neighborhood, planting trees, vegetables, herbs. Will the block and beyond become the farm?

The plant guilds described here are applicable to a broad range of climatic conditions. From Planting Zones 4 through 8, with slight variations, all of these guilds are replicable in rural, suburban, and urban locations. It is important to remember that there is a structure and a process for designing and developing plant guilds that can be applied in any climate, any biome, any region, predicated on what will grow there.

Say you own one-sixth of an acre in town. The footprint of the house is seventeen hundred square feet. If you subtract this from the overall breadth of the land, you are left with small areas for plants. You decide to plant a dense matrix of perennials in the front yard and a mix of perennials and annuals in the side and backyards, stacking as many plant functions as possible, utilizing many vertical as well as horizontal niches. Because of the small footprint what you currently have are basically three diverse plant guilds, front, side, and back. These polycultures contain a variety of edible, medicinal, and utility species that supply the family with a large percentage of basic needs.

In order to make these gardens happen you may have to take down trees to admit sunlight, then chip them and retain the feast of organic matter for the gardens. You dig swales in the front yard to slow down and collect the water so that the plants might drink. You comb the neighborhood for leaves, harvest cardboard from the local dairy bottling plant, compost kitchen scraps, stockpile horse manure from the stables nearby. Wherever organic materials are found you bring them home and set up a sheet mulch system throughout the yard. You install a water collection system and relish the fact that you can use the water from the top of the ridge of your property (the roof) for irrigation. You go to the city council with a report about urban chickens and farming. After two years of watching the report bandied about by infinite committees you now have laying hens: eggs, manure, enjoyment.

All in all, you plant 250 species of plants on this postage stamp of a property, and there is room for 500 more if need be. You forage throughout the neighborhood (we are not delimited by property lines).

The neighbors have taken note of what is happening. Some are simply curious. Others have begun planting themselves. They have become the beneficiaries of the harvest, sharing across fences and alleyways, striking up conversations that usually begin with, "So what is it you're up to now?"

Plant guilds are not limited to a few simple functions. You have ample opportunity to design and develop diverse guilds that focus on specific modalities: animal foods, oils, fibers, medicines, spices, endlessly. You might embed yield functions in a broad services guild, or you can design based on a particular theme that meets basic needs. Experiment, explore, ask yourself, *What do I need for my family's sustenance?* Proceed from here.

You continue every year to insert more species into niches in the understory—the guilds are always in process. You may begin by planting the larger woody species and work under and around them. As these mature you are getting a closer look at their spread and needs. Because you have created a forest environment on a large portion of the property, care is minimal, organic matter is layered in by you and autumn leaf fall, and there is no "weeding," so to speak—simply chop-and-drop.

There are infinite functions and uses for the plants listed in the fifteen plant guilds here. It would be near impossible to include all of them. Instead we've selected prominent features of each to delineate the beginnings of a descriptive language for research and interest. Our research is never-ending. Every plant could potentially meet every animal and human need depending on cultivation practices and preparation after harvest. Go for it!

CASE STUDIES: FIFTEEN PLANT GUILDS

Fruit and Nut Guild

> **Region:** Lower Midwest
>
> **USDA Planting Zone:** 7a
>
> **Biome:** Eastern woodlands
>
> **Scale and setting:** Suburban lot
>
> **Rainfall average:** 50 inches per year
>
> **Time to maturity (succession):** Overstory at 10 years
>
> **Beneficial outcomes of the guild:** Food, medicine, and utility for inhabitants; sharing of nutrients among plants
>
> **Animal relationships:** Flowers blooming throughout the year attract beneficial insects and pollinators; small mammals eat a variety of plant parts

One of our authors implements this polyculture on a suburban property in Southern Illinois, a meeting ground of several ecosystems, including eastern woodlands, the northernmost point of the Gulf Coast, and the Ozark Plateau, within a few miles of the confluence of the Mississippi and Ohio Rivers, which together drain two-thirds of all the surface water in the continental United States. Fifty inches of rain per year fall on this site. A deep sheet mulch, a series of swales, removal of the concrete sidewalk, and a dense understory planting have all helped buffer and make use of this large volume of rainwater. Zone 7a stretches like a band across the continental US, to either side of the Mason-Dixon line. Many of the plants depicted here are applicable in diverse locations, including in planting zones to either side of Zone 7a. It is important that we push the climatic edge, so to speak, and experiment with these polycultures in differing climate regimes.

The mature female ginkgo tree in this guild drops a bumper crop of nuts every year and creates a towering canopy above plants that have been arranged in the understory facing west and receive approximately six to eight hours of sunlight per day. Some small trees, primarily horsechestnut, had to be removed in order to maximize sunlight to the guild. These trees were sent through the chipper and used as woody carbon to help develop this small forest environment. A selection of dwarf Asian pears and hardy almonds serve as the dominant understory trees, with a shrub layer of raspberry and fig; an herbaceous layer of comfrey, fennel, wild ginger, mints, groundnut, licorice, caraway, and sweet violet is interspersed in and around the larger woody species.

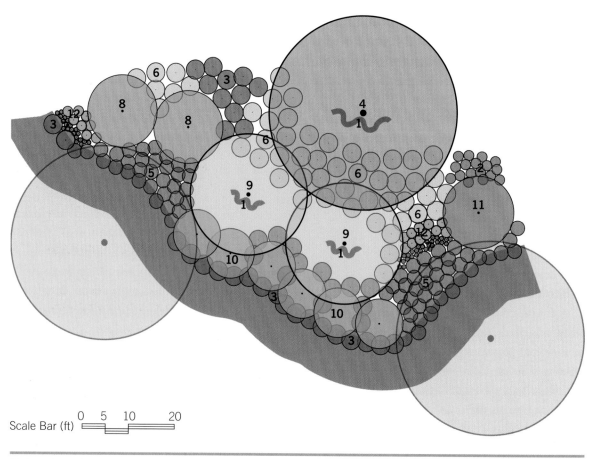

Fruit and Nut Guild.

Several functions (only a small ration) were chosen for this plant guild. Comfrey is the Permaculture plant par excellence, known for its uncanny ability to tap deep into the soil matrix and pull up a wide assortment of minerals (dynamic accumulator). Comfrey is placed in several strategic points in almost any guild. It is used for food; for medicine (it's known as knitbone because it can speed up the healing of sprains and broken bones); to attract beneficial insects; to fix nitrogen and supply it to other plants in the matrix; for chop-and-drop mulch; and so on. How can we enumerate all the functions of this miraculous plant?

An herbaceous layer of mint and several other plants presents a variety of medicinals, edibles, and utility plants for the taking. It would take hours to delineate the many benefits from this dense cauldron of green organisms. These herbaceous species flower at intervals across the growing season and attract pollinators and beneficial insects.

As the fruit and nut trees and shrubs mature, the possibilities for processing into myriad products are endless — not to mention the fact that the family can go out the back and front doors and gorge on it all. Raspberries, figs, Asian pears, almonds offer a stream of fresh food from early spring to late autumn. If we preserve what we harvest for

CASE STUDIES: FIFTEEN PLANT GUILDS

TABLE 7.1 FRUIT AND NUT GUILD PLANT LIST

ID	Common Name	Scientific Name	Plant Type	Height	Spread	Ecological Function	Human Use/Crop
1	Groundnut	*Apios americana*	Vine	5'	3'	Nitrogen Fixer	Container Garden, Food, Medicine
2	Fennel	*Foeniculum vulgare*	Perennial	4'	2'	Insecticide, Insectary, Dynamic Accumulator	Dye, Food, Essential Oil, Insect Repellent, Medicine
3	Spearmint	*Mentha spicata*	Perennial	2'	4'	Aromatic Pest Confuser, Insectary	Container Garden, Food, Essential Oil, Medicine
4	Ginkgo	*Ginkgo biloba*	Deciduous Tree	60'	40'	Soil Builder, Erosion Control	Medicine, Soap
5	Canadian Wild Ginger	*Asarum canadense*	Perennial	1'	3'	Soil Builder, Erosion Control	Food, Medicine
6	Comfrey	*Symphytum officinale*	Perennial	3'	4'	Domestic Animal Forage, Chemical Barrier, Insecticide, Insectary, Mulch Maker, Dynamic Accumulator, Water Purifier	Biomass, Compost, Food, Medicine
7	Violet	*Viola odorata*	Annual	9"	9"	Insectary	Container Garden, Food, Essential Oil, Medicine
8	Hardy Almond	*Prunus dulcis*	Perennial	15'	15'	Mulch Maker, Edge Species, Soil Builder	Nuts, Oil
9	Asian Pear	*Pyrus pyrifolia*	Perennial	25'	25'	Mulch Maker, Insectary	Fruit, Medicine
10	Raspberry	*Rubus* spp.	Perennial	5'	10'	Wildlife Food, Insectary	Fruit, Edible Greens, Medicine, Tea
11	Fig	*Ficus* spp.	Perennial	15'	15'	Insectary, Cools Environment in Hot Places	Fruit, Medicine
12	Caraway	*Carum carvi*	Perennial	2'	2'	Insectary	Essential Oil, Culinary Herb, Medicine, Food
13	Licorice	*Glycyrrhiza glabra*	Perennial	2'	2'	Nitrogen Fixer, Dynamic Accumulator	Medicine, Candy, Flavoring, Food

the winter, we have a 365-day bounty that does not require a trip to any market. We become our own market. And then do not forget to harvest the neighborhood (all those acorns under tire and not in my muffins?).

So why these particular plants in this plant guild? The combination of functions that is placed in relationship here creates a forested environment that requires little care. The livestock (animalcules) in this yard do most of the work. The key is in the building of fertility, and all these plants supply it: They mine the soil for minerals and bring them to the other layers in the soil matrix (comfrey, wild ginger); they supply biomass and organic matter to the soil microbiology at leaf fall in autumn (overstory and understory species, shrubs — all plants for that matter). The flowers that bloom most of the year attract multitudes of beneficial and pollinating insects (all of these guild plants flower), and the seeds they produce feed a diversity of avian life that fertilizes the yard. This plant guild supplies the household with food and medicine, fiber and oil protein, carbohydrates, culinary herbs, and the pure pleasure of hanging out with the plants. This small tidbit of a forest of sustenance is simply a beginning into your own research and experimentation.

Ginkgo, Asian pear, and hardy almond anchor this site and create a dense overstory. The ginkgo, already sixty feet tall, dominates the eastern flank, with almond and Asian pear placed west of the ginkgo on either side so that maximum sunlight may reach them. As the pears and almonds mature, the plants in all of the other vertical layers and in the root zone will create enough biomass and organic matter to feed all members of this polyculture. Partitioning and proper spacing are achieved through conscious placement based on sunlight availability and nutrient need.

Notes on Specific Plants

HARDY ALMOND (*PRUNUS DULCIS*)

Excellent tree for thickets and edges. Flowers bloom in early spring. Pollinated by insects. The tree is self-fertile and does not need a companion. Drops its leaves in the autumn, adding organic matter to the soil matrix. Protein- and oil-rich nut. The oil is used predominantly in massage oil formulas and as a lubricant.

ASIAN PEAR (*PYRUS PYRIFOLIA*)

Fruit contains up to 9 percent sugar. Can be planted in Hardiness Zones 4–9. Grows to about twenty-five feet tall.

COMFREY (*SYMPHYTUM OFFICINALE*)

The dynamic accumulator par excellence! Leaves can be cut three to four times during the growing season and used as nutrient mulch around all plants in a guild (chop-and-drop). The deep taproot opens up clay soil and allows for maximum air and water circulation. As a medicinal, comfrey is used for external and internal treatments. Externally it's known as knitbone and used for broken bones, sprains, bruises, and cuts. It's especially effective internally for pulmonary congestion and dry coughs.

FENNEL (*FOENICULUM VULGARE*)

Nectary for attracting beneficial insects. Tasty, licorice-like tea.

GINKGO (*GINKGO BILOBA*)

Prolific producer of ginkgo "nuts" from the female tree. The slimy and stinky skin of the nut is stripped away, and the nut is roasted for a delectable treat. Chinese women in New York City's Chinatown make a yearly autumn pilgrimage to Central Park to harvest them. The ginkgo leaf is used medicinally to open up blood vessels in the body. Planted in cities because of its resilience, pollution resistance, and ancient heritage as the oldest known deciduous woody species.

RASPBERRY (*RUBUS SPP.*)

Produces not only fruit but also tea from leaves, fibers from stems, and dye from stem and leaves. Acts as a true understory shrub and can spread rapidly.

FIG (*FICUS SPP.*)

Prefers warmer climes, but some varieties will survive temperatures down to 10°F (-12°C) if properly insulated over the cold months. Fruit is well known, eaten fresh or dried. A gentle laxative is created from the fruit syrup.

WILD GINGER (*ASARUM CAUDATUM*)

One of the top ground covers for full-shade situations. The root smells and tastes like culinary ginger and can be used as a substitute. Used in Chinese medicine as a warming agent.

MINTS (*MENTHA SPP.*)

Peppermint is the "hotter" and spearmint the "cooler" mint. Excellent ground covers and herbaceous species. Spread rapidly, but in a diverse polyculture are blocked by other species. Create a water- and soil-holding root mass.

GROUNDNUT (*APIOS AMERICANA*)

Nitrogen fixer. The tuber has the taste of a sweet potato. The tuber will keep in storage through the winter.

LICORICE (*GLYCYRRHIZA GLABRA*)

Nitrogen fixer. Used in herbal formulas to balance all other herbs.

CARAWAY (*CARUM CARVI*)

Nectary for beneficial insects and pollinators. Seeds used in baking. Excellent as tea for warming the body in the cold months, and for removing excessive flatulence.

SWEET VIOLET (*VIOLA ODORATA*)

Candied blossoms are a real treat. A prolific ground cover with shallow roots.

Pawpaw Delight Guild

> **Region:** South of 40° latitude
>
> **USDA Planting Zones:** 7a–9
>
> **Biome:** Eastern woodlands
>
> **Scale and setting:** Suburban, urban, and rural sites
>
> **Rainfall averages:** 30–50 inches per year
>
> **Time to maturity (succession):** 10 years
>
> **Beneficial outcomes of the guild:** Food, medicine, utility; sharing of nutrients among plants
>
> **Animal relationships:** Flowers attract beneficial insects and pollinators; birds eat fruit and drop nutrient-rich manure back into the plant matrix; small mammals eat fruit, buds, bark, and flowers

The pawpaw is endemic to more southerly states, usually below forty degrees latitude, but recently more hardy varieties have come online. An overstory of pawpaw and persimmon, indigenous to North America, offers up fruits not typically consumed by most Americans. The pawpaw spreads rapidly in shade, but needs breaks in the canopy that admit sunlight for maximum fruit ripening. Hot, humid summers are ideal for these two trees. In and around them the vertical layers include a dense shrub layer of bush apricots and cherries, along with elderberry; in hotter climes (typically in Mediterranean and desert biomes) rosemary can grow into a woody shrub. Three plants from the herbaceous layer are spotlighted here: dandelion, red clover, and yucca.

Dandelion

The thesaurus does not list this word as a synonym for *ubiquitous*. Guess Roget did not have to mow his front lawn every week. Or maybe he did, but refused to allow this universal "pest" into his treasure house of words and phrases. Dandelion. Dent-de-lion. Lion's tooth. Bright little suns waking in springtime among the fresh green. A landscaper's nemesis. A homeowner's dilemma. A survivalist's dream. A wildcrafter's boon. Probably the most infamous wildflower in America. But how wild can the dandelion be? It follows human habitation relentlessly, bringing clovers, and sorrels, and plantains (referred to by Native Americans as the "white man's footsteps") along for the ride.

CASE STUDIES: FIFTEEN PLANT GUILDS

The common dandelion (*Taraxacum officinale*) *does* seem to grow everywhere. The familiar rosette of spear-shaped leaves peeps through the dry sands of the desert and pops up at elevations where other plants could hardly breathe. A meadow of solid dandelions is one of nature's most pleasing sights: a saturated sheet of yellow gently rustling in the breeze. And after the delicate white puffball seed heads form, the fields take on an unearthly look!

The yellow flowers of dandelion appear from March through September atop hollow, purplish stems that contain a milky, latex juice. At night and when it rains these flower heads close up; they then reopen at the first signs of clearing and sunlight. The bracts beneath the flower heads are reflexed. The leaves are deeply lobed and jagged, resembling spearheads.

Many flies, bees, and insects come to languish in the blossoms' nectars. It is not unusual, when strolling through a field of dandelions on a warm summer day, to hear a loud drone rising from the thousands of bees swarming over the blossoms. Pick your way carefully to the center of such fields and simply listen and take it all in.

There are many birds, mammals, hoofed browsers, and domesticated farm animals that eat different parts of the dandelion. The grouse, partridge, pheasant, quail, and wild turkey enjoy the seed heads. Songbirds, such as blackbirds, goldfinches (which relish any downy seed!), sparrows, siskins, and towhees eat the seeds. Rabbits, porcupines, ground squirrels, mice, and prairie dogs eat seeds, foliage, and roots. Deer browse on dandelions from spring to fall. Pigs and goats, when left to pasture on the farm, find dandelions highly palatable.

The flower heads of dandelions can be made into fritters. In the autumn, after a massive acorn-gathering expedition, you can grind acorns into flour, add water, coat a few dandelion flowers with this batter, and deep-fry them in vegetable oil.

A delicacy in France, the younger leaves of the common dandelion are an excellent salad ingredient or cooked green. In this country we often avoid incorporating the bitter taste into our meals, but it's an important digestive and liver cleansing flavor. Dandelion greens fill this gap. The flower buds, when steamed or boiled, are especially good. They can also be turned into pickles and fermented with other vegetables. The root is baked until brown, ground, and perked into a coffee-like beverage (add in some roasted chicory root for a richer flavor). Remember that this is not a coffee substitute, but a delicious brew in its own right.

The dandelion is high in protein, calcium, vitamin A, and other important nutrients. The entire plant can be dried, ground into flour, and used as an addition to recipes calling for flour.

Add dandelions to sugar, ginger, orange rind, lemon, and yeast, and you'll produce the famous dandelion wine. This wine makes for a delicious spring tonic. The leaves and roots are a safe diuretic (the leaf being a bit more diuretic). Kidney inflammation and bladder infection respond to the diuretic and cleansing action of dandelion. As a general liver tonic, dandelion root breaks up liver congestion and restores liver function after a bout with hepatitis. It can be drunk freely in decoction to dissolve urinary stones and relieve some forms of constipation. It helps lower blood pressure and has a positive effect on the heart. Dandelion clears obstructions in the gallbladder, pancreas, and spleen. It treats anemia with its high mineral and vitamin content. As a blood purifier it aids in clearing skin diseases, psoriasis, and acne. Dandelion balances blood sugar levels whether you're suffering from hypo- or hyperglycemia. Dandelion is a diamond in nature's medicine chest.

Dandelion is a central plant in biodynamic gardening and farming, and because of its dynamic accumulating abilities it is prized among gardeners

TABLE 7.2 PAWPAW GUILD PLANT LIST

ID	Common Name	Scientific Name	Plant Type	Height	Spread	Ecological Function	Human Use/Crop
1	Bush Cherry	*Prunus* spp.	Perennial Shrub	6'	4'	Wildlife Food, Biomass, Organic Matter	Food, Medicine (Cough Syrup)
2	Bush Apricot	*Prunus* spp.	Perennial Shrub	6'	4'	Wildlife Food	Food, Medicine
3	Elderberry	*Sambucus nigra*	Perennial Shrub	12'	10'	Wildlife Food (Birds), Biomass, Organic Matter	Food (Fruits), Medicine (Flowers and Inner Bark), Tea
4	Persimmon	*Diospyros virginiana*	Perennial Tree	25'	15'	Wildlife Food, Biomass, Organic Matter	Fruit, Medicine, Wood
5	Pawpaw	*Asimina triloba*	Perennial Tree	25'	20'	Edges, Open Woods, Biomass, Organic Matter, Wildlife Food	Fruit, Medicine
6	Yarrow	*Achillea millifolium*	Perennial	2'	5'	Dynamic Accumulator (K, Cu, P), Insectary, Nectary, Ground Cover	Medicine, Biodynamic Compost Activator
7	Comfrey	*Symphytum officinale*	Perennial	4'	5'	Domestic Animal Forage, Chemical Barrier, Insecticide, Insectary, Mulch Maker, Dynamic Accumulator, Water	Biomass, Compost, Food, Medicine
8	Red Clover	*Trifolium* spp.	Perennial	1'	1'	Nitrogen Fixer, Dynamic Accumulator (P), Nectary, Ground Cover, Cover Crop, Insectary (spiders)	Food, Medicine (Alterative), Forage
9	Wild Ginger	*Asarum caudatum*	Perennial	1'	3'	Soil Builder, Erosion Control	Food, Medicine, Culinary Herb
10	Tulip	*Tulipa gesneriana*	Perennial	2'	1'	Early-Spring Insectary	Early-Spring Pollinator and Beneficial Insect Attractor
11	Anise Hyssop	*Agastache foeniculum*	Perennial	3'	1'	Invertebrate Shelter, Insectary	Medicine, Tea, Spice
12	Dandelion	*Taraxacum officinale*	Perennial	1'	1"	Dynamic Accumulator (K, P, Ca, Cu, Fe), Compost Accelerator, Insectary	Food, Medicine, Wine, Latex
13	Feverfew	*Tanacetum parthenium*	Perennial	2'	2'	Dynamic Accumulator, Biomass	Medicine (Migraine Headaches)
14	Oregano	*Origanum* spp.	Perennial	1'	2'	Nectary, Insectary, Ground Cover (Erosion Control)	Tea, Spice, Medicine

CASE STUDIES: FIFTEEN PLANT GUILDS

ID	Common Name	Scientific Name	Plant Type	Height	Spread	Ecological Function	Human Use/Crop
15	Yucca	*Yucca filamentosa*	Perennial	6'	3'	Wildlife Food, Wildlife Shelter	Food, Fruit, Soap, Fiber, Medicine, Basketry
16	Rosemary	*Rosmarinus officinalis*	Perennial	5'	4'	Wildlife Food, Nectary, Insectary	Medicine, Spice, Tea, Incense, Essential Oil
17	Horseradish	*Armoracia rusticana*	Perennial	4'	4'	Biomass, Organic Matter, Dynamic Accumulator	Food, Culinary Herb, Medicine

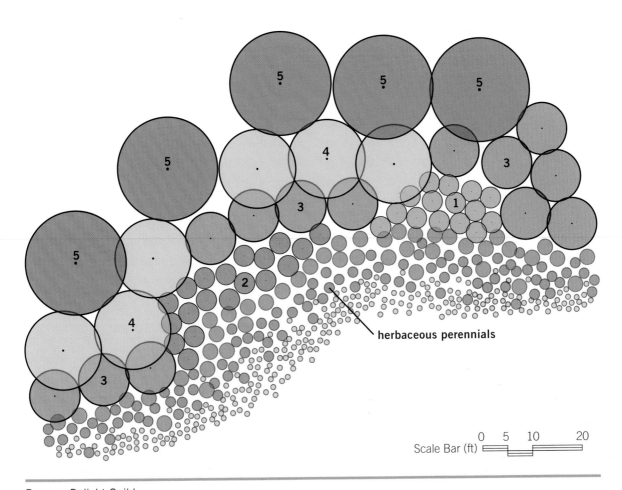

Pawpaw Delight Guild.

and farmers for its ability to mine minerals from deep in the soil profile. It is readily applied to the garden as a mulch or fermented foliar spray.

How much food we have growing right in the front yard (and back)! In recent years the dandelion has been cultivated more frequently for its delectable greens. It can even be found now in many American produce markets. So why does this most wonderful plant still have to absorb the opprobrium of so many lawn doctors and perfectionists? What with the high prices we must pay at the checkout counter for days-old (and weeks-old) produce, it makes good, uncommon sense to harvest a few spring leaves, roots, and flower heads to welcome the abundance of the sun's spring return and celebrate Mother Earth in all her vital immediacy.

Clover

There are seventy-five species of true clover (*Trifolium* spp.) in America. Of the many varieties of clover — including tomcat, clammy, foothill, hop, rabbitfoot, white, and alsike — red clover is most abundantly found in all ecosystems. This durable perennial prefers meadows, grasslands, and lawns, but it can be seen on high mountain plateaus and in the driest deserts. It grows from one to three feet tall with several stems supporting many three-sectioned ovate leaves, each imprinted with a light green V. The flowers range in color from pink to purple. They are round, fragrant, and plentiful.

The red clover is not a favorite of bees, which prefer white clover as the wild source plant for clover honey production. Red clover is a favorite food plant for many other animals, though, game birds and fur-bearing animals being the most frequent visitors. Quail, grouse, wild turkey, partridge, marmot, and woodchuck (especially when alfalfa is in short supply) prefer leaves, seeds, and sometimes the whole plant. Songbirds eat clover seeds. Small animals such as squirrels, mice, and gophers enjoy flower pods and foliage, and deer browse on the entire plant. Clover is also cultivated as forage for range animals and as a nitrogen-fixing cover crop for soil improvement.

The entire red clover plant can be eaten. Because of its abundance it is an excellent wild food source. Use red clover leaves in salads, soups, and stews, or eat them steamed. The raw plant may be a bit hard to digest for some; soaking for a few hours helps. The plant can be dried and ground into nutritious all-purpose flour for cooking and baking. Clover is excellent juiced, contains many vitamins and minerals, and is a good protein source. From city parks to deep wilderness, clover feeds and nourishes many.

Red clover is known medicinally around the world for its quality alterative and blood-cleansing properties. Current cancer research cites its blood-thinning activity, which helps remove accumulated toxins from the entire system. Tonic for convalescents, clover stimulates liver and gallbladder activity, improving appetite and relieving constipation. For fevers and inflammatory conditions, and all debilitating illnesses (such as hepatitis and mononucleosis), clover can be used frequently with no side effects. It is slightly diuretic, and serves as an expectorant for coughs, mucus, and colds. Use it also as a fomentation over rheumatic and arthritic joints, and as a poultice for sores, athlete's foot, rashes, and cancerous growths.

Yucca

Yucca is coming into its own in most American climates. While it's primarily a desert plant, you can now see yucca in yards from south to north. This plant has been and remains primary subsistence for many Native tribes around the country: fibers, food, medicine, soap.

A staple food for several Native tribes, the fruits are fleshy and sweet. The cooked fruit is made into cakes and dried for later use. You can also make a nutritious drink by dissolving the dried fruit in water. Flowers are used in salads. Flowers and flowering stems are excellent in stir-fries. The young stalks taste like asparagus, and young leaves are palatable when steamed.

Medicinally, the pulverized leaves are used as an anti-emetic to prevent vomiting. The sap, or gum, that oozes from the plant acts as a temporary bandage to stop bleeding and keep a wound clean.

Notes on Specific Plants

BUSH CHERRIES (*PRUNUS* SPP.)

A small cherry "tree" topping out at about five feet. Brought from Asia, this is an excellent fruit-bearing shrub for the understory in a polyculture. Produces an abundance of sweet cherries year after year.

BUSH APRICOTS (*PRUNUS* SPP.)

Also from Asia, very similar in its functions to bush cherry, bearing sweet and nourishing apricots.

ELDERBERRY (*SAMBUCUS NIGRA*)

Elderberry is often selected for plant guilds. The birds that drop phosphorous and potassium from their dung on forest gardens love to eat the berries and hang out in the intertwining, tangled branches. One of the most efficacious remedies for colds and chest congestion is created from the elderberry, and the tea infused from the blossoms is delectable. Elderberry is an early successional species that spreads rapidly and produces seemingly infinite berries even in partial to full shade.

PERSIMMON (*DIOSPYROS VIRGINIANA*)

When the persimmons come ripe in the fall (got to get them just right to avoid the pucker), what would anyone rather eat? Not only does the persimmon fruit stand out, but the wood of the tree is hard and durable. Tea from the leaves is high in vitamin C. You can roast the seeds and grind them up for a coffee-like (notice: coffee-like?) drink, and press the seeds for oil (a substance that is in high demand, but rarely produced from farm or garden).

PAWPAW (*ASIMINA TRILOBA*)

This tree is found in open woods and edges. The fruit ripens in autumn and is delectable, somewhat like a cross between banana and papaya. The pawpaw is pollinated by flies, and the flower has a characteristic carcass smell. It grows in the understory, but needs a break in the canopy for full fruit development.

YARROW (*ACHILLEA MILLIFOLIUM*)

Yarrow is hardy in most climates in the continental United States. It can be found in abundance in open fields and meadows. The feathery leaves, when squeezed between the fingers, smell sulfurous. One of the most potent of all plant medicines, it is used for a variety of ailments: colds and flu, digestive issues, external compresses for liver problems. It is also one of the plants used in the biodynamic preps for compost activation.

COMFREY (*SYMPHYTUM OFFICINALE*)

The dynamic accumulator par excellence! Leaves can be cut three to four times during the growing season and used as nutrient mulch around all plants in a guild (chop-and-drop). The deep taproot opens up clay soil and allows for maximum air and water circulation. As a medicinal, comfrey is used for external and internal treatments. Externally it's known as knitbone and used for broken bones, sprains, bruises, and cuts. It's especially effective internally for pulmonary congestion and dry coughs.

RED CLOVER (*TRIFOLIUM* SPP.)

Sometimes referred as a nitrogen scavenger, red clover fixes this most important, very mobile element in nodules on its roots through the mediation of bacteria. Once the roots die, the nitrogen is sloughed off into the soil matrix for immediate use by other plants in a guild. It's used for animal forage, and is an important blood alterative for human beings.

WILD GINGER (*ASARUM CAUDATUM*)

One of the top ground covers for full-shade situations. The root smells and tastes like culinary ginger and can be used as a substitute. Used in Chinese medicine as a warming agent.

TULIP (*TULIPA GESNERIANA*)

A spring bloomer that attracts pollinators and beneficial insects early in the season.

ANISE HYSSOP (*AGASTACHE FOENICULUM*)

Sometimes known as licorice mint, this native of Korea is perennial and offers up some delicious and invigorating tea when brewed. It is medicinal, as most mints are, for stomach and digestive complaints.

DANDELION (*TARAXACUM OFFICINALE*)

Why the dandelion has become the bane of the suburban lawn, Lord knows. For food, medicine, and land repair there are few other plants as nutritious, healing, efficient, prolific, and persistent as this Asian denizen. It has been in the United States for over eight hundred years. Does this make it native?

FEVERFEW (*TANACETUM PARTHENIUM*)

Lots of biomass. One of the best herbal remedies for migraine headache.

OREGANO (*ORIGANUM SPP.*)

Italian food anyone?

YUCCA (*YUCCA FILAMENTOSUM*)

Yucca is utilized for everything from making twine and cordage to weaving baskets, mats, and sandals — not to mention as shampoo, food, medicine, waterproofing (from the gum in the plant), and paintbrushes. Multifarious opportunities in the guise of one plant.

ROSEMARY (*ROSMARINUS OFFICINALIS*)

Used in the bath, the essential oil in this herb stimulates the entire system.

HORSERADISH (*ARMORACIA RUSTICANA*)

The well-known hot and spicy taproot is not simply a condiment, but a powerful sinus congestion reliever.

CASE STUDIES: FIFTEEN PLANT GUILDS

Four Vines Guild

> **Region:** Southern Illinois
>
> **USDA Planting Zone:** 7a
>
> **Biome:** Forest edge, eastern woodlands
>
> **Scale and setting:** Suburban lot, forest edge (ecotone)
>
> **Rainfall average:** 50 inches per year
>
> **Time to maturity (succession):** 5 years
>
> **Beneficial outcomes of the guild:** Food for humans and animals, potent plant medicines, fibers
>
> **Animal relationships:** Insectary, nectary, animal shelter, seclusion

Vines are typically edge species growing in field-to-forest transitional areas (ecotones), climbing shrubs and trees, establishing themselves in vertical niches between herbaceous layers and overstory, and at times climbing above the overstory seeking all-out sunlight. As we structure our plant guilds, we can plant strong scaffold trees (mulberry) to act as trellises or construct trellises, pergolas, or fences that offer stable, vertical frames to support prolific vine growth. Below and in front of the vines, in full sun, a shrub layer of yellow raspberries and roses overshadows herbaceous and ground cover layers that include taprooted species and dynamic accumulators that pull minerals from deep in the soil profile, then drop them in the topsoil, feeding phosphorous, nitrogen, calcium, potassium, and trace minerals to other plants.

This polyculture is an eastern woodlands guild for habitats from Illinois to the Florida panhandle, from the Atlantic Ocean to the Mississippi River, but may also be replicated in many climates and soil types west of the Mississippi. All four of the vines produce abundant, health-giving, luscious fruit at different times in the growing season: schisandra in early summer, grapes in midsummer, akebia in early autumn, passionfruit in late autumn. In essence, along with the mulberry and roses, these four vines act as overstory for numerous herbaceous species.

Notes on Specific Plants

AKEBIA (*AKEBIA QUINATA*)

Akebia produces a sweet and tasty fruit. The vines are exceptional basket-making material, the flowers are made into a tea, and the rinds of the fruit pod are stuffed with meat and rice for consumption.

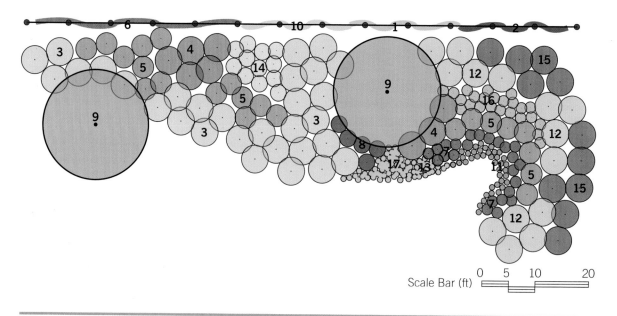

Four Vines Guild.

GRAPES (*VITIS VINIFERA*)

We all know the table grape, of which there are hundreds of varieties, each with a unique taste. Wine gets its flavor from the skin of the grape; it evolves in the fermentation process. The vines are prolific and, when grown against a building, offer a cooling effect during the hot months of summer. When the vines lose their leaves in autumn, this allows sunlight to pass through and warm the wall.

YELLOW RASPBERRIES (*RUBUS* SPP.)

Sweeter than red raspberries! Excellent edge species, and produces extravagantly in full sun.

ROSE (*ROSA* SPP.)

The most praised, poeticized, archetypal flower of them all. Planted simply for beauty, but we cannot forget all the other functions and uses of this majestic flowering plant: food (rose hips), candy (candied rose petals), essential oils.

COMFREY (*SYMPHYTUM OFFICINALE*)

The dynamic accumulator par excellence! Leaves can be cut three to four times during the growing season and used as nutrient mulch around all plants in a guild (chop-and-drop). The deep taproot opens up clay soil and allows for maximum air and water circulation. As a medicinal, comfrey is used for external and internal treatments. Externally it's known as knitbone and used for broken bones, sprains, bruises, and cuts. It's especially effective internally for pulmonary congestion and dry coughs.

PASSIONFLOWER (*PASSIFLORA INCARNATA*)

Tangled vines give birth to one of the most beautiful blooms in the plant kingdom. From the bloom a round, passionate-tasting fruit appears that is worth the effort of simply planting seeds. Prefers a warmer climate. The last perennial plant to germinate in late spring, early summer.

CASE STUDIES: FIFTEEN PLANT GUILDS

TABLE 7.3 FOUR VINES GUILD PLANT LIST

ID	Common Name	Scientific Name	Plant Type	Height	Spread	Ecological Function	Human Use/Crop
1	Akebia	*Akebia quinata*	Vine	5'	20'	Biomass, Organic Matter, Shade	Food, Basket Making
2	Grape	*Vitis vinifera*	Vine	15'	20'	Nectary, Insectary, Wildlife Food, Shade, Cooling Effect	Greens, Fruit, Wine, Medicine
3	Yellow Raspberry	*Rubus* spp.	Perennial Shrub	5'	5'	Wildlife Food	Food, Medicine
4	Rose	*Rosa* spp.	Perennial Shrub	7'	5'	Wildlife Food	Food, Medicine, Candy, Essential Oil
5	Comfrey	*Symphytum officinale*	Perennial	4'	4'	Domestic Animal Forage, Chemical Barrier, Insecticide, Insectary, Mulch Maker, Dynamic Accumulator, Water	Biomass, Compost, Food, Medicine
6	Passionflower	*Passiflora incarnata*	Vine	20'	40'	Biomass, Organic Matter	Food, Medicine
7	St. Johnswort	*Hypericum perforatum*	Perennial	1'	2'	Ground Cover (Erosion), Biomass, Organic Matter	Medicine
8	Hollyhock	*Alcea rosea*	Perennial	4'	3'	Nectary, Insectary	Food, Medicine
9	Mulberry	*Morus nigra*	Tree	30'	20'	Biomass, Wildlife Food	Coppice, Food, Medicine, Bow Wood, Fiber
10	Schisandra	*Schisandra chinensis*	Vine	20'	20'	Biomass, Insectary	Food
11	Bluebead Lily	*Clintonia borealis*	Perennial	2'	1'	Insectary, Nectary	Food
12	Peppermint	*Mentha* spp.	Perennial	2'	5'	Biomass, Soil Stabilization	Food, Medicine, Tea, Essential Oil
13	White Clover	*Trifolium* spp.	Perennial	1'	2'	Nitrogen Fixer, Biomass	Land Repair, Medicine, Food
14	Sage	*Salvia officinalis*	Perennial	3'	3'	Perennial Ground Cover	Tea, Medicine, Spice
15	Spearmint	*Mentha* spp.	Perennial	2'	5'	Root Mass Stabilizes Soil, Dynamic Accumulator	Tea, Medicine
16	Mullein	*Verbascum thapsus*	Biennial	8'	2'	Early Succession Plant for Disturbed Site Repair	Medicine, Oil for Earaches, Leaves for Lung Issues and Toilet Paper
17	Nasturtium	*Tropaeolum majus*	Annual	2'	5'	Dense Root Zone and Growth Good for Small Animal Sheltering	Flowers and Pods for Food

ST. JOHNSWORT (*HYPERICUM PERFORATUM*)

Used in homeopathic medicine for burns and injuries to nerves in finger and toe tips. An oil made from the yellow flowers (turns red) is used for the same purpose.

HOLLYHOCK (*ALCEA ROSEA*)

Similar to marsh mallow in growing habit and human uses. It is demulcent, which helps to loosen and soothe dry coughs, intestinal cramps, and constipation.

MULBERRY (*MORUS NIGRA*)

Mulberries are harvested for as much as a month and a half after they first appear. The tree is sometimes considered a weed; birds ingest the berries endlessly and drop them at various locations, where trees seems to pop out of nowhere. Keep one, two, or a few in your planting regime and enjoy some of the best fruits available. Highly perishable, they should be eaten quickly or processed into jams and jellies, or dried almost immediately upon harvest.

SCHISANDRA (*SCHISANDRA CHINENSIS*)

A classic vine used in Chinese medicine. Bears a tasty fruit.

BLUEBEAD LILY (*CLINTONIA BOREALIS*)

The young leaves of this plant are edible while still only a few inches tall.

PEPPERMINT (*MENTHA SPP.*)

Peppermint is a "hotter" mint than spearmint. Used in teas and as a digestive aid, it's a prolific grower and excellent ground cover.

WHITE CLOVER (*TRIFOLIUM SPP.*)

The nitrogen fixer par excellence, it regenerates and reinvigorates depleted soils faster than any other plant in the kingdom. Plant directly into white clover for maximum nitrogen absorption.

SAGE (*SALVIA OFFICINALIS*)

Used for culinary flavoring for thousands of years; also an important medicinal for sore throat and excessive sweat.

SPEARMINT (*MENTHA SPP.*)

The "cooler" brother to peppermint. A sun tea made from spearmint during a hot summer cools like nothing else.

MULLEIN (*VERBASCUM THAPSUS*)

Mullein shows up especially on disturbed sites and begins the regeneration process of the area quickly. The stalk can grow up to eight feet tall, at the end of which grow several delicate yellow flowers. Mullein is an early successional biennial.

NASTURTIUM (*TROPAEOLUM MAJUS*)

This annual produces peppery-tasting flowers that are used in salads and as garnishes.

CASE STUDIES: FIFTEEN PLANT GUILDS

Annual–Perennial Guild

> **Region:** All of North America
>
> **USDA Planting Zones:** 4–9
>
> **Biome:** Woodlands, open fields, or meadow
>
> **Scale and setting:** At any scale
>
> **Rainfall averages:** 15–50 inches per year
>
> **Time to maturity (succession):** 1–10 years
>
> **Beneficial outcomes of the guild:** Abundance of annual and perennial crops for food, medicine, and utility
>
> **Animal relationships:** Insectary, nectary, shelter for invertebrates and small mammals

A polyculture or integrated forest garden is not all about perennial species. We, of course, have been raised on diets typically centered on annual vegetables, meats, grains, and beans (and McDonald's). Therefore, it would not do justice to our varied palates to eliminate an annual culture from our gardening activities. Although forest environments rely strongly on the mycelial web belowground, which acts as a kind of communication network that translocates minerals, hormones, vitamins, and the like among woody species, and annual cropping environments are enriched by bacterial action, the two come together in dynamic ways. Each year we can plant annual species in and among perennials and vary annual selection accordingly: for vegetable variety, rotation of crops, and the ecological benefits of different species.

The Annual–Perennial Guild integrates typical annual crops with perennials. The core of this combination of species affords us a vast nutritional and utility palette for just about everything we need for our basic necessities.

Notes on Specific Plants

BLUEBERRY (*VACCINIUM CORYMBOSUM*)

The blueberry fruit is abundant in antioxidants and very healing for the eyesight. A perennial shrub with a long life that stays in production for years.

COMFREY (*SYMPHYTUM OFFICINALE*)

The dynamic accumulator par excellence! Leaves can be cut three to four times during the growing season and used as nutrient mulch around all plants in a guild (chop-and-drop). The deep taproot opens up clay soil and allows for maximum air and water circulation. As a medicinal, comfrey is

TABLE 7.4 ANNUAL–PERENNIAL GUILD PLANT LIST

ID	Common Name	Scientific Name	Plant Type	Height	Spread	Ecological Function	Human Use/Crop
1	Blueberry	*Vaccinium corymbosum*	Perennial Shrub	6–12'	8'	Animal Food, Insectary, Nectary	Food, Medicine
2	Comfrey	*Symphytum officinale*	Perennial	4'	4'	Domestic Animal Forage, Chemical Barrier, Insecticide, Insectary, Mulch Maker, Dynamic Accumulator, Water	Biomass, Compost, Food, Medicine
3	Dogwood	*Cornus* spp.	Tree	30'	30'	Dynamic Accumulator (Ca), Biomass	Medicine, Understory Species—Takes Shade Well
4	Chives	*Allium tuberosum*	Perennial	1'	1'	Soil Erosion	Food, Spice, Medicine
5	Parsley	*Petroselinum crispum*	Annual	1'	2'	Soil Erosion	Food, Medicine, Spice, Insect Repellent
6	Peppermint	*Mentha* spp.	Perennial	2'	5'	Biomass, Soil Stabilization	Food, Medicine, Tea, Essential Oil
7	Oregano	*Origanum* spp.	Perennial	1'	2'	Nectary, Insectary, Ground Cover (Erosion Control)	Tea, Spice, Medicine
8	Thyme	*Thymus vulgaris*	Perennial	1'	5'	Erosion Control, Biomass	Medicine, Spice
9	Bamboo	*Phyllostachys* spp.	Perennial	40'	Indefinite	Erosion Control, Animal Shelter, Shade	Food, Medicine, Fiber, Building Material
10	Feverfew	*Tanacetum parthenium*	Perennial	2'	2'	Dynamic Accumulator, Biomass	Medicine (Migraine Headaches)
11	Lemon Balm	*Melissa officinalis*	Perennial	2'	4'	Prolific Biomass Producer	Medicine, Tea
12	Summer Savory	*Satureja hortensis*	Annual	2'	2'	Organic Matter	Spice, Tea, Medicine
13	Basil	*Ocimum basilicum*	Annual	4'	2'	Biomass	Spice, Tea, Medicine
14	Strawberry	*Fragaria virginiana*	Perennial	1'	3'	Biomass, Organic Matter	Food, Medicine
15	Calendula	*Calendula officinalis*	Annual	2'	2'	Biomass	Medicine
16	Cayenne Pepper	*Capsicum annuum*	Annual	4'	3'	Organic Matter	Food, Medicine
17	Cilantro	*Coriandrum sativum*	Annual	4'	2'	Organic Matter	Spice, Medicine
18	Columbine	*Aquilegia* spp.	Perennial	4'	3'	Organic Matter	Food, Flower
19	Sedum	*Sedum* spp.	Perennial	1'	3'	Organic Matter, Prolific Biomass	Cover Crop, Undersown Crop

CASE STUDIES: FIFTEEN PLANT GUILDS

ID	Common Name	Scientific Name	Plant Type	Height	Spread	Ecological Function	Human Use/Crop
20	Dandelion	*Taraxacum officinale*	Perennial	1'	1'	Dynamic Accumulator (K, P, Ca, Cu, Fe), Compost Accelerator, Insectary	Food, Medicine, Wine, Latex
21	White Clover	*Trifolium repens*	Semi-Perennial	1'	3'	Nitrogen Fixer	Cover Crop, Medicine, Food
22	Annual Crops	Various Species	Annual	Varied	Varied	Biomass, Nitrogen Fixers, Etc.	Based on Yearly Choices

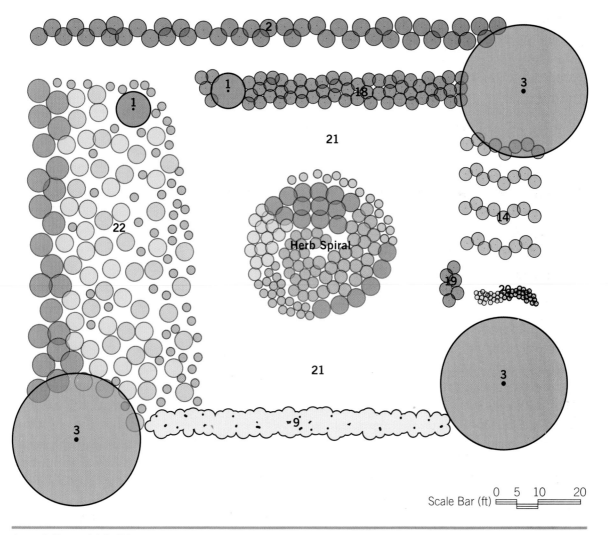

Annual–Perennial Guild.

used for external and internal treatments. Externally it's known as knitbone and used for broken bones, sprains, bruises, and cuts. It's especially effective internally for pulmonary congestion and dry coughs.

DOGWOOD (*CORNUS* SPP.)

The prime dynamic accumulator of calcium. Grow it over a compost bin so that calcium-rich leaves will fall directly on the pile. Dogwood bark has been used in place of quinine for malarial symptoms.

CHIVES (*ALLIUM TUBEROSUM*)

Added to many dishes for a unique garlic-onion taste. If left in the ground, chives are perennial, as are other members of the *Allium* genus.

PARSLEY (*PETROSELINUM CRISPUM*)

Insect repellent if rubbed on uncovered skin and an excellent diuretic for excessive water in the body.

PEPPERMINT (*MENTHA* SPP.)

Peppermint is a "hotter" mint than spearmint. Used in teas and as a digestive aid, it's a prolific grower and excellent ground cover.

OREGANO (*ORIGANUM* SPP.)

Italian food anyone?

THYME (*THYMUS VULGARIS*)

An excellent ground cover, thyme spreads rapidly, and because of its low height surrounds other plants in the guild, thus adding to the biomass in the root cluster and shading the soil from the sun in order to retain maximum water for all plants.

BAMBOO (*PHYLLOSTACHYS* SPP.)

Except in the far north, bamboo is a plant of choice in all polyculture systems. Its soil erosion prevention is well known, especially on steep embankments and slopes. Bamboo has been used as sustenance and utility for many cultures around the world for centuries.

FEVERFEW (*TANACETUM PARTHENIUM*)

Lots of biomass. One of the best herbal remedies for migraine headache.

LEMON BALM (*MELISSA OFFICINALIS*)

Another prolific grower that produces large quantities of biomass for soil and compost enhancement. The prime tea for small children's upset stomachs.

SUMMER SAVORY (*SATUREJA HORTENSIS*)

A culinary spice and beautifully scented plant for potpourris.

BASIL (*OCIMUM BASILICUM*)

Italian food anyone?

STRAWBERRIES (*FRAGARIA VIRGINIANA*)

A perennial ground cover par excellence, yielding copious fruits. The stolons that grow out the sides of the plants are replanted easily.

CALENDULA (*CALENDULA OFFICINALIS*)

The principal herb for topical skin creams and salves.

CAYENNE PEPPER (*CAPSICUM ANNUUM*)

Hot and spicy and a healing herb. Dry peppers and keep them in a cool, dark place for future use in curries and for warming a cold body.

CILANTRO (*CORIANDRUM SATIVUM*)

The leaves and seeds of cilantro are used as culinary herbs.

COLUMBINE (*AQUILEGIA* SPP.)

An unusually shaped flower brightens the garden landscape. The young greens of columbine are edible.

CASE STUDIES: FIFTEEN PLANT GUILDS

SEDUM (*SEDUM* SPP.)

Utilized primarily as a dense ground cover and on living roofs for soil stabilization.

DANDELION (*TARAXACUM OFFICINALE*)

Why the dandelion has become the bane of the suburban lawn, Lord knows. For food, medicine, and land repair there are few other plants as nutritious, healing, efficient, prolific, and persistent as this Asian denizen. It has been in the United States for over eight hundred years. Does this make it native?

WHITE CLOVER (*TRIFOLIUM REPENS*)

The nitrogen fixer par excellence, it regenerates and reinvigorates depleted soils faster than any other plant in the kingdom. Plant directly into white clover for maximum nitrogen absorption.

ANNUAL CROPS SELECTED EVERY YEAR BASED ON NEED AND PREFERENCE

Arugula, assorted Japanese greens, assorted lettuces, sweet potatoes, bell peppers, broccoli, bush beans, cabbage, mesclun, okra, potato, kale, chard, onion, tomato, peas, beans, spinach, beets, zucchini, summer squash, turnips, mustard greens.

Poisonous Plant Guild

> **Region:** Across North America
> **USDA Planting Zones:** 4–9
> **Biome:** Ecotone between field and forest, woodland edges
> **Scale and setting:** From intensive to vast
> **Rainfall averages:** 10–60 inches per year
> **Time to maturity (succession):** 10 years
> **Beneficial outcomes of the guild:** See text
> **Animal relationships:** See text

The focus in the Poisonous Plant Guild is on the human uses of the plants. The relationships set up, in terms of ecological services performed for the grouping, are similar to the function in other plant guilds. The plants of this guild, in combination, support one another in countless ways: sharing nitrogen, hormones, minerals, vitamins, enzymes. There is enough biomass generated in the Poisonous Plant Guild to support an infinite variety of soil creatures that will keep the guild free from stress and healthy.

Though they may appear to be useless in terms of human sustenance, these poisonous (or partially poisonous) plants are used extensively in homeopathic and herbal medicine circles (and for food and animal forage). What put Socrates to sleep (forever) is a potent medicine that has been used as a cure for centuries. It is true that any plant — any substance, for that matter — may be used to heal or injure. We choose the former. Not all of the parts of these plants are toxic to humans, but we need to be sure to "follow the rules" on what we can and cannot ingest to ensure our safety. Either inquire with those who are in the know about these species, or observe astutely and study intensively. (*The authors of this book are not responsible for the use of any of these plants for any purpose.*)

Here is a description of some of the plants from this guild. Because it would be impossible to depict every function of every plant on these pages (the functions are theoretically unlimited), a few of the most salient plants (aren't they all!) are defined for your enjoyment and practical consideration.

Toxicodendron vernix, poison sumac, is a small tree that grows upward of thirty feet tall. In this plant guild poison sumac, along with elderberry (*Sambucus nigra*), towers over the other species. It is more virulent than poison oak and ivy in terms of contact dermatitis and creates swellings, eruptions, and pustules on the skin. What is interesting is that what causes can also cure. Homeopathic toxicodendron relieves skin eruptions, swellings, and pustules — the same symptoms that contact with poison sumac elicits.

What are the benefits of a tree that causes virulent skin disorders? For one, it is a tree and the centerpiece of our plant guild, and it performs ecological functions for the entire guild: It generates biomass through autumn leaf fall, sloughs off organic matter in the root zone, and feeds microorganisms in the soil. It is a node in the mycelial web of the forest. Deer and all classes of livestock browse poison sumac. Birds eat the fruits. It has a high percentage of crude protein along with phosphorus, potassium, and other essential minerals. It is an important tree in the rehabilitation of disturbed lands and is a cog in the wheel of a forest's natural succession.

Some Native tribes used poison sumac to remove warts, to cauterize wounds, and to stop bleeding. There were also times that they would ingest a drink of poison sumac to stop dysentery. (This is not something that we would recommend.)

Poison ivy (*Toxicodendron radicans*) has a similar chemistry to poison sumac. Contact dermatitis sufferers know this plant all too well: The trifoliate, almond-shaped leaves are unmistakable, changing from deep green to bright red in the fall. A "hairy" vine hugs and scales even very large trees, and the oils are still active during the winter months. The white fruits are a favorite winter food of some birds and other animals. Seeds are spread mainly by animals and pass through the digestive tract unchanged.

Poison ivy is typically rampant in transitional ecotones. It is an important vine layer species that dominates sunny edges, transitional areas between forest and field — in ecological parlance, an ecotone. Homeopathic rhus tox is used to stave off the poison ivy infection. Rhus tox is the poison ivy plant substance diluted in a medium several times until the original substance disappears. It leaves an imprint of its actions on the medium, and this is what we ingest. Like cures like.

How often do we get pigeonholed into thinking only one thing about this plant: It causes terrible skin rashes. "I do not want to go near it." "Did you hear about the guy who breathed in the smoke from a poison ivy plant fire?" Although we surely have to pay heed to these warnings, we also need to open a new "leaf" on the topic of all plants, and the possibility that many functions may be very valuable to us.

Poison hemlock (*Conium maculatum*), infamous as the poison that put Socrates into everlasting sleep, has demonstrated effectiveness against asthma, bladder disorders, liver disorders, cancer, and dysfunctions of the pelvic organs when taken homeopathically. Poison hemlock acts as a pioneer species, quickly colonizing disturbed sites and displacing natives during early succession.

Deadly nightshade (*Atropa bella-donna*) is a member of the nightshade (Solanaceae) family, many species of which are used for human consumption (tomato, potato, chile pepper, paprika), for recreational use (tobacco), and as garden ornamentals (petunia). The family is informally known as the nightshade or potato family. Nightshades are known for possessing a diverse range of alkaloids. As far as humans are concerned, these alkaloids can be desirable, toxic, or both, though they presumably evolved because they reduced the tendency of animals to eat the plants. *Atropa bella-donna,* "beautiful lady" in Italian, got this name because Italian women used juice from deadly nightshade berries to dilate their eyes, making them bright and shiny as a beauty aid. Bella-donna is also called devil's berries or death cherries. It is used homeopathically for treatment of fever, common cold, coughs, arthritis, menstrual cramps, headaches, stomach upsets, nerve

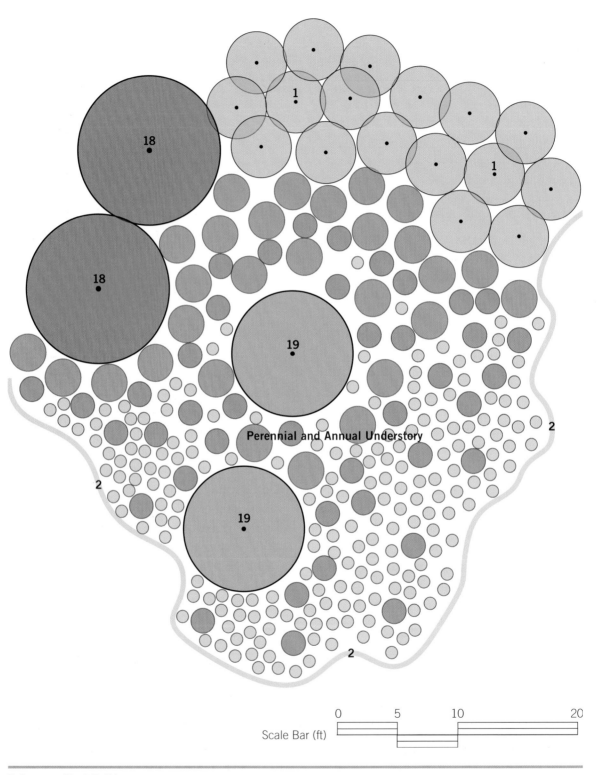

Poisonous Plant Guild.

CASE STUDIES: FIFTEEN PLANT GUILDS

TABLE 7.5 POISONOUS PLANT GUILD PLANT LIST

ID	Common Name	Scientific Name	Plant Type	Height	Spread	Ecological Function	Human Use/Crop
1	Poison Sumac	*Toxicodendron vernix*	Perennial	5'	5'	Early Successional Species, Biomass	Homeopathic Medicine
2	Poison Ivy	*Toxicodendron radicans*	Vine	30'	30'	Edge Species, Protects Forest, Biomass, Bird Food	Homeopathic Medicine
3	Poison Hemlock	*Conium maculatum*	Perennial	6'	2'	Early Successional Species, Biomass	Homeopathic Medicine
4	Deadly Nightshade	*Atropa bella-donna*	Perennial	2'	2'	Shade Species, Biomass	Homeopathic Medicine
5	Pokeberry	*Phytolacca americana*	Perennial	8'	5'	Biomass, Wildlife Food	Food, Dye, Medicine
6	Mayapple	*Podophyllum peltatum*	Perennial	1'	2'	Early-Spring-Flowering Plant, Biomass, Understory Species	Food, Medicine
7	Jimsonweed	*Datura stramonium*	Annual	5'	3'	Biomass	Medicine, Ceremonial Herb
8	Hyacinth	*Hyacinthus orientalis*	Perennial	2'	2'	Biomass, Organic Matter	Medicine, Flower
9	Daffodil	*Narcissus* spp.	Perennial	1'	1'	Early-Spring Attractor for Beneficial Insects and Pollinators	Pollinator for Fruit Trees
10	Castor Bean	*Ricinus communis*	Annual	10'	4'	Prolific Growth and Biomass	Oil, Medicine
11	Larkspur	*Delphinium* spp.	Perennial	3'	3'	Organic Matter	Flower, Medicine
12	Monkshood	*Aconitum napellus*	Perennial	3'	3'	Organic Matter	Homeopathic Medicine
13	Star of Bethlehem	*Ornithogalum umbellatum*	Perennial	6"	6"	Biomass, Ground Cover	Medicine
14	Lily of the Valley	*Convallaria majalis*	Perennial	1'	1'	Biomass, Ground Cover	Medicine
15	Iris	*Iris* spp.	Perennial	4'	2'	Early-Spring Insect Attractor	Medicine
16	Foxglove	*Digitalis* spp.	Annual-Perennial	4'	2'	Biomass	Medicine
17	Rhubarb	*Rheum rhabarbarum*	Perennial	4'	5'	Biomass, Organic Matter, Dynamic Accumulator	Food, Medicine
18	Yew	*Taxus baccata*	Shrub-Tree	20'	15'	Biomass	Bow Wood, Food
19	Elderberry	*Sambucus nigra*	Shrub-Tree	20'	20'	Animal Food, Biomass	Food, Medicine, Utility
20	Jack in the Pulpit	*Arisaema triphyllum*	Perennial	2'	1'	Biomass, Early-Spring Bloomer	Medicine

pain, inflammation of the brain or spinal cord, depression, insomnia, flu, boils, and abscesses.

Pokeberry (*Phytolacca americana*) is a tall, beautiful plant whose purple berries stain the fingers when squished. Although most parts of this plant are highly toxic, its young spring greens have been eaten for centuries. The shoots are boiled, drained, and boiled again to make poke salad, the traditional rural dish of the southern United States. You can even find poke salad in cans in southern markets. This plant contains chemicals called "pokeweed mitogens" that are being studied for use in treatments of autoimmune diseases including AIDS and rheumatoid arthritis. The berries of poke are used for dye production.

Mayapple and Jack in the pulpit are understory (shade-loving) plants that grow in patches and large stands deep in the woods from early spring on. The "apples" are found on plants with two leaves. When they turn a deep yellow, they are ready for eating and can be consumed raw; cooked and made into jelly, jam, or marmalade; used as a pie filling; or dried.

> *Go and catch a falling star,*
> *Get with child a mandrake root,*
> *Tell me where all past years are,*
> *Or who cleft the devil's foot...*

The mystical, magical American mandrake root (another name for the mayapple) is a powerful herbal medicine. We say this with caution because it is a strong liver and intestinal stimulant that only an experienced practitioner should recommend if needed.

Jack in the pulpit is used medicinally as a skin ointment or poultice, applied topically. If you're using the plant internally for any reason, its toxins must be removed through drying, roasting, or leaching. The root, in particular, can be peeled, ground, dried, and roasted to make a bread or cereal that has a chocolate-like flavor. The root is thinly sliced into chips that are then roasted into edible, chocolate-flavored wafers. Again, please be cognizant of the potential toxic effects of this plant. It's a beautifully sculpted member of the botanical realm and deserves recognition simply for that.

Hyacinth, daffodil, and iris bloom early in the spring and attract beneficial insects just as wildflowers bloom on the forest floor after the winter.

Even though we've included suggestions for uses of these poisonous plants, it is very important that anyone wishing to use them work with a qualified practitioner in the medical profession. Our intention in writing specifically about these plants is simply to reveal the many uses of plants not commonly on our radar as useful species.

Notes on Specific Plants

POISON SUMAC (*TOXICODENDRON VERNIX*)

Homeopathic medicine for virulent skin rashes.

POISON IVY (*TOXICODENDRON RADICANS*)

Homeopathic medicine for virulent skin rashes.

POISON HEMLOCK (*CONIUM MACULATUM*)

The plant that infamously poisoned the philosopher Socrates. Used homeopathically.

DEADLY NIGHTSHADE (*ATROPA BELLA-DONNA*)

Used homeopathically for the nervous system, congestion, furious excitement, twitching, convulsions, and pain. It has a marked action on the

CASE STUDIES: FIFTEEN PLANT GUILDS

vascular system, skin, and glands. Bella-donna always is associated with hot, red skin, flushed face, glaring eyes, throbbing carotids, and an excited mental state. In the practice of homeopathy the symptom picture of the patient is matched with the symptoms caused by the medicine to be administered.

POKEBERRY (*PHYTOLACCA AMERICANA*)

Poke salad is a delicacy of early spring. The young shoots and small leaves are boiled in three or more changes of water and eaten like asparagus. Sold in cans in the Deep South.

MAYAPPLE (*PODOPHYLLUM PELTATUM*)

When the apple turns yellow it is ready for eating. You will see the apple in plants with two leaves growing in the crotch on the stems bearing the leaves. All other parts of this plant are poisonous, and the apple must be ripe before you ingest it.

JIMSONWEED (*DATURA STRAMONIUM*)

Associated primarily with ritual uses in Native American tribal ceremonies.

HYACINTH (*HYACINTHUS ORIENTALIS*)

An exquisite flower planted for copious biomass and stunning beauty.

DAFFODIL (*NARCISSUS SPP.*)

An early-spring flowering plant that attracts pollinators and beneficial insects.

CASTOR BEAN (*RICINUS COMMUNIS*)

Used in a compress and placed over the liver and digestive system for relief of upset stomach and liver ailments.

LARKSPUR (*DELPHINIUM SPP.*)

Flower.

MONKSHOOD (*ACONITUM NAPELLUS*)

Used in homeopathic medicine for sudden shock or onset of illness.

STAR OF BETHLEHEM (*ORNITHOGALUM UMBELLATUM*)

Found on lawns all over the United States. Some varieties of this plant are edible and medicinal, and some are poisonous.

LILY OF THE VALLEY (*CONVALLARIA MAJALIS*)

Used in modern medicine as a heart remedy.

IRIS (*IRIS SPP.*)

Another spring ephemeral that draws in pollinators and beneficials.

FOXGLOVE (*DIGITALIS SPP.*)

Another plant used for heart issues by modern medicine, known in medical circles as digitalis.

RHUBARB (*RHEUM RHABARBARUM*)

Strawberry-rhubarb pie anyone? Only the stalks are edible. Beware!

YEW (*TAXUS BACCATA*)

The best wood for making bows in the United States, hands down. Native peoples traded Pacific yew staves for osage orange (*Maclura pomifera*) staves, another superlative bow-making wood, the trees of which are indigenous to Oklahoma and thereabouts.

ELDERBERRY (*SAMBUCUS NIGRA*)

Elderberry is often selected for plant guilds. The birds that drop phosphorus and potassium from their dung on forest gardens love to eat the berries and hang out in the intertwining, tangled branches. One of the most efficacious remedies for colds and chest congestion is created from the elderberry, and the tea infused from the blossoms is delectable. Elderberry is an early successional species that spreads rapidly and produces seemingly infinite berries even in partial to full shade.

JACK IN THE PULPIT (*ARISAEMA TRIPHYLLUM*)

The spathe that hides the flowers and, later on, the bunch of red berries is worth having around for its beauty alone.

CASE STUDIES: FIFTEEN PLANT GUILDS

Asian Pear Polyculture: Dry-Land Design for Acidic Alluvial Soil

> **Region:** Texas, southern states
>
> **USDA Planting Zone:** 8
>
> **Biome:** Gulf Coast prairie and marsh
>
> **Scale and setting:** Rural homestead
>
> **Rainfall average:** 48-plus inches per year
>
> **Time to maturity (succession):** Canopy 10 years, shrubs and perennials 3–5 years
>
> **Beneficial outcomes of the guild:** Food, medicine, and utility for inhabitants
>
> **Animal relationships:** Flowers blooming throughout the year attract beneficial insects and pollinators

The southern central United States has had chronic drought conditions. The soil is dry, and many trees are dying. The East Texas location for this polyculture has calcareous (seafloor) soil with an extremely acid 5.3 pH. It called for drought-hardy plants sited in a thick, cooling group for retaining soil moisture. Implementation needed heavy mulch with micro-drip irrigation on each plant until winter weather and precipitation arrived. You could potentially leave the irrigation in for another year or until the plants are well established in season three.

This is a progressive and dynamic polyculture. The plant niches change over time to accommodate the overstory and buffer extreme weather conditions from hurricane to drought. Thanks to pruning, tree spread is unlikely to achieve full breadth until the other plants are well established. Dense planting is crucial to create enough transition from the dry and hot surroundings to the cooler, moist interior of the polyculture. The seven-layer vertical forest structure applies here, as does a lateral layered effect from outlying grasses to perennial forbs and the interior's small shrubs.

Wherever the site, the structure of the plant resource guild and polyculture mimics a forest's ecosystem. You must follow the structure to build a vertical partitioning of resources and a dynamic cycling of nutrients and materials. Ecological function must be assessed and supplied to each plant. This doesn't require a lot of plants; still, greater diversity of species and plant families will ensure a healthy agro-ecosystem.

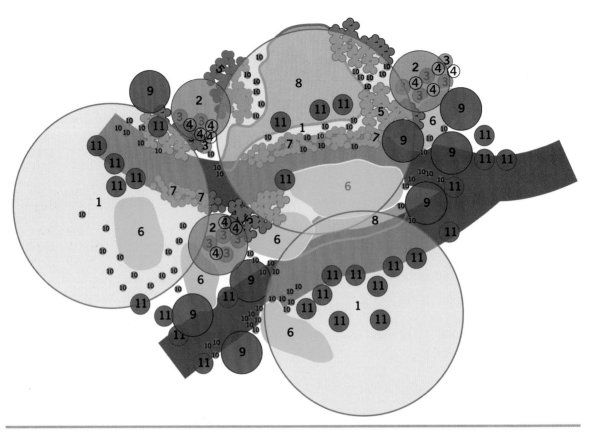

Asian Pear Polyculture.

Notes on Specific Plants

ASIAN PEAR (*PYRUS* SPP.)

Asian pear is the anchor plant for this guild. It is the intended long-term overstory plant with a height and spread of forty feet. It may be small and only a couple of feet wide when it is planted, but over time it will grow large enough to create its own microclimate, buffering the heat and storing moisture on an arid windswept plain.

While the Asian pears grow, the other plants will be creating soil and organic material. They will be expanding and increasing the soil organisms around the trees. With each season the understory plants will spread beneath the pear trees, increasing the available water capacity and cooling the soil. Within a couple of years of installation, elderberry will be fruiting and growing across the landscape, providing a food source for beneficial insects and birds, not to mention habitat.

ELDERBERRY (*SAMBUCUS NIGRA* L. SSP. *CANADENSIS*)

Elderberry is drought-tolerant and makes a good nurse plant for the quince and pear. It flowers in June and July.

COMFREY (*SYMPHYTUM OFFICINALE*)

Comfrey mines the soil for minerals and replenishes the surface with organic material, protecting it from erosion. At the same time it feeds organisms that make nutrients available to trees.

TABLE 7.11 ASIAN PEAR POLYCULTURE GUILD PLANT LIST

ID	Common Name	Scientific Name	Plant Type	Height	Spread	Ecological Function	Human Use/Crop
1	Asian Pear	*Pyrus pyrifolia*	Deciduous Tree	40'	40'	Bird Habitat, Nurse Plant	Food, Organic Materials
2	Elderberry	*Sambucus nigra* L. ssp. *canadensis*	Deciduous Shrub	12'	12'	Wildlife Habitat, Insecticide, Insectary, Mulch Maker, Nurse Plant, Soil Cultivator, Water Purifier, Wildlife Food	Dye, Food, Insect Repellent, Medicine
3	Russian Sage	*Perovskia atriplicifolia*	Perennial	60"	36"	Dynamic Accumulator	Food, Aromatics/Fragrance, Medicine
4	Pale Purple Coneflower	*Echinacea angustifolia*	Perennial	48"	30"	Domestic Animal Forage, Insecticide, Insectary, Dynamic Accumulator, Reclamator, Wildlife Food	Cut Flower, Dried Flower, Medicine
5	White Clover	*Trifolium repens*	Perennial	8"	18"	Domestic Animal Forage, Insecticide, Insectary, Mulch Maker, Nitrogen Fixer, Nurse Plant, Erosion Control, Ground Cover	Food, Medicine
6	Bachelor's Button	*Centaurea montana*	Perennial	18"	12"	Insectary	
7	Strawberry	*Fragaria x ananassa*	Perennial	12"	18"	Dynamic Accumulator	Food
8	Miner's Lettuce	*Claytonia perfoliata*	Annual	12"	12"	Wildlife Food	Food, Medicine
9	Quince	*Cydonia oblonga*	Deciduous Tree	25'	25'	Hedge	Food, Medicine
10	Chicory	*Cichorium intybus*	Perennial	48"	24"	Biomass, Compost	Dye, Food, Medicine
11	Comfrey	*Symphytum officinale*	Perennial	36"	48"	Biomass, Compost, Domestic Animal Forage, Chemical Barrier, Insecticide, Insectary, Mulch Maker, Dynamic Accumulator, Water Purifier	Food, Medicine

CHICORY (*CICHORIUM INTYBUS*)

Chicory, another taprooted plant, is so slender it hardly casts a shadow, but its pencil-thin taproot can go down six feet bringing up phosphorus and minerals to the leaves. Roasted chicory root is used as a coffee substitute and flavoring for chocolate. Birds are attracted to its seeds and bring more nutrients to the polyculture.

RUSSIAN SAGE (*PEROVSKIA ATRIPLICIFOLIA*)

Russian sage is drought-tolerant thanks to its deep-feeding taproot. The flowers are edible and used in salads. On the landscape Russian sage complements purple coneflower.

BACHELOR'S BUTTON (*CENTAUREA MONTANA*)

Bachelor's buttons tolerate drought conditions. As the canopy expands, they accept the partial shade of the growing pear and elderberry while providing a summer source of nectar for hummingbirds and butterflies. They spread via rhizomes.

STRAWBERRY (*FRAGARIA* SPP.)

Strawberries bloom spring through fall and advance across the ground with rooting stolons. Their young leaves are edible and accumulate minerals that break down into the soil. All consumers cherish the fruit. Some strawberries are available that do not spread with stolons and stay in a limited space.

MINER'S LETTUCE (*CLAYTONIA PERFOLIATA*)

Miner's lettuce is a perennial leafy green, high in vitamin C. It likes partial shade, but can thrive in full shade. Young shaded stalks, flowers, and leaves can be eaten as a salad. Leaves turn bitter with age and direct sun. This makes a good ground cover in shaded areas. Reseeding may be necessary over time.

QUINCE (*CYDONIA OBLONGA*)

Quince is a small pear-like fruit. Depending on how you prune the plant, it can be as dense as a hedge. It needs a cold period if it's to flower, and it's not drought-tolerant. This plant may need microdrip irrigation to get established.

WHITE CLOVER (*TRIFOLIUM REPENS*)

White clover is the final understory plant, covering all the vacant soil spaces. It will supply nitrogen and organic material to the soil. During hot periods it will act as an insulation layer keeping the ground cool. It is drought-sensitive, but has a fourteen-inch rooting depth.

CASE STUDIES: FIFTEEN PLANT GUILDS

Ginseng/Sugar Maple Polyculture Guild: Woodland Medicinal Polyculture

Region: Northern United States

USDA Planting Zone: 3

Biome: Deciduous woodland

Scale and setting: Rural homestead

Rainfall average: 26-plus inches per year

Time to maturity (succession): Canopy (existing) 10 years, shrubs and perennials 4–12 years

Beneficial outcomes of the guild: Food, medicine, and utility for inhabitants

Animal relationships: Flowers blooming throughout the year attract beneficial insects and pollinators

The goal of this polyculture is to augment agricultural forest woodland with ecologically functional plants and highly diverse sources of harvestable crops. The harvestable products from this forest garden will include food and medicinal crops as well as increased ecological services and natural capital.[1] Not all the plants are perennials; some may need to be planted annually if they are not self-seeding.

Sources online from Cornell University and Ohio State[2] have discussed the incorporation of ginseng and ginger into existing hardwood forests as a way to expand revenue-generating acreage in farming. The research focuses on developing secondary and low-maintenance crops within an existing ecology requiring little input or maintenance.

The first layer is the tall-tree canopy, the overstory, the highest and tallest sun-loving plants that begin to create shade. The canopy trees are the long-term overstory and for this project have been well established for many years in healthy woodland. Care must be taken not to disturb the ecological services previously existing in the planting area or to cause stress to the trees involved. Long-term overstory trees may consist of nut-producing deciduous trees and/or conifers such as Korean nut pine (*Pinus koraiensis*). In this case the major overstory tree is sugar maple, which provides excellent ecological services for the production of ginseng.[3] Overstory canopy can be as low as five feet or as high as eighty. The positions of the plants in the forest layers are relative

INTEGRATED FOREST GARDENING

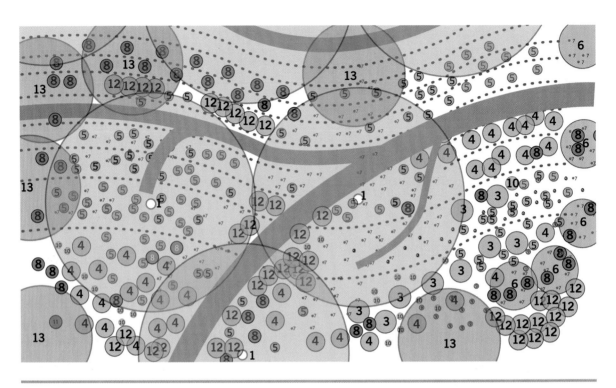

Ginseng/Sugar Maple Polyculture Guild.

1	*Acer saccharum*, Sugar Maple
2	*Artemisia annua*, Sweet Wormwood
3	*Ziziphus jujuba*, Jujube or Red Date
4	*Morus rubra*, (Natural) Red Mulberry
5	*Glycyrrhiza uralensis*, Chinese Licorice (P, NF)
6	*Phellodendron amurense*, Amur Corktree
7	*Coptis trifolia*, Goldthread
8	*Vigna unguiculata*, Cow Pea (N)
9	*Salvia multiorrhiza*, Red Sage
10	*Akebia quinata*, Chocolate Vine

Associated plant symbols.

1	Sugar Maple
2	*Artemisia annua* - L. (A)
3	*Ziziphus jujuba*
4	*Morus rubra*
5	*Glycyrrhiza uralensis* (P, NF)
6	*Phellodendron amurense*
7	*Coptis trifolia*
8	*Vigna unguiculata*
9	*Salvia multiorrhiza*
10	*Akebia quinata*

Using Microsoft Excel, you can diagram the plants in relative scale. Care should be taken when setting up the chart parameters.

CASE STUDIES: FIFTEEN PLANT GUILDS

TABLE 7.12 GINSENG/SUGAR MAPLE POLYCULTURE GUILD PLANT LIST

ID	Common Name	Scientific Name	Plant Type	Height	Spread	Ecological Function	Human Use/Crop
1	Sugar Maple	*Acer saccharum*	Deciduous Tree	75'	50'	Wildlife Habitat, Reclamator, Soil Builder, Windbreak, Erosion Control, Wildlife Food	Food, Wood
2	Sweet Wormwood	*Artemisia annua*	Annual	9'	3'	Moist, Dry	Medicine, Aromatherapy
3	Jujube	*Ziziphus jujuba*	Deciduous Tree	32'	23'		Food
4	Mulberry	*Morus rubra* (Shaped)	Shrub	4'	5'	Wildlife Habitat, Wildlife Food	Medicine
5	Wild Licorice	*Glycyrrhiza lepidota*	Perennial	36"	18"	Nitrogen Fixer	Food, Fiber, Medicine
6	Amur Corktree	*Phellodendron amurense*	Deciduous Tree	45'	30'	Nitrogen Fixer, Dynamic Accumulator, Reclamator, Erosion Control, Wildlife Food	Food, Medicine
7	Goldthread	*Coptis trifolia*	Perennial	12"	6"	Insecticide	Medicine, Wood
8	American Hogpeanut	*Amphicarpaea bracteata*	Vine	36"	36"	Wildlife Food, Nitrogen Fixer	Dye, Food, Medicine
9	Sage	*Salvia officinalis*	Perennial	24"	24"	Aromatic Pest Confuser, Mulch Maker	Compost, Food, Essential Oil, Medicine
10	Akebia	*Akebia quinata*	Vine	30"	10'	Nectary	Food, Aromatic Medicine
11	Black Currant	*Ribes nigrum*	Deciduous Shrub	6'	3'	Container Garden, Dye, Food, Essential Oil	Container Garden, Dye, Food, Essential Oil
12	Chinese Rhubarb	*Rheum officinale*	Shrub	6'	5'	Insecticide, Mulch Maker	Container Garden, Food, Medicine
13	Pagoda Dogwood	*Cornus alternifolia*	Deciduous Tree	25'	30'	Wildlife Food	Biomass, Dried Flower

to height, though ground-level layers have fewer extremes in mature height. Some large shrubs can be pruned to a tree form.

Building a forced polyculture of any kind requires a considerable amount of planning. You'll also need many sources of information as you research your niche and available species. Keeping the final plant list to a minimum for the initial placement of the ecologically functioning species will help you build a functional structure. After the initial nine to twelve species are in place, including the overstory, you can add a number of shrubs, companion perennials, and a ground cover species.

When you first learn about a new plant, you may be so excited to have it in your landscape immediately that you look for a source of seeds and/or transplants. But remember that all plants grow in ecology of companion plants. The niche in which a species grows is filled with other plants that support it and the soil ecology. Taking a plant out of its natural environment and isolating it in pots or planting beds removes it from the supporting mechanisms that facilitate its growth. This is followed by reduced yields, disease, and reduction of the phytonutrients it might supply to its consumer. Instead cultivate these plants within a supporting ecology where all ecological functions are present and filled by plants with those functional roles that may seem to reduce the yield — except that the aggregate yields of all the associated plants far exceed the yield of any single one.

Notes on Specific Plants

SUGAR MAPLE (ACER SACCHARUM)

Sugar maple supplies one of the world's healthiest and most sought-after foods, maple syrup. Boiled sap reduced to syrup has seventeen calories per teaspoon and is rich in vitamin B_2 and manganese. Native Americans reduced the sap to an almost dry sugar for long-term storage and ease of use.

SWEET WORMWOOD (ARTEMISIA ANNUA)

Sweet wormwood is the first Chinese medicinal herb. Its aroma deters insects and is said to alleviate malaria symptoms.

JUJUBE (ZIZIPHUS JUJUBA)

Jujube (the word means "big date") is used in cooking and eaten dried or raw. The dried fruit stores well and is filled with rice in Chinese recipes. Leaves can be eaten, but are considered famine food; dried they are used as a wound dressing. The plant prefers stony or gravelly soil.

MULBERRY (MORUS RUBRA)

Mulberry will offer many harvests. The fruit is sweet, while twigs and barks have medicinal uses. It has a deep taproot for drought tolerance.

WILD LICORICE (GLYCYRRHIZA LEPIDOTA)

This nitrogen fixer grows in moist soils and along roadsides. The root is chewed and used to ease mouth pain, stomach ailments, and coughs. Leaves are used for poultice.

AMUR CORKTREE (PHELLODENDRON AMURENSE)

Amur corktree is one of the top fifty Chinese medicinal plants. Its bark is a strong kidney detoxifier that must be used with professional guidance. The bark can be harvested and dried for storage.

GOLDTHREAD (COPTIS TRIFOLIA)

Goldthread or coptis is a low-growing, shade-loving ground cover. The entire plant is used for teas and a root-beer-like beverage. It is considered mildly toxic and should be used with caution and moderation.

AMERICAN HOGPEANUT (AMPHICARPAEA BRACTEATA)

American hogpeanut is an understory vine, usually planted near trees. Its seeds are a snack some people find similar to peanuts. It's super tolerant of all soils, tolerates a broad pH, serves as a nitrogen fixer, and can become a spreading nuisance if not contained.

SAGE (SALVIA OFFICINALIS)

Sage grows perennially over a wide range of habitats from woodland to prairie. It is a full-sun, drought-tolerant aromatic perennial. Leaves are used in smudge sticks for room purification and as a pest repellent.

CASE STUDIES: FIFTEEN PLANT GUILDS

AKEBIA (*AKEBIA QUINATA*)

Akebia fruit is likened to tapioca, while the leaves are used in tea. Stems are used in Chinese medicinal treatments, but the plant must be managed and can spread quickly, taking over a large area. Harvest it frequently, or have small livestock browse it as forage.

BLACK CURRANT (*RIBES NIGRUM*)

Black currants are high in vitamin C and are used in pies and juices. The plant's seed oil is high in vitamin E and used in skin lotions. The leaves can be used to make a yellow dye; the fruit creates blue and violet dyes.

CHINESE RHUBARB (*RHEUM OFFICINALE*)

Chinese rhubarb makes a dense ground cover that deters rabbits. The stems are eaten raw or cooked into tangy pies and jams. An insecticide from the leaves kills aphids, spider mites, and leaf miners.

PAGODA DOGWOOD (*CORNUS ALTERNIFOLIA*)

Pagoda dogwood or green osier is an important wildlife habitat and food plant. Leaves, stems, and fruit supply nutrition to birds, deer, rabbits, beavers, and bears. The plant attracts birds away from orchard fruits and is accustomed to moist or wet soil.

Boreal Forest Berry Guild

Region: Northern United States and Canada

USDA Planting Zones: 2–3b

Biome: Boreal and coniferous forest

Scale and setting: Urban or rural homestead

Rainfall average: 26-plus inches per year

Time to maturity (succession): Canopy shrubs 3–5 years

Beneficial outcomes of the guild: Food, medicine, and utility for inhabitants; soil health

Animal relationships: Flowers blooming throughout the year attract beneficial insects and pollinators; each plant supplies important forage for birds, deer, and bears

This simple resource guild partitions the soil resource and allows for sun and air movement among the plants. Distance between similar species decreases disease transmission and adds stress to any insect pests. The plants are layered in a three-level structure for sunlight with the blueberry being the canopy, lingonberry the understory, and the bunchberry the ground cover.

Flowering times, fruit bearing, and seasonal effects are all staged over weeks with an aggregate harvest exceeding what a monoculture of plants may have produced. The bunchberry also allows access to the other fruit and can be disturbed occasionally for harvests. All these plants use acidic soils. The grouping also makes a good understory in woodland design.

Using an elevation view can give you a preview of the composition after installation. Drawing the guild on grid paper helps. Just keep in mind the front and back positions of the plants.

Notes on Specific Plants

NORTHBLUE BLUEBERRY (*VACCINIUM ANGUSTIFOLIUM*)

Northblue blueberry is the overstory plant in this polyculture, accepting full sun to partial shade. This cultivar is self-fertile with dark blueberries and intense fall leaf colors.

LINGONBERRY (*VACCINIUM VITIS-IDAEA*)

Lingonberry is the understory plant, growing to sixteen inches in height. It has many common names, one being cowberry. Fruits ripen in late

CASE STUDIES: FIFTEEN PLANT GUILDS

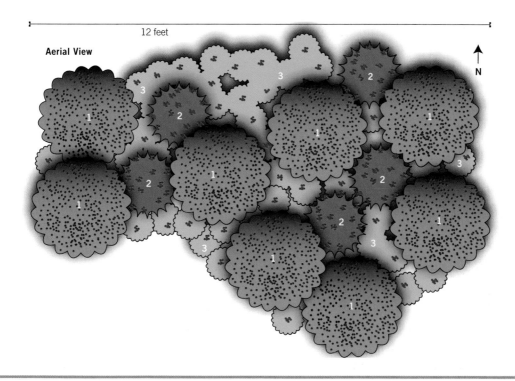

Boreal Forest Berry Guild.

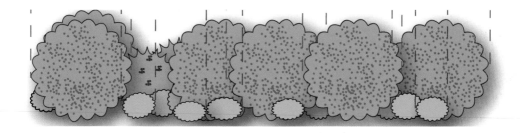

Elevation process view.

TABLE 7.13 SIMPLE BERRY GUILD PLANT LIST

ID	Common Name	Scientific Name	Plant Type	Height	Spread	Ecological Function	Human Use/Crop
1	Northblue Blueberry	*Vaccinium angustifolium*	Deciduous Shrub	3'	3'	Erosion Control, Wildlife Food	Container Garden, Food, Medicine
2	Lingonberry	*Vaccinium vitis-idaea*	Evergreen Shrub	5"	2'	Insectary	Dye, Food, Medicine
3	Bunchberry	*Cornus canadensis*	Deciduous Shrub	5"	2'	Water Purifier, Wildlife Food	Food, Pectin

summer and autumn; they can be dried or eaten fresh, but are most often used in pies or jams and taste best after a frosty night. Although blueberry and bunchberry have a broad soil pH range, lingonberry requires an acidic soil (with a pH near 5) and will withstand cold to -40ºF (-40°C).

BUNCHBERRY (*CORNUS CANADENSIS*)

Bunchberry is a short woody shrub that as a ground cover spreads shallow fibrous roots. It is an important wildlife forage plant for deer and sharp-tailed grouse. It is well adapted to shade as an understory plant. Each berry has two large oval seeds; the pulp tastes like a mild apple. Bunchberry is known for neutralizing acid rain. The leaves are rich in calcium.

CASE STUDIES: FIFTEEN PLANT GUILDS

Salsa Garden Guild

> **Region:** All of North America
>
> **USDA Planting Zones:** 3–9
>
> **Biome:** Multiple biome use
>
> **Scale and setting:** Kitchen garden
>
> **Rainfall averages:** 12–35 inches per year
>
> **Time to maturity (succession):** 120 days average
>
> **Beneficial outcomes of the guild:** Multiple fungicides, insecticide, insect repellent, beneficial insectary, flavor enhancer
>
> **Animal relationships:** Pollinators, bees, butterflies, wasps, beetles

The Salsa Garden Guild is designed as a modified keyhole bed. Plants are installed in close proximity but with diverse positions. Some of the plants are incompatible with others, but many can share their ecological services and functions. Tomato is one of the compatible plants of the Nightshade family. In this case all the plants in the guild benefit from fungicide, aromatic repellents, and pest confusers. Sunlight and nutrients are partitioned, because the plants are spread in a diverse pattern. Resources are partitioned via shallow roots, deep fibrous roots, rhizomes, and taproots. Plants are positioned to take advantage of aromatic characteristics. A tomato hornworm that attacks one tomato plant will have to use extra effort to get to the others, which leaves time for you to notice. Next to the tomato is basil. Basil is said to add flavor to the tomatoes and is an aromatic pest confuser along with onions, nasturtiums, collards, and borage.

Allowing space between the plants brings in more light and creates a microclimate for moisture. Comfrey and asparagus occupy the center island of the garden. This allows for the comfrey to be harvested numerous times during the growing season and used as mulch under the garden plants. Asparagus will grow unhindered by the comfrey and has its own nematocide (for nematodes).

Use quarter-inch grid paper to sketch the polyculture planting pattern or draw sections to visualize architecture and root patterns. There are many ways to position the plants in the garden bed based on the direction of the sun and previous plantings. Since you don't want to repeat any one species in the same position, it's wise to design the planting scheme on a number of different

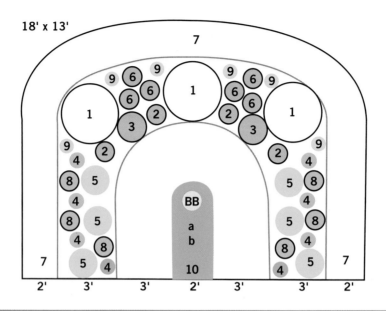

Salsa Garden Guild.

TABLE 7.14 SALSA GARDEN PLANT LIST

ID	Species	Form	Ht x W	Root	Uses & Functions	Spacing	# Plants
1	Tomato (A)	UR	36"x36"	Fibrous Deep	Aromatic Pest Repellent, Fungicide	30"	3
2	Basil	UR	18"x12"	Fibrous Shallow	Flavor Enhancer, Aromatic Pest Confuser	10"	4
3	Borage	Mound	24"x18"	Tap	Aromatic Pest Repellent, Insecticide	16"	2
4	Collard	UR	36"x10"	Tap	Aromatic Pest Repellent	10"	2
5	Sea Kale (A)	UR	30"x18"	Fibrous Deep	Insectary, Mulch Maker, Soil Builder	16"	2
6	Peppers	UR	14"x16"	Fibrous Shallow	Insecticide	12"	2
7	Nasturtium	Mound	14"x12"	Rhizome	Aromatic Pest Repellent	8"	20
8	Onions	UR	24"x9"	Fibrous Shallow	Aromatic Pest Confuser, Fungicide	9"	6
9	Cilantro	UR	24"x18"	Tap	Insectary	24"	8
10a	Comfrey	UR	30"x48"	Tap	Chemical Barrier, Dynamic Accumulator, Mulch	24"/3'	8
10b	Asparagus	UR	54"x24"	Tap 10'	Nematocide	24"	12

Note: Based on a planting guide in *Edible Forest Gardens* by Dave Jacke.

pieces of paper. Some gardeners have a four- to five-year rotation of the plants and have mapped out the crop positions for each year. This garden bed is just one example of designing to increase the diversity of the plants in their positions and reduce pest damage.

CASE STUDIES: FIFTEEN PLANT GUILDS

Dwarf Cherry Tree Polyculture

Region: Northern United States

USDA Planting Zone: 3

Biome: Deciduous woodland

Scale and setting: Rural homestead

Rainfall averages: 26-plus inches per year

Time to maturity (succession): 8 years (pruned); shrubs and perennials 3–12 years

Beneficial outcomes of the guild: Food, medicine, and utility for inhabitants

Animal relationships: Flowers blooming throughout the year; attracts beneficial insects and pollinators; field sanitation with chickens

Fruit trees are popular targets for pests. Their succulent fruit is easily pierced and provides a nutritious environment for insect eggs and larvae. Trees can be susceptible to multiple pests at different times of the season. As an alternative to continual spraying, this guild uses aromatic plants to mask target fruit and confuse pests, which makes locating the fruit more difficult. Additionally, beneficial habitat near the anchor tree hosts pest predators and supplies alternative food sources when pest numbers are low.

In a redundant ecological system, a few select plants provide multiple functions and services. In table 7.15, *Baptisia australis, Allium schoenoprasum,* and *Dalea purpurea* provide insectary services and soil-building functions. Five of the nine plants are a source of human food and medicine. Plant placement is guided by the functions contributed to the guild. Ground covers and mulching plants are dispersed for even distribution of organic material.

Notes on Specific Plants

CHERRY (*PRUNUS* SPP.)

North star cherry, a grafted dwarf tree, is the anchor plant in this polyculture. It has mid- to late-summer fruit and is susceptible to tent worms in northern latitudes, fruit flies in the south. Bird netting protects the harvest, and winter pruning shapes the tree for access and airflow.

WILD BLUE INDIGO (*BAPTISIA AUSTRALIS*)

Wild blue indigo supplies nitrogen to the soil through a mycorrhizal association. Early-summer flowers can be cut for sale or enjoyment. The tiny transplant will flourish into a huge herbaceous perennial display of color. This is a great ground cover and habitat plant, though it has a short bloom time for pollinators. Buy the plants small by the tray and pot for future use.

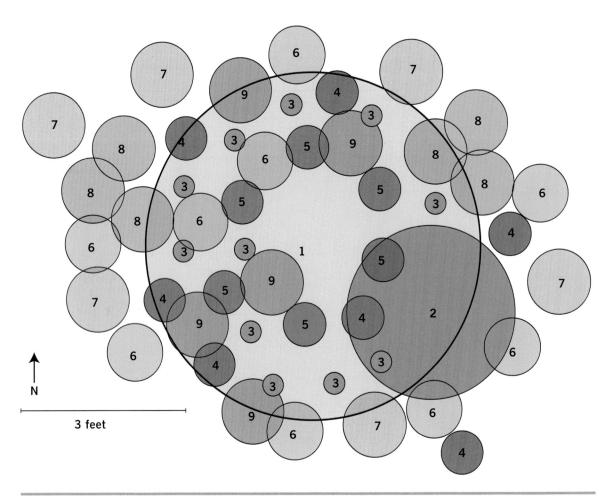

Dwarf Cherry Tree Polyculture Guild.

MEADOW BLAZING STAR (*LIATRIS LIGULISTYLIS*)

Meadow blazing star has late-summer blooms to fill in the pollen gap after spring and summer flowers. It's a tall plant that can be spread throughout the planting area. The dried flowers can be harvested for arrangements; they persist on the stem across seasons.

CHIVES (*ALLIUM* SPP.)

Chives, the system's aromatic pest confusers, belong to the *Allium* (onion) genus and can be harvested throughout the growing season. They have many ecological benefits.

DAFFODIL (*NARCISSUS* SPP.)

Daffodils are placed in a circle around the tree trunk to keep rodents from girdling the tree trunk (chewing off the bark all the way around the tree). They're among the first blooms in spring, with an increasing number each year.

FRENCH SORREL (*RUMEX ACETOSA*)

French sorrel can be harvested with the chives for salads. It is a deep-rooted green perennial with a lemony flavor. Pick only the young leaves; they're used as an accent in foods. Older leaves are not palatable. Left uncut, the plant will set seed for new plantings.

CASE STUDIES: FIFTEEN PLANT GUILDS

TABLE 7.15 DWARF CHERRY TREE POLYCULTURE GUILD PLANT LIST

ID	Common Name	Scientific Name	Type	Height	Spread	Ecological Function	Human Uses
1	North Star Cherry	*Prunus* spp.	Fruit Tree	8–10'	6–8'	Hedge, Insectary, Wildlife Food	Dye, Food, Medicine
2	Wild Blue Indigo	*Baptisia australis*	Deciduous Shrub	4'	3'	Insectary, Mulch Maker, Nitrogen Fixer, Dynamic Accumulator, Soil Builder, Erosion Control	Food, Medicine
3	Meadow Blazing Star	*Liatris ligulistylis*	Hardy Perennial	36"	6"	Insectary	Dried Flowers
4	Chives	*Allium schoenoprasum*	Hardy Perennial	18"	15"	Aromatic Pest Confuser, Insectary, Dynamic Accumulator, Soil Builder, Erosion Control	Food
5	Daffodils	*Narcissus* spp.	Bulb	18"	12"	Insecticide, Rodent Deterrent	Cut Flowers, Dye
6	French Sorrel	*Rumex acetosa*	Perennial	20–24"	10–12"	Insectary	Food, Medicine, Oil, Dye
7	Veronica	*Veronica spicata*	Perennial	24"	8–18"	Hummingbird and Butterfly Attractor, Rabbit-Resistant	Food, Cut Flowers, Medicine
8	Purple Prairie Clover	*Dalea purpurea*	Perennial	1–2.5'	4–18"	Domestic Animal Forage, Insectary, Nitrogen Fixer, Dynamic Accumulator, Reclamator, Erosion Control, Wildlife Food	Container Garden, Food, Insect Repellent, Medicine
9	Creeping Thyme	*Thymus serpyllum* 'Coccineus'	Perennial	4"	18"	Ground Cover	

VERONICA (*VERONICA SPICATA*)

Veronica has shallow roots and needs loamy moist soil, but is sun- and partial-shade-tolerant, making a good ground cover This rabbit-resistant barrier plant grows in a dense pattern, blooming four weeks or more.

PURPLE PRAIRIE CLOVER (*DALEA PURPUREA*)

Purple prairie clover is a spreading ground cover and nitrogen fixer. The leaves are edible.

CREEPING THYME (*THYMUS* SPP.)

Creeping thyme advances across the soil in a carpet fashion. Drought-tolerant, it can look pretty sad when it's dry, but it will bounce back.

RUDDOCK GUILDS

South Milwaukee, Wisconsin, is a small suburban city along the western shores of Lake Michigan. The climate is moderated by proximity to the lake, keeping summer and winter temperatures less extreme than in the counties to the west, allowing an extended frost-free growing season of several more weeks. It is in the USDA Hardiness Zone 5 at about forty-three degrees latitude and receives an average precipitation of 34.8 inches.

Early land survey records from the 1830s record woody vegetation then as being primarily oak, ash, basswood, willow, cottonwood, and maples, with an understory of hazels. The eastern woodland northern deciduous forest biome is dominant but there are incursions of black walnut, hickory, and beech along the lake bluffs and creek, and white cedar in the shaded ravines near the lake.

The Ruddock homesite was first occupied in the 1870s and was part of a farm orchard. In 1984, when they purchased the property, trees on the site were a large silver maple, an older plum, and a young green ash. A few pasture roses and a local garlic variety called settler's were all that remained of the original garden. Bryce and Debby Ruddock have named the site the Spirit Tree Urban Food Forest.

Since they came to the partially empty sixth-of-an-acre slate of grass and a few trees, they have altered the landscape, removing lawns and planting additional trees and perennials, as well as a vegetable garden. Five of their guilds will be portrayed in the following section, but they are not the entire picture, just snapshots of the larger landscape. All five fade into the others at the boundaries. These transitions are soft and not clearly defined, enabling much biological sharing across the edges.

Many species of animals move energy in the forms of seeds and fertility from one guild to another. Rabbits feed and share across the site while ground-foraging bird species such as sparrows, juncos, and thrushes mix the layers of soil detritus as they search for food. Squirrels, chipmunks, and mice serve as gardeners when they bury seeds and nuts, constantly digging some up and moving them elsewhere. Along with grub-digging searches by skunks and raccoons, this opens up bare soil patches for seeds to germinate.

Mallard ducks eat wild rice seedlings in the pond so that any excess has been removed, enabling the remaining plants to grow with less competition for resources. The ducks' wastes help nourish the rice. Every year the ducks return in spring and remain until the wild rice reaches a foot in height before leaving to raise their young.

In the woodlands, trees and shrubs are places of refuge for bird species where they can hide from the Cooper' hawk that stops by every few days looking for an easy meal. The bird wastes in turn feed the woodland plants. The site has been used for Permaculture course trainings, especially in regard to rainwater collection. The educational possibilities that the integrated forest and pond plantings provide are almost endless. Neighbors from around the city stop by and ask about the plantings; after a few minutes of conversation, ideas, like windblown seeds, are sent forth, hopefully to take root in another location and build new ecological relationships at that site.

The Ruddocks have learned so much from the simple observation of nature, and their children and grandchildren have grown up seeing and being active partners in the guilds' designs, learning nature's processes on the way. When all is said and done, can anyone ever ask for more?

CASE STUDIES: FIFTEEN PLANT GUILDS

Cider Guild

> **Region:** Southeastern Wisconsin near Lake Michigan
> **USDA Planting Zone:** 5a
> **Biome:** Eastern woodlands northern deciduous forest
> **Scale and setting:** Suburban lot
> **Rainfall average:** 34.8 inches
> **Time to maturity (succession):** Overstory at 5–20 years depending on fruit tree rootstock selected
> **Beneficial outcomes of the guild:** Shared resources for plants and food; medicines for humans
> **Animal relationships:** Pollen and nectar for insects and birds; habitat for predator species; decoy food for rabbits

Planting an apple or pear tree to be used for juice/cider production maximizes yields from the guild. A standard-sized tree, although taking up more space than a smaller one, has the advantage of casting greater shade beneath its canopy. The cooling effect is increased. It also may be pruned to a semi-dwarf size without adversely affecting its vigorous and hardy rootstock.

Notes on Specific Plants

APPLE (*MALUS* SPP.), PEAR (*PYRUS* SPP.)

An apple or pear tree, on standard rootstock, is the anchoring canopy species here with a projected maturity of twenty years until reaching a height of twenty-five feet and heavy fruit production. Gradual closure of the canopy by the spreading tree will require you to replace more sun-tolerant species — such as the herbs and strawberries — with shade-tolerant woodland strawberries and currant species fruits. The design can be implemented on sites of varying sizes and soil types by changing the choice of rootstock for the fruit tree, which means it can be of dwarf or semi-dwarf size; you can also opt for increased disease resistance and ability to anchor against strong winds in sandier or wetter soils. It is important to follow the general outline of functions with a fruit tree polyculture, using the different niche layers to maximize yield while at the same time providing for the resilience needed to maintain it.

Apple tree species that are suitable for sweet cider or the high-tannin apples that are useful for

INTEGRATED FOREST GARDENING

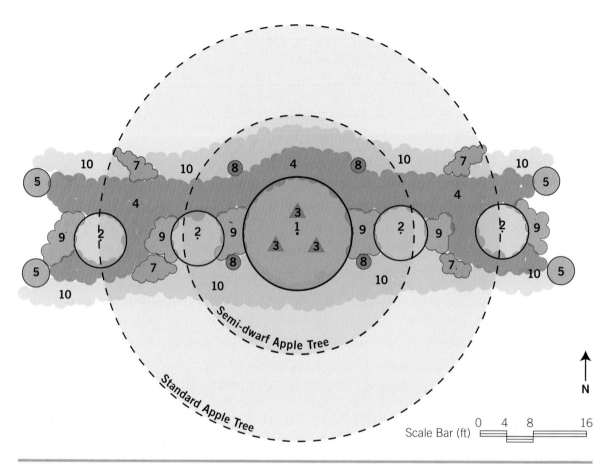

Cider Guild.

hard cider are preferred cultivars. For pear species, use perry pear types, which once again are hard and tannin-rich.

CURRANTS (*RIBES* SPP.)

Black, red, or white currants are all useful beneath the spreading apple or pear tree crowns. Each type offers different culinary and medicinal uses.

STRAWBERRY (*FRAGARIA* SPP.)

Strawberries will grow well until the canopy closes over, at which time you can plant shade-tolerant cultivars of woodland strawberries.

CULINARY HERBS

These herbs can confound the olfactory senses of apple pest species and give food new flavors. The white Dutch clover edge plantings and sweet pea vines both attract insect pollinators and set atmospheric nitrogen.

CASE STUDIES: FIFTEEN PLANT GUILDS

TABLE 7.6 CIDER GUILD PLANT LIST

ID	Common Name	Scientific Name	Plant Type	Height	Spread	Ecological Function	Human Use/ Crop
1	Apple	*Malus* spp.	Deciduous Tree	8–40'	8–40'	Wildlife Food, Insectary, Nectary	Food
2	Red, White, or Black Currants	*Ribes* spp.	Deciduous Woody Shrub	3–6'	3–8'	Early-Season Insect Pollen	Food
3	Pea	*Pisum sativum*	Annual Vine	2–6'	6"	Nectar	Food
4	Musk Strawberry	*Fragaria moschata*	Perennial	12"	8"	Animal Food, Insectary, Ground Cover	Food, Medicine
5	Lemon Balm	*Melissa officinalis*	Clumping Perennial	18"	2'	Soil Stabilizer, Insectary	Medicine, Flavoring
6	Thyme	*Thymus* spp.	Perennial	1'	6"	Ground Cover, Insectary, Pest Repellent and Confuser	Food, Medicine
7	Oregano	*Origanum vulgare*	Perennial	2'	8"	Ground Cover, Insectary, Pest Confuser	Food, Medicine
8	Onions and Garlic	*Allium* spp.	Biennial or Perennial	4'	12"	Pest Repellent	Food, Medicine
9	Baikal Skullcap	*Scutellaria baicalensis*	Ground Cover	12"	12"	Insectary	Traditional Chinese Medicine
10	White Dutch Clover	*Trifolium repens*	Ground Cover	8"	12"	Erosion Control, Nitrogen Fixer, Rabbit Decoy Plant	Edible Flowers, Honey Source

INTEGRATED FOREST GARDENING

Pawpaw Patch Guild: Floodplain and Rich Bottomland Polyculture

Region: Southeast Wisconsin near Lake Michigan

USDA Planting Zone: 5a

Biome: Eastern woodland northern deciduous forest

Scale and setting: Urban and suburban

Rainfall average: 34.8 inches

Time to maturity (succession): 10 years for overstory pawpaw species

Beneficial outcomes of the guild: Carbon sequestration, food and medicinals, temperature modification

Animal relationships: Insectary, nectary, food, shelter, habitat

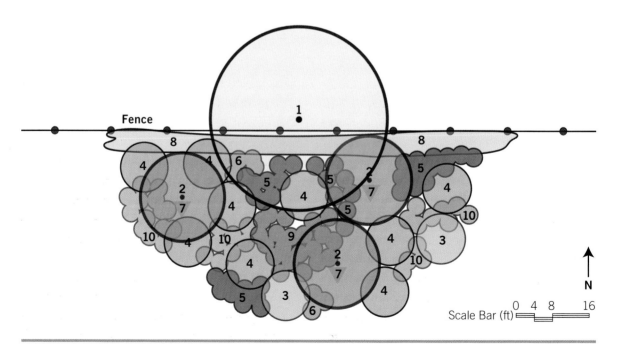

Pawpaw Patch Guild.

CASE STUDIES: FIFTEEN PLANT GUILDS

TABLE 7.9 PAWPAW PATCH GUILD PLANT LIST

ID	Common Name	Scientific Name	Plant Type	Height	Spread	Ecological Function	Human Use/Crop
1	Sugar Maple	Acer saccharum	Deciduous Tree	98'	30'	Insectary, Nectary, Food, Shelter, Habitat	Food, Medicine, Lumber
2	Pawpaw	Asimina triloba	Deciduous Tree	16–30'	15'	Insectary, Food	Food, Insecticide, Dye, Fiber
3	Spicebush	Lindera benzoin	Shrub	10'	8'	Insectary, Nectary, Food	Food
4	Ciwujia	Eleutherococcus senticosus	Shrub	8'	8'	Insectary	Medicine
5	Currants	Ribes spp.	Shrub	5'	3'	Insectary, Nectary, Food	Food, Medicine
6	Wild Senna	Senna hebecarpa	Perennial	5'	3'	Insectary, Nectary, Nitrogen Fixer	Medicine
7	Groundnut	Apios americana	Perennial Vine	50'	2'	Insectary, Nitrogen Fixer	Food
8	Snakeroot	Cimicifuga racemosa	Perennial	3'	1'	Insectary, Nectary	Medicine
9	Solomon's Seal	Polygonatum biflorum	Perennial	3'	3'	Insectary, Nectary, Food	Food, Medicine
10	Wild Ginger	Asarum canadense	Perennial Ground Cover	6"	3'	Insectary, Nectary	Food, Medicine, Fiber

Pawpaws are a clonal understory species with the largest fruits of any native North American fruit tree species and are hardy in USDA Zones 5 to 9. Because they are not self-pollinating, you'll need to plant at least two varieties. Sugar maple is used as a nurse species to provide shade for the young pawpaw seedlings. Spicebush is a shrub species that also needs at least two for pollination — one male, one female. Low currant shrubs give a summer fruit yield while wild senna and vining groundnuts fix nitrogen for the trees. The roots of snakeroot plants yield medicinal estrogens. Wild ginger carpets the soil along with fallen leaves from the previous season. Planting a few Solomon's seal roots can lead to a harvest of edible shoots after a few years — the plants will quickly form a large colony three feet across.

You can substitute oak or hickory as the nurse canopy tree in areas of oak savanna biome. Southward, try pecan as the overstory. Farther north in Zone 4 areas of the Midwest, try planting a butternut (*Juglans cinerea*) overstory with a hazel species shrub layer and currants beneath. Ground species could include spikenard (*Aralia* species) along with wild ginger.

The larger tree acts as a nurse crop for the pawpaws while they are getting established by protecting them from intense sunlight; it also serves as a source of edible nuts and oils, wildlife food, edible leaves in the case of the beech, and wood. It will be at least twenty years before the overstory species can shade the young pawpaws, so it's best to use this polyculture beneath an existing medium to large tree for a quick start. Otherwise you can design for the long term, setting out the larger-growing tree species that you want and waiting for it to achieve a size and canopy spread that will shade a pawpaw grove. The patch of native fruit trees and accompanying partners provides food for humans and wildlife, shelter for wildlife, soil nutrients, temperature mitigation in the shade, enhanced leaf transpiration (which cools temperatures), and wood and medicinal products.

Pawpaws are a relic of days gone by when settlers harvested them from wild trees. They are also tolerant of juglone, the allelopathic chemical found in all walnut family trees. Using them as the understory planting on sites where low juglone-producing walnut family species such as hickory and pecan are growing will give a diverse yield where apples and pears may not do well. There are many improved cultivars with larger and tastier fruits than the wild-sourced seedlings. They are grafted onto seedling rootstock for wide soil adaptability.

Notes on Specific Plants

SPICEBUSH (*LINDERA BENZOIN*)

Spicebush provides more than just berries to season food with. It also has leaves and stems that can be used to make a tasty hot tea.

GROUNDNUT (*APIOS AMERICANA*)

The groundnut species all have beautiful flowers and form edible tubers, with highly nutritive starches.

WILD GINGER (*ASARUM CANADENSE*)

Wild ginger is used similarly to the spice ginger of commerce.

Persimmon Wood Guild

> **Region:** Southeastern Wisconsin near Lake Michigan
>
> **USDA Planting Zone:** 5a
>
> **Biome:** Eastern woodland northern deciduous forest
>
> **Scale and setting:** Urban and suburban
>
> **Rainfall average:** 34.8 inches
>
> **Time to maturity (succession):** 15 years for canopy closure and fruit yields
>
> **Beneficial outcomes of the guild:** Food, carbon sequestration, shared resources for all species
>
> **Animal relationships:** Insectary, nectary, habitat, food

The canopy species, American persimmon, was selected to allow food for northward-migrating insect species of butterfly displaced by warming weather in the southern forests. Native persimmon seedlings are the easiest way in which to begin this guild. Homegrown seedlings are easy to start and have a higher degree of survival than bareroot transplants from a nursery; their taproots adjust faster. They are shade- and drought-tolerant, and are relatively disease-free. The mature trees will be sixteen to sixty-six feet tall with a rounded or conical crown and may be coppiced to allow for a shorter, shrubby growth. Growth is slow, with flowering beginning at about eight years of age and maximum fruiting from twenty-five years old onward.

Notes on Specific Plants

PERSIMMON (*DIOSPYROS VIRGINIANA*)

Persimmon is the canopy species in a shaded situation beneath an existing nurse tree that can be removed later. This nurse tree can be an ash, maple, oak, tulip poplar, or any other. When a persimmon gets older, it is tolerant to full sun. It could be grown beneath a larger nut tree canopy such as hickory in a large guild setting. As the trees get taller, add some nitrogen-fixing wood vetch in the understory. Substitutions for American persimmon can be made with other *Diospyros* species such as Oriental persimmon (*D. kaki*), Texas persimmon (*D. texana*), and date persimmon (*D. lotus*), which are native to

TABLE 7.7 PERSIMMON WOOD GUILD PLANT LIST

ID	Common Name	Scientific Name	Plant Type	Height	Spread	Ecological Function	Human Uses/Crop
1	Green Ash	*Fraxinus pennsylvanica*	Deciduous Tree	66'	30'	Carbon Sequestration, Insectary, Habitat	Lumber, Food, Medicine
2	American Persimmon	*Diospyros virginiana*	Deciduous Tree	16–66'	15–30'	Carbon Sequestration, Wildlife Food	Lumber, Food, Insecticide
3	Raspberry	*Rubus idaeus*	Shrub	5'	1'	Soil Stabilizer, Wildlife Food, Insectary	Food, Medicine
4	Thimbleberry	*Rubus parviflorus*	Shrub	5'	1'	Soil Stabilizer, Wildlife Food, Insectary	Food
5	Groundnut	*Apios americana*	Vine	50'	1'	Nitrogen Fixer, Wildlife Food	Food
6	Wood Vetch	*Vicia caroliniana*	Perennial	3'	1'	Nitrogen Fixer, Wildlife Food	Medicine
7	Tennessee Coneflower	*Echinacea tennesseensis*	Perennial	3'	2'	Wildlife Food	Medicine
8	Tulips	*Tulipa* spp.	Perennial Bulb	2'	1'	Wildlife Food, Nectary	Cut Flowers
9	Daffodils	*Narcissus* spp.	Perennial Bulb	2'	2'	Nectary	Cut Flowers
10	Oregano	*Origanum vulgare*	Perennial Herb	2'	3'	Insectary, Nectary, Soil Stabilizer	Food, Medicine
11	Mints	*Mentha* spp.	Perennial Herb	1'	2'	Insectary, Nectary, Soil Stabilizer	Food, Medicine
12	Woodland Strawberries	*Fragaria vesca*	Perennial Ground Cover	1'	6"	Insectary, Nectary, Food	Food, Medicine

warmer and drier biomes such as the southwestern states. There are no persimmon species reliably hardy in Zone 4 and less. Try substituting an apple or pear hardy to Zone 3 or 4.

RASPBERRY (*RUBUS* SPP.), STRAWBERRY (*FRAGARIA* SPP.)

Beneath the persimmon is dappled light that is perfect for raspberries and woodland strawberries with a spring bulb assortment for earliest bloom. The cane fruits and strawberries will begin to bear in a year or two. An evergreen ground cover of raspberry may be substituted for the strawberries; it will grow only a foot tall (*Rubus hayata-koidzumii*) and is hardy in Zones 6 through 10. For colder regions the arctic raspberry (*R. arcticus*) is a suitable ground cover. It is circumpolar in distribution in northern regions. Raspberries have a shallow and running root system that does not directly compete with the persimmon for nutrients.

CASE STUDIES: FIFTEEN PLANT GUILDS

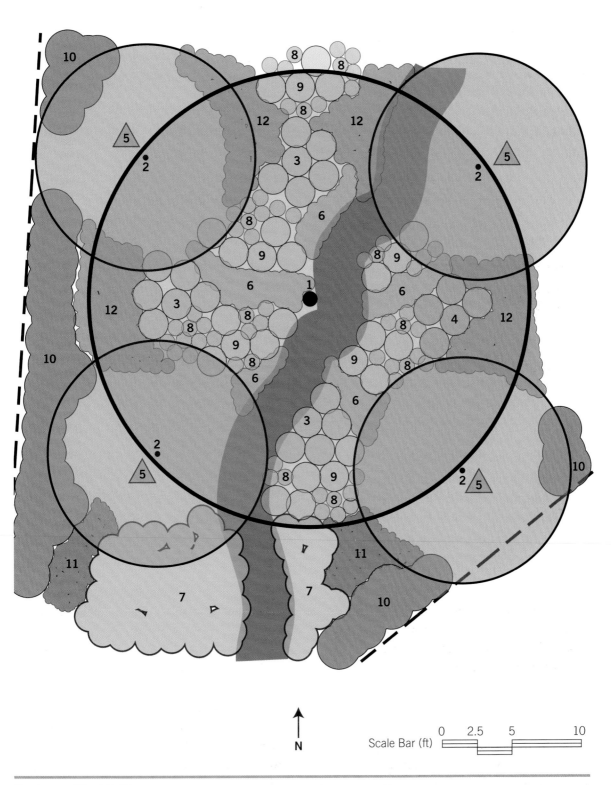

Persimmon Wood Guild.

OREGANO (*ORIGANUM* SPP.), RUE (*RUTA GRAVEOLENS*), MINT (*MENTHA* SPP.)

Along the edges are sunny niches to plant herbs such as oregano and rue. In the shaded areas mint is the ground cover.

CONEFLOWER (*ECHINACEA* SPP.)

A few summer-blooming coneflowers can provide visual interest during the hot and dry months.

SPRING BULBS

Spring bulbs provide cut flowers, serve as early pollen sources for pollinators, and are a storehouse for nutrients in the leaves. When the plants go dormant, phosphorus and other minerals are released for uptake by later-blooming species.

CASE STUDIES: FIFTEEN PLANT GUILDS

Wild Rice Pond Guild

> **Region:** Southeast Wisconsin near Lake Michigan
>
> **USDA Planting Zone:** 5a
>
> **Biome:** Eastern woodland northern deciduous forest
>
> **Scale and setting:** Urban and suburban
>
> **Rainfall Average:** 34.8 inches
>
> **Time to maturity (succession):** First year
>
> **Beneficial outcomes of the guild:** Restoration of endangered ecosystem, education, climate modifier, food, medicine
>
> **Animal relationships:** Insectary, nectary, pollen, habitat, water

Traditionally most of the Great Lakes states were primary habitat for wild rice, but this changed with massive agricultural inputs and the draining and rechanneling of local waterways. This polyculture is an effort to restore some of the lost relationships that were once integral to how the local ecosystems functioned.

Wild rice is a self-sowing annual aquatic grass that requires submergence in water to germinate after several months and a cold period. It is an important food and cultural resource for indigenous peoples of the northern tier of states. Migrating waterfowl feed on it both as a grain and for the young spring shoots. Planting the non-native watercress in this guild helps to balance the pond-sized nutrient system by converting ammonia compounds to soluble nitrates that other plants can use. Fish and bird wastes in the pond provide those nutrients for the watercress. Deeper water areas are places for arrowhead and cattails to grow, while near-shore shallows are excellent spots for the calamus or sweetflag plants.

Areas of rocky outcrops surrounding the pond are niches in which to plant culinary and medicinal herbs needing full sun and reflective heat. These attract pollinators, which in turn draw amphibians and bird species also attracted by available water.

Wild rice is able to grow throughout most of the contiguous United States and much of eastern Canada. A hot-weather-adapted variety (*Zizania texana*) is native to Texas and an endangered species so would be an excellent choice for southern areas. This polyculture is adaptable to sites from twenty-five square feet up to many acres in size so is applicable to both small yard ponds and farms, as

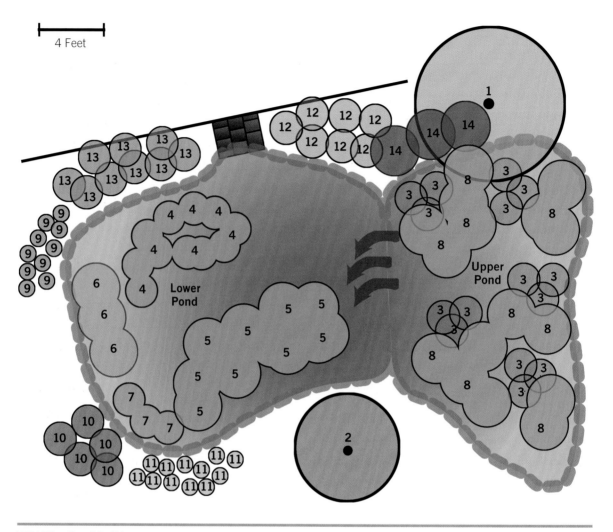

Wild Rice Pond Guild.

well as natural area restoration projects. Additional nutrients are provided by animal species and runoff to the pond, which serves as a nutrient sink.

Wild rice provides food for wildlife and humans. Yields rise depending on planting area, but expect a pound of seed or more from a seventy-five-square-foot planting. Ducks will visit the pond in spring and feed on emerging rice sprouts, thinning the excess plants for you. Arrowhead species yield a starchy tuber — ready in autumn every year — that is also a food source for humans. Sweetflag plants look like smaller cattails but with aromatic edible, medicinal roots.

Cattails provide both visual interest and foodstuffs in the form of edible pollen, seeds, stems, and starchy roots. Small birds are also drawn to the mature seed heads as a food source. The herbs planted along the rocky edges outside the water benefit from the extended season provided by the pond water and stone walls functioning as a heat

CASE STUDIES: FIFTEEN PLANT GUILDS

TABLE 7.8 WILD RICE POND GUILD PLANT LIST

ID	Common Name	Scientific Name	Plant Type	Height	Spread	Ecological Function	Human Use/Crop
1	Juneberry	*Amelanchier canadensis*	Deciduous Shrub	20'	15'	Insectary, Nectary, Wildlife Food	Food, Medicine
2	Shrub Rose	*Rosa rugosa*	Shrub	8'	5'	Insectary, Nectary, Wildlife Food	Food, Medicine, Cosmetic
3	Northern Wild Rice	*Zizania palustris*	Annual Aquatic Grass	5'	2'	Insectary, Wildlife Food, Shelter	Food, Fiber
4	Arrowhead	*Sagittaria latifolia*	Perennial Aquatic	4'	2'	Insectary, Nectary, Wildlife Food	Food, Medicine, Hunting Bait
5	Cattails	*Typha* spp.	Perennial Aquatic	6'	3'	Insectary, Habitat, Wildlife Food	Food, Fiber, Medicine
6	Sweetflag	*Acorus calamus*	Perennial Aquatic	2'	3'	Insectary	Food, Medicine, Dye Mordant
7	Marsh Marigold	*Caltha palustris*	Perennial Aquatic	2'	2'	Insectary, Nectary	Food, Medicine
8	Watercress	*Nasturtium officinale*	Perennial Aquatic	1'	3'	Insectary, Nectary	Food, Medicine
9	Blue Lobelia	*Lobelia siphilitica*	Perennial	2'	1'	Insectary, Nectary	Medicine, Ceremonial
10	Blueflag Iris	*Iris versicolor*	Perennial Rhizome	2'	2'	Insectary and Nectary	Medicine, Fiber, Charm
11	Ramps	*Allium tricoccum*	Perennial Bulb	1'	1'	Wildlife Food	Food, Medicine, Accumulator
12	Oregano	*Origanum vulgare*	Perennial Herbal Ground Cover	2'	2'	Insectary, Nectary	Food, Medicine
13	Lavender	*Lavandula* spp.	Perennial Herbal Ground Cover	3'	2'	Insectary, Nectary	Food, Medicine
14	Dwarf Catnip	*Nepeta mussinii*	Perennial Ground Cover	1'	3'	Insectary, Nectary, Shelter	Medicine

sink. This in turn draws many pollinators and other insects to the water, including dragonflies and beetles. Mosquito control is provided by amphibian species such as frogs and small insect-eating birds, as well as fish in the pond itself.

An additional function of this guild is education: It well illustrates functional relationships among species and restoration techniques for reclamation of damaged wetlands. However you implement the pond, the species composition should be supportive of not just one another but the surrounding ecosystem as well. More species, including mammals large and small, birds, amphibians, and insects, will all enter into the design.

Notes on Specific Plants

WILD RICE (*ZIZANIA PALUSTRIS*)

Wild rice needs slightly flowing water to thrive, so you will need a recirculation system in a small pond. You can provide this with either a pump or an aerator and scale up accordingly if the pond is larger. Natural areas will provide their own flow controls to waterways lower in the watershed.

CATTAILS (*TYPHA* SPP.)

Cattails can become invasive in a small pond after just a few years, so be sure to begin harvesting them the third year on a regular basis.

SWEETFLAG (*ACORUS CALAMUS*)

Sweetflag can also become invasive if allowed to persist unimpeded. Harvest it yearly, replanting a few of the rhizomes.

WATERCRESS (*NASTURTIUM OFFICINALE*)

Watercress will need to be replanted every spring season unless your climate is warmer than Zone 6. Simply purchase the cress at the grocery store and toss individual stems into the pond. It will root at the leaf nodes.

OREGANO (*ORIGANUM* SPP.), LAVENDER (*LAVANDULA* SPP.), CATNIP (*NEPETA* SPP.)

Herbs such as oregano, lavender, and catnip will spread both vegetatively and by seed, so be sure to begin harvest every summer and repeat when needed to control their spread.

CASE STUDIES: FIFTEEN PLANT GUILDS

Hedge Wall Guild

> **Region:** Southeast Wisconsin near Lake Michigan
>
> **USDA Planting Zone:** 5a
>
> **Biome:** Eastern woodland northern deciduous forest
>
> **Scale and setting:** Urban and suburban
>
> **Rainfall average:** 34.8 inches
>
> **Time to maturity (succession):** 3 years
>
> **Beneficial outcomes of the guild:** Carbon sequestration, temperature modification, wood, food, medicine
>
> **Animal relationships:** Insectary, nectary, habitat, food

This location called for drought-hardy and durable species, as well as ones that would form both a visual and a sound barrier from traffic in a side alley. We chose species that could be coppiced and pruned to conform to municipal standards for plants along a right-of-way.

Perennial herbaceous *Silphium* plants serve as nurses the first few years after woody plants are added, giving shade from the hot sun and drawing pollinators. Hemlocks were planted early on, and in their dry and acidic shade we planted lingonberries. Other species include sea berries, pea shrub, willow, arborvitae, and forsythia. Magnolia vines and Dutchman's pipe vines clamber over the fence and into the tall shrub layer. Ground cover plants include mints and daylilies.

In Zone 6 and warmer areas you can substitute hardy citrus for some of the shrub species. In colder zones try haskaps (edible *Lonicera* species) and bush plums. Rose species (especially *Rosa villosa*) make good alternatives, with an upright growth and large fruit hips. Functions of the species in the guild include screening and filtering of both visual and sound annoyances, wildlife food and habitat, food and medicines for humans, water runoff filtration during storm events, and biodiversity conservation.

Notes on Specific Plants

COMPASS PLANT, CUP PLANT (*SILPHIUM* SPP.)

Compass plant, cup plant, and other *Silphium* species grow up to eight feet tall and provide a screen from view, large amounts of pollen- and nectar-rich flowers for native pollinators, and seeds for goldfinches. They provide shade for the newly planted woody shrub species and can be removed at a later time when the other species are established.

INTEGRATED FOREST GARDENING

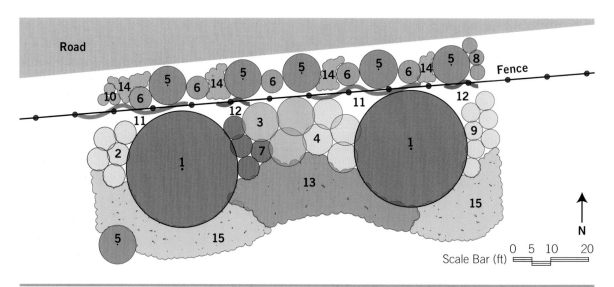

Hedge Wall Guild.

HEMLOCK (*TSUGA CANADENSIS*)

Hemlocks can be hedged and kept to manageable heights of from six to thirty feet by pruning excess growth. They provide a year-round screen and a winter refuge for birds.

SEA BERRY (*HIPPOPHAE RHAMNOIDES*), PEA SHRUB (*CARAGANA ARBORESCENS*)

Fruiting crops of sea berry will need little extra water once established and provide vitamin-rich berries to be juiced, and as does pea shrub. Both set atmospheric nitrogen that can be harvested for other plants in the guild.

ARBORVITAE (*THUJA OCCIDENTALIS* L. VAR *NIGRA*)

The arborvitae, or white cedar, is a medicinal tree that also screens views and dust while providing animal shelter and a frame on which to grow magnolia vines and Dutchman's pipe.

LINGONBERRY (*VACCINIUM VITIS-IDAEA*)

Lingonberries are a dry-soil ground cover related to the cranberry and can grow in acidic soil beneath evergreens.

MINT (*MENTHA* SPP.)

Different mints, especially invasive types such as peppermint and spearmint, work well here: Their expansion to one direction is stopped by an alley, and to the other by the dense shade of the hedge.

DAYLILY (*HEMEROCALLIS* SPP.)

Daylilies are edge plants that rarely get out of hand for the same reasons.

CASE STUDIES: FIFTEEN PLANT GUILDS

TABLE 7.10 HEDGE WALL GUILD PLANT LIST

ID	Common Name	Scientific Name	Plant Type	Height	Spread	Ecological Function	Human Use/Crop
1	Eastern Hemlock	*Tsuga canadensis*	Evergreen Tree	80'	30'	Food, Shelter, Habitat	Food, Medicine Lumber, Tanning Agents
2	Arborvitae	*Thuja occidentalis* L. var *nigra*	Evergreen Tree	35–60'	6'	Habitat, Shelter, Food	Food, Medicine, Lumber
3	Forsythia	*Forsythia* spp.	Shrub	10'	10'	Habitat, Shelter	Medicine, Ornamental, Cut Flowers
4	Basket Willow	*Salix* spp.	Tall Shrub	20'	8'	Insectary, Habitat	Fiber, Medicine, Artist Charcoal
5	Pea Shrub	*Caragana arborescens*	Tall Shrub	20'	10'	Nitrogen Fixer, Insectary	Livestock Feed, Food for Humans
6	Sea Berry	*Hippophae rhamnoides*	Shrub	12'	6'	Insectary, Food, Nitrogen Fixer	Food, Medicine
7	Apple Rose	*Rosa villosa*	Shrub	10'	6'	Insectary, Food	Food, Medicine, Cosmetic
8	Prinsepia	*Prinsepia sinensis*	Shrub	6'	4'	Insectary, Nectary, Food	Food
9	Beach Plum	*Prunus maritima*	Shrub	3–8'	3–8'	Insectary, Nectary, Food	Food
10	Cup Plant	*Silphium perfoliatum*	Perennial	8'	3'	Insectary, Nectary, Food, Habitat, Water	Medicine, Fiber, Visual Barrier
11	Magnolia Vine	*Schisandra chinensis*	Perennial Vine	30'	6"	Insectary, Food	Food, Medicine
12	Dutchman's Pipe	*Aristolochia durior*	Perennial Vine	8'	12	Insectary, Habitat	Medicine, Visual Barrier
13	Daylily	*Hemerocallis* spp.	Perennial	2'	3'	Erosion Control	Food, Medicine, Fiber
14	Mints	*Mentha* spp.	Perennial	1'	2'	Erosion Control, Insectary, Ground Cover	Food, Medicine, Pest Repellent
15	Lingonberries	*Vaccinium vitis-idaea*	Shrub	12"	24"	Insectary, Erosion Control, Food	Food, Medicine

Moving Forward . . .

If this book was a thousand pages, we could go on and on about the potential uses of all these plants. What we've written here instead is but a small serving of encouragement for you, the reader and guild builder, to do your research and begin to uncover veil after veil of the form and function that each plant supplies to you as a human being, and how that plant can serve the biological genius of a particular site and region. What kinds of relationships can we design into our plant matrices that will help create high yields, and provide for all the creatures that help do the work for us? How deeply can we tap into this never-ending resource, the basis for all life, the primary producers of this world?

In a garden the change of seasons penetrates deep into the bones. Those of us who work the land spend most of our days out in the elements. Our bodies are like tuning forks tracking the heat, humidity, rain clouds, and winds; the first frosts of autumn and the winter chill. When the snow quietly blankets the land we know intimately that the tracks imprinted on the pure white landscape will soon melt into spring and hasten the seeds to their fruition.

A year in the garden is a year of constant change. The microcosm of the natural world is unpredictable. But we can always rely on the greater cycles of the seasons. We know that the sun will beat its path across an arc that is predictable—though what the weather will bring, we can only guess. We attempt to read the signs, we lay out our plans, and we proceed with our work, but we must keep all our senses open, our minds clear. We must stay present to the changes in air pressure, the shapes of the clouds, the levels of humidity, the movement of water and wind. To become efficient cultivators of the soil and caretakers of plants requires single-mindedness, focus, and patience. We are part and parcel of the natural ebb and flow. What may appear chaotic in the natural world has an underlying logic all its own.

In the greater context, this year is no more significant than any other year. It is simply that we who work the land become more aware of the intimate metamorphosis through time and the more intimate metamorphosis of the way all life is in constant communication. Through observation we come to see the subtleties of the land and what we need to do in order to raise yields and the overall abundance the land can provide. Abundance is not simply about increasing crop yields. It is about reaping the infinite resources of our hearts, minds, and bodies in sustainable and harmonious ways. It is about enjoying the fruits of our work with the larger community and aligning ourselves with an ethical basis for all we do. The land is a unity, everything working with everything else. There is no waste in the natural order of things. The economy of nature is such that life and death will always continue. Everything is food and sustenance for everything else, and we, as caretakers of the land, must consciously see to it that this ongoing process of death and renewal is not interfered with. We cannot "grow" anything. We can only nurse what is already there by consistently balancing all the elements and providing the platform for the Grace of Life to work its magic. An astute Permaculture practitioner utilizes observation as the essential foundation of gardening practice.

Plants, our main topic throughout this book, are the primary consumers. Without them the secondary and tertiary consumers cannot live and thrive. Plants grow together in striking ways. Rarely, if ever, do we see a lone plant, an isolated being without a community of other beings surrounding it. And so we leave you with our final mantra: *Plant, observe, grow . . .*

ACKNOWLEDGMENTS

WAYNE WEISEMAN

To Bill Mollison, who carried the torch and "got it done." For all of my teachers through the years, too many to name and enumerate. We come into this life to learn how to observe and experience, and I have been given every opportunity to do so. For all of my students, hey, I have learned most everything I know in the realm of Permaculture from you. To my parents, for tolerating what many might conceive of as indiscretion. Most of all, to my girls, wife and daughter, and Mother Earth, all gifts of the spirit.

I would like to thank the editors at Chelsea Green for helping us to see things through to the "tedious" end. And to my partners in writing this tome, Dan and Bryce, thanks for hanging with me through this experience. A life of learning is what it is all about, and you are the consummate life-long learners.

And to the plants, the primary producers on this earth, without whom we would not be writing this book. It is you who give us the sustenance to express our love.

DANIEL HALSEY

It is amazing to think that in 2003 I picked up my first copy of *Permaculture* magazine and in a little over ten years, my life, career, and living system have completely changed; not only into abundance, but into understanding the beauty of nature's self-sufficient and resilient prosperity. This would never have happened except for the guidance of my first Permaculture teachers: Paula Westmoreland, Bruce Blair, and Guy Trombley, and their teachers and those who spread Permaculture from the voice of Bill Mollison and David Holmgren.

A special thank you to my wife Ginny, who has been tirelessly supportive as I continue my horticulture design, study, and application on our landscape.

BRYCE RUDDOCK

Nothing has been possible without the efforts of our ancestors. Whatever I have accomplished is the result of their inspiration and support. Gratitude to my great-grandfather Elmer Ruddock, a Michigan logger who knew the old growth trees; to Harold Ruddock, my grandfather, who worked with wood; to my grandmother Helen Ruddock, who taught us the ways of plants and insects; and to my parents, Bryce and Kathleen Ruddock, who took their children back to the woods every time possible so that that circle of awareness could be complete.

Thank you to Bill Wilson of Midwest Permaculture for seeing a potential in me and encouraging its growth. Thanks to Jesse Tinges for his help with illustrating. An extraordinary gratitude to my animal world teachers, who have taught me so much from simple observation of their interactions between themselves and other species. Most of all a heartfelt thank you to my life's partner, Debby Ruddock, for all the patience and love she has provided over the years.

APPENDIX

AERIAL TO ELEVATION TECHNIQUE

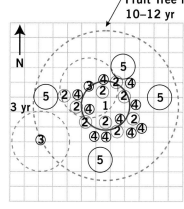
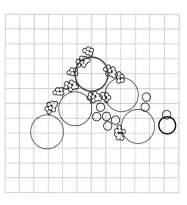
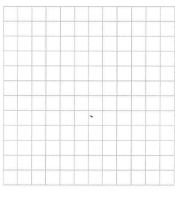
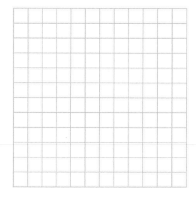
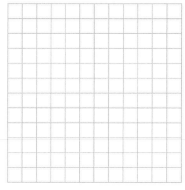
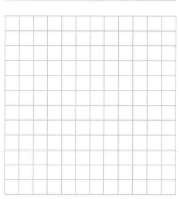

Orchard Polyculture Honey Crisp

1 Honey Crisp   1
2 Alpine Strawberry 10
3 Wild Indigo   2
4 Daffodil   9
5 Optional Comfrey   4

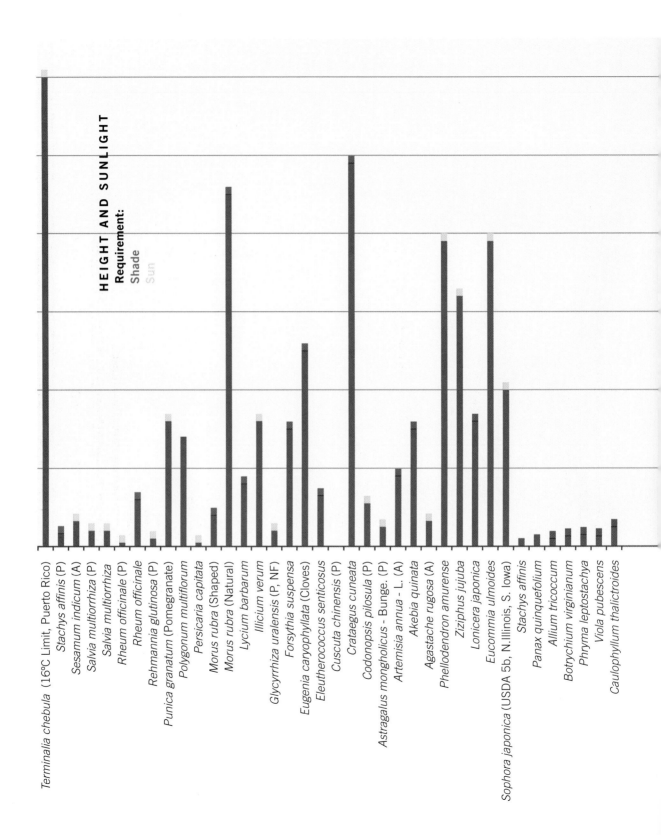

APPENDIX

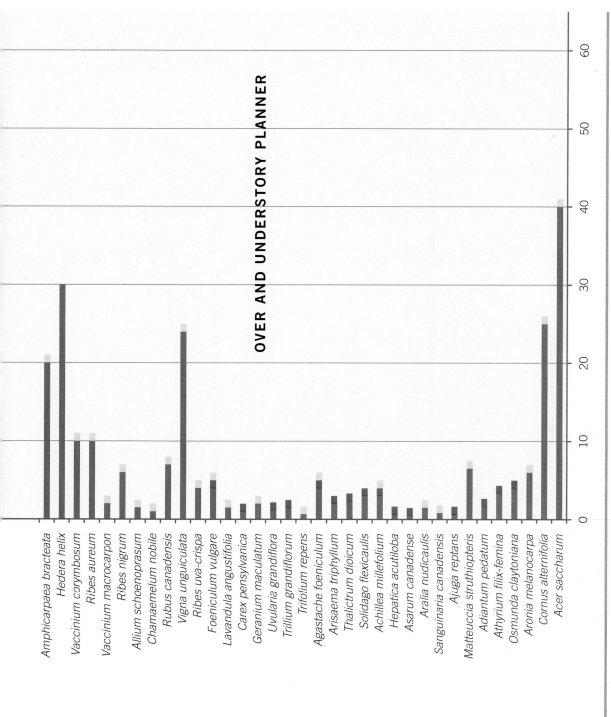

OVER AND UNDERSTORY PLANNER

This chart shows the sunlight needs and heights of plants. Structure and placement start to appear as you come to understand plant characteristics.

INTEGRATED FOREST GARDENING

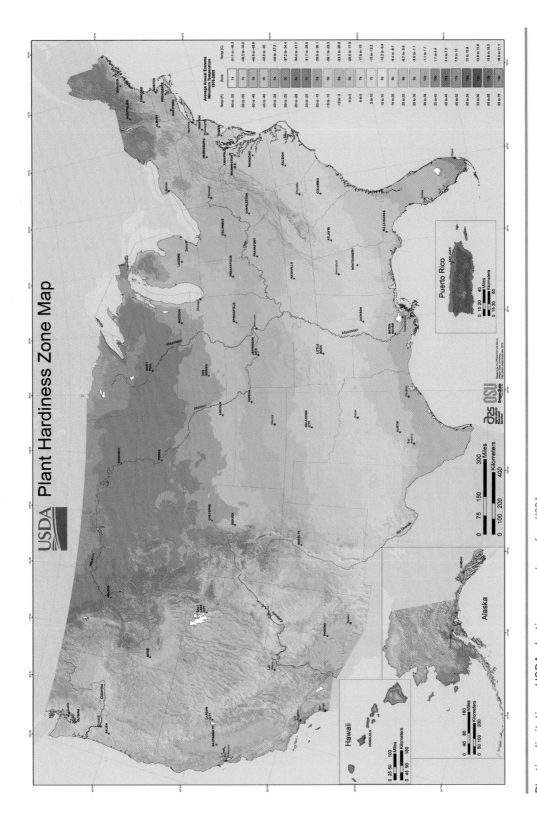

Planting limitations: USDA planting zones. Image from USDA

NOTES

Chapter 1

1. Yeomans, Ken B. and P.A. Water for Every Farm - Yeomans Keyline Plan. 2009. Keyline Designs Publisher.

Chapter 2

1. Cooperative competition occurs when two or more organisms live to the best advantage of each rather than competing to dominate the growing niche. Many organisms need the same resources, but can share the resources and even trade their products for the products of other plants or organisms. Mutualism is a form of cooperative competition where two organisms combine their efforts so that both can thrive.
2. Simard, S. W., M. D. Jones, and M. Durall. 2002. "Carbon and Nutrient Fluxes Within and Between Mycorrhizal Plants." In *Mycorrhizal Ecology*. Ecological Studies 157. Springer-Verlag, Berlin.
3. Hart, Robert. 1991. *Forest Gardening*. Chelsea Green Publishing.
4. Mollison, Bill. 1988. *Permaculture: A Designer's Manual*. Tagari.
5. Mollison, B. C. 1990. *Permaculture: A Practical Guide for a Sustainable Future*. Island Press.

Chapter 3

1. Natural Capital Plant Database (permacultureplantdata.com), a web-based search engine for finding plants for ecological function, tolerance, and characteristics.
2. Nabhan, Gary Paul. 2013. *Growing Food in a Hotter, Drier Land*. Chelsea Green Publishing.
3. Gatti, Roberto Cazolla. June 2011. "The Biodiversity Niches Differentiation Theory: A New Way to Consider Relations Between Species and Niches in the Economy of Nature." *Economology Journal* 1, no. 1.

Chapter 5

1. Seed, John. 1985. "Anthropocentrism." In *Deep Ecology* by Bill Devall and George Sessions. Gibbs M. Smith, pp. 243–46.
2. Henderson, Carol L. 1987. *Landscaping for Wildlife*. Minnesota Department of Natural Resources, p. 61.
3. Sanford, Dr. Malcolm. 1998. "Pollination, the Forgotten Agricultural Input." In *Proceedings of the Florida Agricultural Conference and Trade Show* (Lakeland, FL, September 29–30, 1998) by Ferguson et al., pp. 45–47.
4. Vaughn, Mace, and Scott Hoffman. February 2007. "Enhancing Nest Sites for Native Bee Crop Pollinators." *Agroforestry Notes* 34.

5. Thomas, Susan R., et al. 2002. "Botanical Diversity of Beetle Banks, Effects of Age and Comparison with Conventional Arable Field Margins in Southern UK." *Agriculture, Ecosystems and Environment* 93, pp. 403–12.
6. Collins, K. L., N. D. Boatman, and J. M. Holland. August 2003. "A 5-Year Comparison of Overwintering Polyphagous Predator Densities Within a Beetle Bank and Two Conventional Hedge Banks." *Annals of Applied Biology* 143, no. 1, pp. 63–71.
7. Elevitch, Craig, and Kim Wilkinson. 1998–99. "A Guide to Orchard Alley Cropping for Fertility, Mulch and Soil Conservation." Version 1/99 (posted at www.echocommunity.org).
8. Irwin, Kris, and Jerry Bratton. August 1996. "Outdoor Living Barn: A Specialized Windbreak." *Agroforestry Notes* 2.
9. Bishopp, Troy. April 2012. "Passionate About Silvopasture." *Inside Agroforestry* 40, no. 2.
10. Holmgren, David. 2002. "Permaculture: Principles and Pathways Beyond Sustainability." Holmgren Design Services, Hepburn, Victoria, Australia, p. xix.
11. Schoenberger, Michele. 2005. "Agroforestry: Working Trees for Sequestering Carbon on Ag Lands." University of Nebraska–Lincoln Publications Paper 11, pp. 1–14.
12. Pulido, A. F., et al. 1999. "The Traditional Extensive Free Range Pig Farm: A Sustainable or an Endangered Production System." Escuela de Ingeniarías Agararias, Universidad de Extremadura, Ctra. Cáceres s/n, 06006. Badajoz, Spain.
13. Straight, Richard. September 2011. "A New Way in Iowa(y)." *Inside Agroforestry* 19, no. 2, p. 3.
14. Jarchow, Meghan E., and Matt Liebman. 2011. "Incorporating Prairies into Multifunctional Landscapes." *PMR* 1007, Iowa State University.
15. Ampt, Peter, and Susan Doornbos. September 2011. "Communities in Landscapes Project: Benchmark Study of Innovators, Draft Report." University of Sydney, NSW, Australia.

Chapter 6

1. Aldo Leopold (January 11, 1887–April 21, 1948) was an American author, scientist, ecologist, forester, and environmentalist. He was a professor at the University of Wisconsin and is best known for his book *A Sand County Almanac* (1949), which has sold more than two million copies.
2. Calculating swale capacity: Measure from the bottom of the spillway to the bottom of the swale in feet. Divide that number by two for average depth (assuming the swale is concave). Measure the level width of the swale at spillway height and multiply by the average depth. Multiply that number by the length of the swale. This is the cubic-foot capacity of your swale. To convert to gallons, divide the cubic capacity by 7.48.
3. A CSA (Community Supported Agriculture) grower is one who sells shares of the harvest to subscribers at the beginning of the season.
4. Johnson, Gary. 2013. Adapting to Urban Ecosystem Changes, Preparing Minnesota for Climate Change: A Conference on Climate Adaptation. University of Minnesota, November 7, 2013.

Chapter 7

1. Hosack, D. 2008. *Organic Production and Marketing of Forest Medicinals: Building and Supporting a Learning Community Among*

NOTES

Growers. Rural Action, PO Box 157, Trimble, OH. Sustainable Agriculture Research and Education (SARE) program, funded by the US Department of Agriculture; Burkhart, E. 2007. *Opportunities from Ginseng Husbandry in Pennsylvania.* Pennsylvania State University, 112 Agricultural Administration Building, University Park, PA.

2. Donohue, Colin. 2000. *Sustaining Farms and Biodiversity Through Woodland Cultivation of High-Value Crops.* Rural Action Sustainable Forestry; Hankins, Andy. 2000. *Producing and Marketing Wild Simulated Ginseng in Forest and Agroforestry Systems.* Virginia Cooperative Extension publication 354-312 (posted at ext.vt.edu/pubs/forestry/354-312/354-312.html); Adam, Katherine L. October 2004. *Ginseng, Goldenseal and Other Native Roots.* National Sustainable Agriculture Information Service publication IP115 (posted at http://attra.ncat.org/attra-pub/ginsgold.html); Carroll, Chip, and Dave Apsley. 2013. *Growing American Ginseng in Ohio: An Introduction.* Ohio State University Extension Fact Sheet F-56-13 (posted at http://ohioline.osu.edu/for-fact/pdf/0056.pdf).

3. Campbell, C., L. Staats, and B. Beyfuss. 1998. *Ginseng Research Projects.* Uihlein Sugar Maple Field Station, Cornell Cooperative Extension; Burkhart, 2007.

RESOURCES

Agriculture, Rudolf Steiner, Biodynamic Agricultural Association, 1984.

Agroecology: The Scientific Basis of Alternative Agriculture, Miguel Altieri, HarperCollins, 1989.

Animal Communities in Temperate North America, Victor E. Shelford, University of Chicago Press, 1913.

Apple Grower, The, Michael Philips, Chelsea Green Publishing, 1998.

Biotechnical Slope Protection and Erosion Control, Donald H. Gray and Andrew T. Leiser, Van Nostrand Reinhold, 1982.

Black Walnut in a New Century, Proceedings of the 6th Walnut Council Research Symposium, edited by Michler et al., USDA, 2004.

Book of Bamboo, David Farrelly, Sierra Club Books, 1984.

Botany for Gardeners, Brian Capon, Timber Press, 1990.

Bugs in the System, May Berenbaum, Addison-Wesley Publishing, 1995.

Bugs, Slugs, and Other Thugs, Rhonda Massingham Hart, Storey Publishing, 1991.

Changes in the Land, William Cronon, Hill and Wang, 2003.

Chinese Medicinal Herb Farm, The, Peg Schaefer, Chelsea Green Publishing, 2011.

Climate and Agriculture, Jen-Hu Chang, Aldine Publishing, 1968.

Common Sense Pest Control, William Olkowski, et al., Taunton Press, 1991.

Common Weeds of the United States, Agricultural Research Service of USDA, Dover Publications, 1971.

Companion Planting, Richard Bird, Quarto Publishing, 1990.

Complete Trees of North America, The, Thomas S. Elias, Outdoor Life/Nature Books, 1980.

Cornucopia: A Sourcebook of Edible Plants, Stephen Facciola, Kampong Publications, 1990.

Country Life, Paul Heiney, DK Publishing, 1998.

Creating a Forest Garden: Working with Nature to Grow Edible Crops, Martin Crawford, UIT Cambridge, 2010.

Culture and Horticulture, Wolf D. Stohrl, Bio-Dynamic Literature, 1979.

Design for Human Ecosystems: Landscape, Land Use, and Natural Resources, John Lyle, Island Press, 1999.

Dirt: The Ecstatic Skin of the Earth, William B. Logan, Riverhead Books, 1995.

Eco-Farm, Charles Walters and C. J. Fenzau, Acres USA, 1996.

Ecology of North America, The, Victor E. Shelford, University of Illinois Press, Urbana, 1963.

Ecoregion-Based Design for Sustainability, Robert G. Bailey, Springer Books, 2002.

Ecosystem Geography: From Ecoregions to Sites, Robert G. Bailey, Springer, 2002.

Edible Forest Gardens, Volumes I and II, Dave Jacke, Chelsea Green Publishing, 2005.

Edible Forest, The, Robert Hart and A. J. Douglas.

Edible Wild Plants, Oliver Perry Medsger, MacMillan, 1939.

Effects of Climatic Variability and Change on Forest Ecosystems, edited by James M. Vose et al., USDA, December 2012.

Elemental Geosystems, Robert W. Christopherson, Prentice Hall, 2007.

Encyclopedia of Garden Plants, Christopher Brickell, MacMillan, 1989.

Encyclopedia of Organic Farming and Gardening, Rodale Press.

Enduring Seeds, Gary Paul Nabhan, University of Arizona Press, 1989.

Factors of Soil Formation, Hans Jenny, Dover Publications, 2011.

Farmers of Forty Centuries, F. H. King, Rodale Press, 1911.

Farming in Nature's Image, Judy Soule and Jon Piper, Island Press, 1992.

Farming with the Wild, Daniel Imhoff, Sierra Club Books, 2003.

Field and Roadside, John Eastman, Stackpole, 2003.

Five Acres and Independence: A Handbook for Small Farm Management, M. G. Kains, Greenberg, 1935.

Forest and Thicket, John Eastman, Stackpole, 1992.

Forest Ecology, Burton V. Barnes et al., John Wiley and Sons, 1980.

Forest Ecosystems, David Perry, Johns Hopkins University Press, 2008.

Forest Forensics, Tom Wessels, Countryman Press, 2012.

Forest Gardening, Robert Hart, Chelsea Green Publishing, 1996.

Gaia's Garden: A Guide to Home-Scale Permaculture, Toby Hemenway, Chelsea Green Publishing, 2001.

Gardening for the Future of the Earth, Shapiro and Harrison, Bantam Books, 2000.

Gardens of the New Spain, William Dunmire, University of Texas Press, 2004.

Grass Productivity, Andre Voisin, Island Press, 1998.

Green Woodwork, Mike Abbott, Guild of Master Craftsman Publications, 1992.

Growing Food in a Hotter, Drier Land: Lessons from Desert Farmers on Adapting to Climate Uncertainty, Gary Paul Nabhan, Chelsea Green Publishing, 2013.

Growing Unusual Vegetables, Simon Hickmott, Eco-Logic Books, 1988.

Handbook for Garden Designers, Rosemary Alexander, Ward Lock, 1994.

History of World Agriculture, Marcel Mazoyer and Laurence Roudart, Monthly Review Press, 2006.

How Indians Use Wild Plants for Food, Medicine, and Crafts, Frances Densmore, Dover Publications, 1974.

How to Grow More Vegetables, John Jeavons, Ten Speed Press, 2002.

How to Make a Forest Garden, 3rd edition, Patrick Whitefield, Permanent Publications, 2002.

ICS Handbook for Farms, International Textbook Company, 1912.

Insects and Gardens, Eric Grissell, Timber Press, 2001.

Introduction to North American Beetles, Charles A. Papp, Entomography Publications, 1984.

Invasive Plant Medicine, Timothy Lee Scott, Healing Arts Press, 2010.

Land Mosaics, Richard Forman, Cambridge University Press, 2004.

Landscape Ecology, R. T. Foreman, Wiley, 1986.

Landscape Graphics, Grant Reid, Watson-Guptill, 2002.

Landscaping with Fruit: Strawberry Ground Covers, Blueberry Hedges, Grape Arbors, and 39 Other Luscious Fruits to Make Your Yard an Edible Paradise, Lee Reich, Storey Publishing, 2009.

Let It Rot, Stu Campbell, Storey Communications, 1990.

Living Landscape, The, Patrick Whitefield, Permanent Publications, 2009.

Mainline Farming for the Twenty-First Century, Dan Skow, Acres USA, 1991.

Manual of Woody Landscape Plants, Michael A. Dirr, Stipes, 1975.

Monocultures of the Mind, Vendana Shiva, Zed Books, 1993.

RESOURCES

Mulch Book, The, Stu Campbell, Storey Communications, 1991.

Mycelium Running, Paul Stamets, Ten Speed Press, 2006.

Native Harvests: American Indian Wild Foods and Recipes, E. Barrie Kavasch, Dover Publications, 2005.

Natural Way of Farming, The, Masanobu Fukuoka, Bookventure, 1985.

Nature and Classification of Soils, The, Nyle Brady, Prentice Hall, 1996.

Nature and Properties of Soils, The, Nyle C. Brady, Prentice Hall, 2007.

New Roots for Agriculture, Wes Jackson, University of Nebraska Press, 1980.

North American Agroforestry: An Integrated Science and Practice, H. E. Garret, American Society of Agronomy, 2009.

Nourishment Home Grown, A. F. Beddoes, Vital-to-Life Books, 2004.

Once and Future Forest, The, Leslie Jones Sauer, Island Press, 1998.

One Straw Revolution, Masanobu Fukuoka, Other Indian Press, 1978.

Orchard Almanac, Page and Smillie, Spraysaver Publications, 1986.

Organic Method Primer, The, Bargyla Rateaver and Gylver Rateaver, self-published, 1973.

Perennial Vegetables, Eric Toensmeier, Chelsea Green Publishing, 2007.

Permaculture: A Designer's Manual, Bill Mollison, Tagari, 1988.

Peterson Field Guide to Eastern Forests, John C. Kricher, Houghton Mifflin, 1998.

Peterson Field Guides to Trees, Shrubs, Animals, Birds, Forests, et cetera, Houghton Mifflin.

Pilgrim at Tinker Creek, Annie Dillard, Harper, 2007.

Plant, The, 3rd edition, Gerbert Grohman, Biodynamic Association, 2009.

Plant and Planet, Anthony Huxley, Viking Press, 1974.

Plants for Man, Robert W. Schery, Prentice Hall, 1972.

Plowman's Folly, Edward Faulkner, University of Oklahoma Press, 2012.

Principles of Ecological Landscape Design, Travis Beck, Island Press, 2013.

Reaching for the Sun: How Plants Work, John King, Cambridge University Press, 2011.

Real Dirt, The, Miranda Smith, Northeast Region Sustainable Agriculture Research and Education Program, 1998.

Regenerative Design for Sustainable Development, John Lyly, Wiley Publishers, 1996.

Restoration Agriculture, Mark Shepard, Acres USA, 2013.

Restoration Forestry: An International Guide to Sustainable Forestry Practices, edited by Michael Pilarski, Kivaki Press, 1994.

Restoration of Land, The, A. D. Bradshaw and M. J. Chadwick, University of California Press, 1980.

Rocky Mountain Horticulture, George Kelly, Pruett Press, 1967.

Rodale Guide to Composting, The, Jerry Minnich and Marjorie Hunt, Rodale Press, 1979.

Roots Demystified, Robert Kourik, Metamorphic Press, 2008.

RT Permaculture, Neil Bertrando, Reno, NV.

Ruth Stout No-Work Garden Book, The, Ruth Stout, Rodale Press, 1971.

Save Three Lives, Robert Rodale, Sierra Club Books, 1993.

Science in Agriculture, Arden Anderson, Acres USA, 1992.

Secret Life of Plants, The, Peter Tompkins and Christopher Bird, Harper, 1973.

Secret of Life, The, Georges Lakhovsky, W. Heinemann, 1939.

Secrets of the Soil, Peter Tompkins and Christopher Bird, Harper and Row Publishers, 1989.

Seed to Seed, Susan Ashworth, Seed Saver Publications, 1991.
Seeds of Change, Kenny Ausubel, Harper, 1994.
Seeds of Peace, Sulak Sivaraksa, Parallax Press, 1992.
Site Engineering for Landscape Architects, Steven Strom, Wiley, 2013.
Soil Conservation Handbook, from Kenya.
Solviva, Anna Edey, Trailblazer Press, 1998.
Soul of the Soil, Grace Gershunney, agAccess, 1995.
Sunset Western Garden Book, editors of *Sunset* magazine, Oxmoor House, 1994 and later editions for high-altitude zones.
Sustainable Landscape Construction: A Guide to Green Building Outdoors, William Thompson and Kim Sorvig, Island Press, 2007.
Swamp and Bog, John Eastman, Stackpole, 1995.
Teaming with Microbes, Jeff Lowernfels and Wayne Lewis, Timber Press, 2010.
Temperate Zone Pomology, Melvin N. Westwood, W. H. Freeman, 1978.
Tending the Wild, M. Kat Anderson, University of California Press, 2006.
Tools and Devices for Coppice Crafts, F. Lambert, National Federation of Young Farmer's Clubs, 1957.
Tools for Agriculture, Ianca Bruthers, Intermediate Technology Publications, 1992.
Traditional Woodland Crafts, Raymond Tabor, Batsford, 2003.
Tree Crops: A Permanent Agriculture, J. Russell Smith, Island Press, 1987.
Uncommon Fruits for Every Garden, Lee Reich, Timber Press, 2008.
Unlearn, Rewild: Earth Skills, Ideas and Inspiration for the Future Primitive, Miles Olson, New Society Publishers, 2012.
Urban Watershed Forestry Manual, Karen Capiella, Tom Schueler, and Tiffany Wright, USDA Forest Service, July 2005.
Uses of Wild Plants, The, Frank Tozer, Green Man Publishing, 2007.
Weeds and What They Tell, Ehrenfried Pfeiffer, Rodale Press, 1954.
Woodland Year, The, Ben Law, Permanent Publications, 2009.
You Can Farm, Joe Salatin, Polyface, 1998.

Websites

AccuWeather: accuweather.com
Agroforestry Net: agroforestry.net/
Apios Institute: apiosinstitute.org/
Barefoot Beekeeper: biobees.com
Biomimicry 3.8: biomimicry.net
Botanical Image Data Bases: plantbiology.siu.edu/faculty/nickrent/BotImages.html
Botany Database: http://nativeplants.ku.edu/research/ethnobotany-database-2
CGIAR: cgiar.org
Dave's Garden: http://davesgarden.com
Edible Landscaping: ediblelandscaping.com
Future Scenarios: futurescenarios.org
Holistic Management International: holisticmanagement.org
Index Herbarium: http://sciweb.nybg.org/science2/IndexHerbariorum.asp
Midwest Organic and Sustainable Education Service: mosesorganic.org
Natural Capital Plant Database: permacultureplantdata.com
Northwest Center for Alternatives to Pesticides: pesticide.org
Oikos Tree Crops: oikostreecrops.com
One Green World: onegreenworld.com
Perennial Vegetables: http://perennialvegetables.org/perennial-vegetables-for-each-climate-type
Plant Maps: plantmaps.com/index.php
Plant Methods: plantmethods.com
Plants for a Future: pfaf.org

RESOURCES

Plants for Stormwater Design: pca.state.mn.us/index.php/water/water-types-and-programs/stormwater/stormwater-management/plants-for-stormwater-design.html

Pollen Viewer: ncdc.noaa.gov/paleo/pollen/viewer/webviewer.html

Prairiesource Wildflower Reference Guide: prairiesource.com/wrg.htm

Raintree Nursery: raintreenursery.com

Ready, Set, Grow: dnr.state.mn.us/young_naturalists/seeds/index.html

Richters Herbs: richters.com

Seeds of Change: seedsofchange.com

Self Nutrition Data: http://nutritiondata.self.com

Small Farm Success: smallfarmsuccess.info

South Australian Research and Development Institute: sardi.sa.gov.au/home

SouthWoods Forest Gardens: http://southwoodsforestgardens.blogspot.com

Sustainable Urban Landscape Information Series, University of Minnesota: sustland.umn.edu/index.html

Temperate Climate Permaculture: http://tcpermaculture.com/site/plant-index

Trap Cropping: oisat.org/control_methods/cultural__practices/trap_cropping.html

Tree Index: fs.fed.us/database/feis/plants/tree

Tropical Fruit and Nut Plants: hawaiiantropicalplants.com/fruit.html

USDA National Nutrient Database: nal.usda.gov/fnic/foodcomp/search

USDA NRCS Soils: http://soils.usda.gov

USDA Plants Database: http://plants.usda.gov

Web Soil Survey: http://websoilsurvey.nrcs.usda.gov/app/WebSoilSurvey.aspx

INDEX

A

abiotic ecological conditions, 45, 54
acacia (*Acacia* spp.), 100, 107
accelerated succession, 44, 62, 67, 99
accent plants, 67
access to plants, 59, 60, 62, 70, 100, 202
    for harvest, 80
    in scale of permanence, 74
    maintenance of paths, 63
Acer spp. *See* maple (*Acer* spp.)
Achillea spp. *See* yarrow (*Achillea* spp.)
acid soil, 103, 106, 116, 118, 166, 247
Aconitum napellus (monkshood), 243, 245
acorns, 162
    as animal feed, 160
    flour, 32, 155, 157–58
Acorus calamus (sweetflag), 277–78
Actinidia spp. (hardy kiwi), 90, 119
Adiantum pedatum (maidenhair fern), 120
adversity, using, 72
Aesculus hippocastanum (horsechestnut), 111
aesthetic standard, 60, 61, 66–67
Agaricus bisporus (crimini mushroom), 130
Agastache spp. *See* hyssop (*Agastache* spp.)
agricultural toxins, 108–12
agroforestry techniques, 186–92
air layering, 144
air-cleansing plants, 126–27
Ajuga spp. (bugleweed), 109
    A. reptans (carpet), 110
akebia (*Akebia quinata*), 231, 233, 252–53, 255
Alcea rosea (hollyhock), 233–34
alder (*Alnus* spp.), 28, 99, 108
    mountain (*A. tenuifolia*), 111
    red (*A. rubra*), 111
alfalfa, 187
alkaline soil, 103, 106–7, 116
allelopathic plants, 121
alley cropping, 186–87, 191, 192
Alliaria petiolata (garlic mustard), 121, 143
Allium spp., 267. *See also* onion
    A. cernuum (nodding wild onion), 138–39
    A. schoenoprasum (chives), 110, 261–63
    A. tricoccum (ramps), 277
    A. tuberosum (chives), 236, 238
almond, hardy, 180, 219–22
Alnus spp. *See* alder (*Alnus* spp.)

Aloe spp., 117
Aloysia citriodora (lemon verbena), 127
Amanita spp. (fly agaric mushroom), 131
amaranth (*Amaranthus* spp.), 32, 119, 150, 184
Amelanchier spp. *See* serviceberry (*Amelanchier* spp.)
American hogpeanut, 253–54
ammonia, 127
Amorpha canescens (leadplant), 88, 212
Amphicarpaea bracteata (American hogpeanut), 253–54
amur corktree, 252–54
Anemone canadensis (Canada anemone), 138–39
anemone, Canada, 138–39
angelica, 138–39
animals, 40, 192–99. *See also* birds; decomposers/detritovores; insects; livestock
    agroforestry, 191
    carnivores or predators, 195–96
    excluding, 74
    food sources for, 69, 75, 154, 164, 168–69, 171–72, 175–77, 184, 192–99, 264 (*See also* silvopasture (silviculture))
    integration of, 20, 58, 175–99
    omnivores, 193–95
    on marginalized land, 71
    plant propagation, role in, 143, 156, 195, 264
    prairies, 191
    redundancy, 101
    shelter for, 58, 177–79, 264
Annona reticulata (custard apple), 182
annual plants, 12, 98–99, 160. *See also specific plants*
    annual-perennial guild, 235–39
    as barriers, 121
    as cover crop, 101, 102
    as starter plants, 135, 136
    in dry, sandy soils, 105
    salt-tolerant, 117
annual polycultures, 39
Antennaria rosea (pink pussytoes), 110
Anthurium andraeanum (flamingo-lily), 127
Apios americana. *See* groundnut
Apis spp. (bees), 179–81

Apocynum cannabinum (Indian hemp), 138–39
apple, crab apple (*Malus* spp.), 28–29, 32, 76, 85, 105, 147, 265–67
    access to, 59
    fire-resistant, 111
    grafting, 145
    honeycrisp (*M. pumila*), 59, 88, 212
    orchard, creating, 70–71
    pollination, 140, 180
    sunlight for, 136
    volunteer sprouts, 18–19
apricot, 51, 140
    bush, 226, 229
    Prunus armeniaca, 146
aquic soil. *See* wet soil
Aquilegia spp., 109. *See* columbine
Aralia spp., 269
arborvitae, 279–81
Arctium lappa (burdock), 32, 101
Arctostaphylos uva-ursi (kinnikinnick), 110
Areca spp. (date palm), 107
aridic soil, 105
Arisaema triphyllum. *See* Jack in the pulpit (*Arisaema triphyllum*)
Aristolochia durior (Dutchman's pipe), 182, 279, 281
Armeria maritima (sea thrift), 110
Armillaria spp. (fungi), 130, 133
Armoracia rusticana. *See* horseradish
Aronia spp. (chokeberry), 106
arrowhead, 276–77
Artemisia annua (sweet wormwood), 252–54
Asarum spp. *See* wild ginger (*Asarum* spp.)
Asclepias spp. (milkweed), 182
    A. incarnata (swamp), 138–39
asexual plant propagation, 144–47
ash (*Fraxinus* spp.), 18, 105, 108, 145
    green (*F. pennsylvanica),* 111, 272
    white (*F. americana),* 111, 155, 169
Asimina triloba. *See* pawpaw (*Asimina triloba*)
asparagus, 15, 129, 259–60
aspect, 45. *See also* sun, aspect of land to
aspen, 18
    quaking, 111, 129
aster (*Aster* spp.), 90

New England *(A. novae-angliae)*, 88, 138–39, 212
sky blue *(A. azureus)*, 88, 212
smooth *(A. laevis)*, 88
Astralagus spp. (milk vetches), 103
Atropa bella-donna (deadly nightshade), 241–45
Aubrieta deltoidiea (rock cress), 109, 110
Aurinia saxatilis (basket-of-gold), 110
autumn olive *(Elaeaganus umbellata)*, 119, 122, 133, 144
azalea, western, 111

B

bachelor's button, 249–50
backyard orchard guild, 82–91, 136
baikal skullcap, 267
balance in nature, 42
bamboo, 15, 129, 236, 238
bamboo palm, 127
bananas, 130
Baptisia spp. *See* indigo *(Baptisia* spp.)
basil, 142, 236, 238, 259–60
basket willow, 281
basket-of-gold, 110
basswood, American, 168
bay laurel, 127
bayberry, 128, 133
beach plum, 281
beans, 28, 128, 140, 148, 149, 183, 192, 239. *See also* fava beans
  bush snap, 98
  pole, 118, 119, 134
beardtongue, 110, 182
  hairy, 138–39
beargrass, 182
beebalm, spotted, 138–39
beech *(Fagus* spp.), 18, 118
  American *(F. grandifolia)*, 171
  European *(F. sylvatica)*, 111
bees, 32, 58, 137–38, 150, 179–82, 183. *See also* insects
  colony collapse disorder, 179, 183
  solitary, 180, 181, 183–84
beet (beetroot), 32, 129, 239
beetle banks, 185
beetles, 182–83. *See also* insects
beneficial insects, 39, 58, 91, 137–40
benzene ring aromatic hydrocarbons, 127
bergenia, heartleaf *(Bergenia cordifolia)*, 110
berms. *See* raised beds; swale-and-berm systems
berries, 32
Betula spp. (birch), 111
big bluestem, 13–14
bindweed, 119
  field, 130

biochar, 112
biomass, 57, 101, 149, 187
  energy flow, 40
  trees, created by, 152, 166, 171
biomes, 18, 94–99, 104–5
  Southern Illinois, 216, 219–22, 231–34
  Wisconsin, 96, 105, 264–81
biotic conditions, 40, 45
birch, 18, 111
birds, 39, 195, 264
  attracting, 91, 137, 150, 176–77
  diversity of, 71
  pest control and, 58
  pollination by, 140, 183, 184
  seed dispersal by, 58, 143–44, 156, 195
blackberry, 168
blackeyed Susan, 106
blanket flower, 110
bleeding heart, 115
bloom times, 137
bluebeard lily, 233–34
blueberry, 54, 168, 180, 235–36
  northblue, 256–57
  pH for, 106, 116, 118
Bombus spp. (bumblebee), 180–81
boneset, 138–39
borage *(Borago officinalis)*, 98, 103, 163, 259–60
Boraginaceae, 103
boreal forest berry guild, 256–58
boron, 103
bottomland guild, 268–70
boundaries, 74, 203
boxelder. *See* maple *(Acer* spp.)
bramble berries. *See Rubus* spp. (raspberry)
Brassicas spp. (mustards, cabbage), 102, 103, 106, 108, 119, 184, 239
  B. oleracea var. *acephala* (kale), 141
broadleaf lady palm, 127
broccoli, 15, 239
Bromeliaceae, 116
buckthorn, 120, 121, 143
buckwheat *(Eriogonum* spp.), 48, 102, 103, 121, 184
budget, 205, 209–13
buffaloberry, 133
bugleweed, 109
  carpet, 110
buildings. *See* structures
bulbs, 120, 129, 130, 161–62
bunchberry, 256–58
burdock, 32, 101
burning bush, dwarf, 111
butterflies, 32, 58, 140, 150, 182
butternut, 269

C

cabbage, 106, 108, 184, 239

Cacalia atriplicifolia (pale Indian plantain), 138–39
cacao, 182
cacti, 122, 182. *See also* saguaro cactus
calcium, 28, 34, 102, 103
calendula *(Calendula officinalis)*, 236, 238
Calochortus spp. (mariposa lily), 182
Caltha palustris (marsh marigold), 277
Calycanthus floridus (Carolina allspice), 182
CAM (crassulacean acid metabolism) plants, 116–17
camas *(Camassia quamash)*, 136–37
Campsis radicans (trumpet vine), 110
Canada thistle, 33
canopy, 13, 17
  agroforestry, 191
  design, role in, 42, 52, 85
  maturity of, 74, 80, 99
  planning chart, 286–87
  sunlight/shade from, 50, 52–53, 135–37
Capsicum annuum (cayenne pepper), 236, 238
Caragana arborescens (pea shrub), 280–81
caraway, 219, 221, 223
carbon, 32, 34, 102, 103, 112, 137, 152
carbon sequestration, 100, 189, 191
Carex spp. (sedges), 109, 110
carnation, garden, 110
Carnegia spp. *See* saguaro cactus
carnivores, 195–96
Carolina allspice, 182
carrots, 54, 128–29, 185
carrying capacity, 3, 42, 68
Carum carvi (caraway), 219, 221, 223
Carver, George Washington, 124–26
Carya spp. *See* hickory *(Carya* spp.)
Caryopteris × clandonensis (blue-mist spirea), 111
Castanea spp. *See* chestnut *(Castanea* spp.)
castor bean, 243, 245
catalpa, western *(Catalpa speciosa)*, 111
catastrophic occurrences, 107–8
catch and store, 23, 68, 100
catchment systems, 23–26
catnip, 149
  dwarf, 277
cattails, 276–78
cayenne pepper, 236, 238
ceanothus *(Ceanothus* spp.)
  C. gloriosus (Point Reyes ceanothus), 110
  C. prostratus (mahala mat), 110
cedar, white, 121
Celtis occidentalis (common hackberry), 111
Centaurea montana (bachelor's button), 249–50
Cerastium tomentosum (snow-in-summer), 110
Cercis canadensis (eastern redbud), 111, 168
Chamaecyparis thyoides (white cedar), 121
Chamaedorea seifrizii (bamboo palm), 127

INDEX

change, response to, 72, 98, 100, 204
chayote, 129
Chenopodium quinoa (quinoa), 148, 150
cherry, 140, 145, 180
    black, 155, 161
    bush, 226, 229
    cornelian, 32
    dwarf cherry tree polyculture, 261–63
    Evans Bali, 88, 212
    in overstory, 86
    North Star dwarf, 261–63
chestnut (*Castanea* spp.), 145–46, 163, 167, 190
chickens, 22–23, 70
chickpeas, 160
chickweed, 103, 120
chicory, 32, 33, 88, 90, 106, 137, 212, 249
Chinese licorice, 252
chives, 110, 236, 238, 261–63
Chlorophytum comosum (spider plant), 127
chocolate vine. See akebia (*Akebia quinata*)
chokeberry, 106
chokecherry, 111
    Canada red, 111
chop-and-drop, 101, 103, 121, 163, 188, 192, 205, 217, 220
chrysanthemum, florist's mum (*Chrysanthemum × morifolium*), 127
Cichorium intybus. See chicory
cider guild, 265–67
cilantro (*Coriandrum sativum*), 98, 236, 238, 260
Cimicifuga racemosa (snakeroot), 269
Cistus purpureus (orchid rockrose), 110
Citrus × meyeri (dwarf lemon), 127
ciwujia (eleuthero, Siberian ginseng), 119, 190, 269
Claytonia perfoliata (miner's lettuce), 249–50
Cleome serrulata (Rocky Mountain bee plant), 149
climate, 7, 16–17, 72, 73, 94, 100
climate change, 17–20, 18, 137, 192, 211
Clintonia borealis (bluebeard lily), 233–34
clones, 147
clover, 28, 102
    purple prairie (*Dalea purpurea*), 261, 263
    sweet, yellow and white (*Melilotus* spp.), 119
clover (*Trifolium* spp.), 28, 102, 149
    crimson, 102, 149
    dwarf, 119
    red, 224, 226, 228–30
    T. repens (white, Dutch white), 120, 121, 233, 237, 239, 249–50, 266–67
clumped plant distribution, 146
collard, 259–60
columbine, 109, 110, 236, 238

comfrey (*Symphytum officinale*), 88, 137, 150, 212
    access to, 59
    as dynamic accumulator, 28, 101, 103
    as nurse plant, 122–23
    in case study guilds, 219–22, 226, 229, 232–33, 235–38, 248–49, 259–60
    in dwarf fruit tree guild, 118
    in oak guild, 161–63
    propagation from, 204
Communities in Landscapes Project innovation study, 191
companion planting, 30, 39, 98
compass plant, 279
compost, 21–23, 57, 64, 69–70, 109, 171, 203, 205
concept lines, 65–67
coneflower (*Echinacea* spp.), 90, 274
    pale purple (*E. angustifolia*), 249
    purple (*E. purpurea*), 88, 106, 110, 212
    Tennessee (*E. tennesseensis*), 272
coneflower (*Ratibida* spp.)
    prairie (*R. columnifera*), 110
    yellow (*R. pinnata*), 138–39
coniferous trees, 109, 111, 118, 132–33. See also specific trees
Conium maculatum (poison hemlock), 124, 241–44
conservation, 68
consumption, limited, 18, 68, 100
contraction, 141
Convallaria majalus (lily of the valley), 130, 243, 245
Convolvulus spp. (bindweed), 119
    C. arvensis (field), 130
cooperative competition, 48–50
coppicing, 145, 159–60, 188
Coptis trifolia (goldthread), 252–54
coralbells, 110
coreopsis (*Coreopsis* spp.), 110
    sand coreopsis/lanceleaf tickseed (*C. lanceolata*), 138–39
Coriandrum sativum. See cilantro (*Coriandrum sativum*)
corm, 130
corn (*Zea mays*), 116, 124, 128–29, 140, 148–50, 183–84, 192
Cornus spp. See dogwood (*Cornus* spp.)
corporate farms, 3
Corylus spp. See hazel (hazelnut) (*Corylus* spp.)
cost effectiveness, 60, 61, 64–65
cotoneaster (*Cotoneaster* spp.), 109
    cranberry (*C. apiculatus*), 110
cotton, 127, 180
cottonwood, 18, 119

cover crops, 21, 101–4, 108, 133. See also ground covers
cow parsnip, 138–39
cow pea, 58, 252
crab apple. See apple, crab apple (*Malus* spp.)
cranberry, American, highbush, 111, 119, 163
cranesbill, grayleaf, 110
Crataegus spp. See hawthorn (*Crataegus* spp.)
creeping charlie. See ground ivy
creeping holly, 110
Crocus spp., 135
crop rotation, 39, 57, 98
cross-pollination, 140, 147, 179, 191
cultural change, 69
Culver's root, 138–39
cup plant, 121, 138–39, 145, 150, 177, 279, 281
Cupressaceae, 118
currant (*Ribes* spp.), 82, 111, 140, 144, 162, 163, 266–67, 269
    black (*R. nigrum*), 88, 204, 212, 253, 255, 266
    flowering, 111
    propagation from, 204
Cuscuta spp. (dodder), 129
custard apple, 182
cuttings, propagation from, 204–5
Cydonia oblonga (quince), 145, 249–50
cypress, 18, 118
Cypripedioideae, 145

D

daffodil (*Narcissus* spp.), 59, 88, 102, 106, 129, 130, 135–36, 213, 243–45, 262–63, 272
dahlia (*Dahlia* spp.), 130, 145
daisy, Barberton, 127
Dalea purpurea (purple prairie clover), 261, 263
dandelion (*Taraxacum officinale*), 28, 32, 47, 103, 106, 120, 224–28, 230, 237, 239
daphne, Carol Mackie (*Daphne × burkwoodii*), 110
dappled shade. See shade, partial
dark earth, 112
date palm, 107
Datura stramonium (jimsonweed), 243, 245
Daucus carota (carrots, Queen Anne's lace), 54, 128–29, 143, 185
daylily, 110, 128, 279–81
dead nettle, 110
deadly nightshade, 243–45
deciduous trees, 109, 111, 118, 132–33. See also specific trees
decomposers/detritovores, 32, 35, 40, 47, 69, 192, 196–99
deer, 184

dehesas, 160–62, 190
Delosperma (iceplant)
    D. cooperi (purple), 110
    D. nubigenum (yellow), 110
delphinium (*Delphinium* varieties), 110
Delphinium spp. (larkspur), 243, 245
Dendrobium spp., 127
design. *See* Permaculture design; plant guild design
detritovores. *See* decomposers/detritovores
Dianthus (garden carnation, pinks), 110
Dicentra spectabilis (bleeding heart), 115
dicots, 128
Digitalis spp. (foxglove), 243, 245
dill, 161
Diospyros spp. *See* persimmon (*Diospyros* spp.)
diversity, 39, 57, 67, 70–71, 81, 100, 191
    alpha, 148–49
    beta, 149–50
    functional, 99
    importance of, 29–31, 148–50
    of yield, 22, 30
    prairies, 191–92
    resilience and, 15, 17, 71
    resource depletion, avoiding, 42
dodder, 129
dogwood (*Cornus* spp.), 109, 188, 236, 238
    bunchberry (*C. canadensis*), 256–58
    flowering (*C. florida*), 111
    pagoda (*C. alternifolia*), 253, 255
    redosier (*C. sericea*), 111
dormancy, 141
Dracaena marginata (dragon plant), 127
dragon plant, 127
drainage, 106
drought, 107, 112
    animal food sources and, 175–76
    diversity, value of, 17, 71, 148
    plants for, 57, 106, 115, 137, 148, 247
dry conditions, 50. *See also* drought
dry soils, 105–7, 156, 247
Dutchman's pipe, 182, 279, 281
dynamic accumulators, 28, 52, 82, 91, 101–3, 108, 150. *See also* nitrogen-fixing plants

E

earthworks, 23–26
eastern woodlands, 80, 96, 105–6, 216
    plant guilds for, 219–34, 264–81
    trees in, 118, 154, 159
Echinacea spp. *See* coneflower (*Echinacea* spp.)
Echium vulgare (viper's bugloss), 103
ecological conditions, assessment of, 21, 45
ecological functions. *See* functions of plants
ecological services, 20, 23, 58, 69, 91, 98, 99, 137, 177. *See also* ecologically functional landscape; functions of plants
enhancement of, 19, 39, 44, 61, 67, 81, 204
integrated, 34
reduction of, 67, 81
ecologically functional landscape, 60–63, 70–71, 74–75. *See also* functional spaces
ectomycorrhizal fungi, 131–33
edges, 52–53, 65, 71–72, 100, 163, 182, 185, 231
eggplant, 109, 128, 140, 180
Eichhornia crassipes (water hyacinth), 108
Elaeaganus umbellata. See autumn olive (*Elaeaganus umbellata*)
elderberry (*Sambucus nigra*), 140, 182, 226, 229, 240, 243, 245–46, 248–49
Eleocharis dulcis (water chestnut), 130
eleuthero (*Eleutherococcus senticosus*), 119, 190, 269
elevation grid, 76, 285
elm, 18, 169
endomycorrhizal fungi, 131–33
energy flow/transfer, 40, 149, 191
environmental adaptation, 37
environmental soundness, 60–61, 64
Epilobium angustifolium (fireweed), 110
equipment, 60, 62, 69, 191
Equisetum arvense (scouring rush), 128
Ericaceae, 106
Erigonum spp. *See* buckwheat (*Eriogonum* spp.)
erosion, 64, 74, 100, 104, 112, 191
Erythronium spp.
    E. americanum (dogtooth violet), 102, 115
    E. americanum (trout lily), 162
Euonymus alatus 'Compactus' (dwarf burning bush), 111
Eupatorium perfoliatum (boneset), 138–39
evening primrose, 110, 184
external stimulus, 141

F

Fagus spp. *See* beech (*Fagus* spp.)
fava beans, 98, 102, 160
feedback, 68, 100
fences, 74, 184, 203
fennel, 161, 219, 221–22
fernleaf buckthorn, 111
ferns, 115, 119–20, 129
    Boston, 127
    maidenhair, 120
fertility in system, 47
fertilizer, 47, 69, 191
feverfew (*Tanacetum parthenium*), 226, 230, 236, 238
fibrous roots, 33, 39, 47, 53, 106, 115, 128
Ficus spp. *See* fig (*Ficus* spp.)
fig (*Ficus* spp.), 32, 51, 106, 127, 182, 219–23
    rubber plant (*F. elastica*), 127
    weeping (*F. benjamina*), 127
Filipendula ulmaria (meadowsweet), 128, 138–39
fir, 18, 134, 179
fire
    seeds exposed to, 144, 166
    soil, effect on, 107
fire-resistant plants, 109–12, 166
fireweed, 110
flamingo-lily, 127
flax, 127
    blue, 110
flies, 182
floodplains
    forests, 105
    guild for, 268–70
floods, 71, 107–8, 148
Foeniculum vulgare (fennel), 219, 221–22
forest farming, 190–91
forget-me-not, 103
formaldehyde, 127
forsythia (*Forsythia* spp.), 279, 281
    F. suspensa, 119
four o'clocks, 129, 182
four vines guild, 231–34
foxglove, 243, 245
Fragaria spp. *See* strawberry, wild strawberry (*Fragaria* spp.)
Frankia alni, 133
Fraxinus spp. *See* ash (*Fraxinus* spp.)
freeze damage, 114
freeze-and-thaw process, 104
French sorrel, 88, 212, 262–63
frost, 114, 137
fruit and nut guild, 219–23
fruit cocktail trees, 145
fruit set, 137–40
fruit trees, 29, 147, 152. *See also* orchards; *specific trees*
    agroforestry, 191
    dwarf/semi-dwarf, 29, 118, 119
    niche location and, 114
    pollination of, 180
    salt-tolerant, 117
    sunlight for, 136
fruiting shrubs, 25, 29, 114. *See also specific shrubs*
Fuchsia spp., 182
full shade. *See* shade
full sun. *See* sunlight
functional relationship, 3, 70. *See also* ecological services; ecologically functional landscape; functions of plants
functional spaces, 9, 28, 43, 62, 68–70, 80–81, 83, 100
functional structures, 28, 253

INDEX

functions of plants, 79, 99–101, 124–28. *See also specific functions,* e.g. pest control
    agroforestry, 186–92
    backyard orchard, 82, 85, 89, 212–13
    case study guilds, 221, 226–27, 233, 236–37, 243, 249, 253, 260, 263, 267, 269, 272, 277, 281
    hickory, 168–69
    maple, 171–72
    multiple species, 39, 42
    oak, 156–60
    pine, 164–66
    placement in garden by, 59
    shared functions, 27, 77, 101
    stacking functions, 22, 76, 119, 177, 189, 217
    structure of guild defined by, 27
fungi, 102–3, 109, 130–33, 150, 158–59, 164, 171, 189

G

Gaia's Garden, 27
Gaillardia var. (blanket flower), 110
Ganoderma lucidum (reishi mushroom), 190
garlic, 15, 267
garlic chives, 15
garlic mustard, 121, 143
Gaultheria spp.
    G. procumbens (wintergreen), 115
    G. shallon (salal), 110
genetic mutations, 20
Geranium cinereum (grayleaf cranesbill), 110
Gerbera jamesonii (Barberton daisy), 127
germination, 142
ginger, 129. *See also* wild ginger
ginkgo (*Ginkgo biloba*), 219, 221–22
ginseng, 185, 190, 251
ginseng/sugar maple polyculture guild, 251–55
Gladiolus spp., 130, 182
Glechoma hederacea (ground ivy, creeping charlie), 119, 120
Gleditsia triacanthos (thornless honeylocust), 111
    G. var. *inermis,* 133
Glycyrrhiza spp. *See* licorice (*Glycyrrhiza* spp.)
GMOs, 147
golden Alexanders, 138–39
goldenrod, 184
    Riddell's, 138–39
goldenseal, 190
goldthread, 252–54
Good King Henry, 15, 32
gooseberry (*Ribes* spp.), 118, 144, 162, 163, 168
Gossypium spp. (cotton), 127
grafting, 71, 145–47, 201, 204

grains, annual, 160. *See also specific grains*
grapes (*Vitis* spp.), 90, 119, 144, 160, 184
    muscadine (*V. rotundifolia*), 168
    riverbank (*V. riparia*), 119
    V. vinifera, 231–33
grasses, 46–47
    as animal habitat, 177
    as cover crop, 102
    as forage, 187, 189
    as ground cover, 119–21
    as predator refuge, 185
    fungi and, 133
    roots of, 128–29
    salt-tolerant, 117
grazers, 20, 103, 122, 187, 191–93
    prairies, 191
    rotational grazing, 103, 160, 189–90, 192
green manure. *See* cover crop
greenbrier, common, 168
greens, 239. *See also* lettuce
grid models, 76, 83, 285
Grifola frondosa (maitake mushroom), 150, 190
Grohmann, Gerbert, 94
ground covers, 77, 82, 119–21. *See also* cover crop
    as temporary plants, 136–37
    diversity added by, 150
    fire-resistant, 109, 110
    placement in garden of, 83, 86
    silviculture, 189
    succession acceleration, 99
    sunlight for, 50
    weed suppressing, 120
    woodland, 115
ground ivy, 119
groundnut (*Apios americana*), 15, 88, 90, 219, 221, 223, 269–70, 272
groundnut, vine, 135, 144
guilds. *See* plant guilds
gum, 18
Gymnocladus dioicus (Kentucky coffee tree), 111

H

hackberry, common, 111
Hamamelis virginiana (common witchhazel), 168
Handbook on Propagation, 145
hardpan soil, 108
hardwood trees, 20
Hart, Robert, 13, 50, 52
harvest, 60, 68, 82, 98, 100, 191, 204
    access for, 80
    maintenance and, 61, 63
    over-yield and, 44, 64
    returning to system, 70

haskaps, 279
hawthorn (*Crataegus* spp.), 111, 145, 168, 179
hazel (hazelnut) (*Corylus* spp.), 22, 32, 118, 145, 160–63, 168
heat, 50–52, 114
heaths, 106
Hedera helix (English ivy), 127
hedge wall guild, 279–81
hedgerows, 121, 126, 178, 185–86
heirloom varieties, 147
Helianthemum nummularium (sun rose), 110
Helianthus spp. *See* sunflower (*Helianthus* spp.)
heliotrope (*Heliotropium* spp.), 103
Hemenway, Toby, 27
Hemerocallis spp. (daylily), 110, 128, 279–81
hemlock, eastern, 279–81
hemlock, poison, 124, 241–44
hens and chickens, 110
Heracleum maximum (cow parsnip), 138–39
herbaceous plants, 3, 13–15, 22, 77, 118, 119, 121. *See also specific plants*
herbicides, 185
herbs, 29, 106, 118, 127, 136, 161, 185, 191, 266–67. *See also specific herbs*
Heuchera sanguinea (coralbells), 110
hickory (*Carya* spp.), 18, 152, 155, 163, 166–69, 269
    bitternut (*C. cordiformis*), 269
    mockernut (*C. tomentosa*), 269
    pecan (*C. illinoinensis*), 125, 166, 269
    pignut (*C. glabra*), 269
    shagbark (*C. ovata*), 166–69, 168
    shellbark (*C. laciniosa*), 269
    uses of, 168–69
Hippophae rhamnoides (sea berry), 133, 279–81
hollyhock, 233–34
Holmgren, David, 2, 68, 188
Holodiscus discolor (oceanspray), 111
homeopathic medicine, 124, 234, 240–46
honeylocust, thornless, 111, 133
honeysuckle, 110
hophornbeam, 168
horsechestnut, 111
horsemint, 138–39
horseradish (*Armoracia rusticana*), 15, 162, 163, 179, 227, 230
hosta lily (*Hosta* spp.), 110
huckleberry, 168
human use/crop. *See* functions of plants
hyacinth (*Hyacinthus orientalis*), 243–45
hybrid plants, 142, 147, 148
Hydrangea spp., 108
Hydrastis canadensis (goldenseal), 190
Hymenoptera spp. (Mexican wingless bees), 181

Hypericum perforatum (St. Johnswort), 233–34
hyssop (*Agastache* spp.)
  anise (*A. foeniculum*), 88, 212, 226, 230
  yellow giant (*A. neptoides*), 138–39

I

iceplant, purple/yellow, 110
implementation of design, 201–14. *See also* maintenance
  budget, 205, 209–13
  checklist, 205–9
  long-term, 204–5, 211
  planner for, 206–7
  planting, 203, 204, 213
  sequence of, 203–4
  site preparation, 202–3
  steps for, 204
  time line for, 202, 205
Indian hemp, 138–39
indigo (*Baptisia* spp.)
  white wild (*B. alba*), 59, 88, 212
  wild blue (*B. australis*), 261, 263
insectaries/insect habitat, 35, 65, 82, 90, 149–50, 177, 187, 261
insects, 58, 71, 140, 149, 171, 195–99. *See also* beneficial insects; pest control; pollinators
integrated forest gardening
  benefits of, 3
  definition, 11–13, 15
  overview, 1–36
  seven-layer model, 13, 49–50, 52, 82, 85, 91, 99, 115
integrated living system, 46–47
integration, 70, 100
intensive planting, 21
interplanting, 38–39
invasive species, 20, 31, 42
Ipomoea spp. (sweet potato), 129, 239
  I. batatas (common), 148
Ipomopsis spp., 182
iris (*Iris* spp.), 120, 129, 182, 243–45
  blueflag (*I. versicolor*), 106, 277
  hybrids (tall bearded iris), 110
  native dwarf (*I. cristata*), 164
  stinking gladwyn (*I. foetidissima*), 182
ironweed, Missouri, 138–39
irrigation, 26, 63, 107, 117
ivy, English, 127

J

Jack in the pulpit (*Arisaema triphyllum*), 130, 182, 243, 246
Japanese greens, 239
Jerusalem artichoke, 15
jicama, 129
jimsonweed, 243, 245
jujube, 252–54

juneberry, 277
juniper, creeping (*Juniperus horizontalis*), 109

K

kale (*Brassicas oleracea* var. *acephala*), 15, 141, 239
Kentucky coffee tree, 111
key plants, 65
keyline plowing, 108
keyline water-harvesting techniques, 26
kinnikinnick, 110
kiwi, hardy, 90, 119
Kniphofia uvaria (torch lily, red-hot poker), 110

L

Lactuca sativa. *See* lettuce
lady's slipper orchid, 145
lamb's ear, 110
Lamium spp. (dead nettle), 110
land cress, 15
landform, 7, 11, 32, 52, 69–70, 73–74
landrace plants, 147, 148
larch, western, 109, 111
Larix spp.
  L. occidentalis (western larch), 109, 111
  tamarack, 118
larkspur, 243, 245
late figwort, 138–39
Laurus nobilis (bay laurel), 127
lavender (*Lavandula* spp.), 106, 110, 185, 277
Law of Return, 99
leadplant, 88, 212
leaf rollup/curling, 141
legumes, 28, 102, 103, 122, 133–34, 141. *See also specific plants*
lemon balm, 236, 238, 267
lemon trees, dwarf, 127
Lentinula edodes (shiitake mushroom), 130, 158–59, 190
Lepidopterae, 182, 184. *See also* butterflies
lettuce, 76, 98, 140, 141, 239
Liatris liguilistylis (meadow blazing star), 262–63
licorice (*Glycyrrhiza* spp.)
  Chinese (*G. uralensis*), 252
  G. glabra, 219, 221, 223
  G. lepidota (wild), 253–54
lifestyle, healthy, sustainable, 20–21
lilac, 106, 111
lily (*Lilium* spp.), 130, 182
lily of the valley, 130, 243, 245
lily, bluebeard, 233–34
lilyturf, 127
limestone, 104, 116
linden, 103
Lindera benzoin (spicebush), 269–70

lingonberry, 118, 256–58, 279–81
Linum spp. (flax)
  L. perenne (blue), 110
  L. usitatissimum, 127
Liquidambar styraciflua (American sweetgum), 105, 111, 168
Liriope spicata (lilyturf), 127
livestock, 45, 63, 71, 221, 241, 255. *See also* grazers; silvopasture (silviculture)
  agroforestry, 186–91
  containing/shelters, 74, 177–78, 187, 189
  humans as, 67, 75, 80
  organic material processed by, 69
  wastes of, 103, 108, 185
lobelia, blue (*Lobelia siphilitica*), 138–39, 277
locust (*Robinia pseudoacacia*), 99
  black, 28, 103, 133, 169
  purple robe, 111
Lonicera spp. (honeysuckle, haskaps), 110, 279
lupine (*Lupinus* spp.), 88, 110, 160, 213
Lycopersicon esculentum. *See* tomato (*Lycopersicon esculentum*)
Lysichiton americanus (skunk cabbage), 182

M

Maclura pomifera (osage orange), 169
Magnolia spp., 182
magnolia vine. *See* schisandra
mahala mat, 110
Mahonia spp.
  M. aquifolium (Oregon grapeholly), 110
  M. repens (creeping holly), 110
maintenance, 57, 60, 63–64, 204, 213–14
maize, 148, 183. *See also* corn
mammals, 183. *See also* animals
mandrake root (mayapple), 243–45
mangrove, 129
manure, 22, 69
maple (*Acer* spp.), 18, 105, 118, 152, 154, 169–72
  amur (*A. ginnala*), 111
  bigleaf (*A. macrophyllum*), 111, 170
  boxelder/ashleaf (*A. negundo*), 171–72, 177
  Norway (*A. platanoides*), 170
  oriental varieties, 170–71
  red (*A. rubrum*), 111, 168, 170
  Rocky Mountain (*A. glabrum*), 111, 170–71
  silver (*A. saccharinum*), 170
  striped (*A. pensylvanicum*), 170
  sugar (*A. saccharum*), 21–22, 163, 168, 170–72, 177, 252–54, 269
  vine (*A. circinatum*), 111, 171
maps, 82, 83
  mind maps, 76, 78–79, 115

INDEX

soil survey, 105
marginalized land, use of, 71–72, 100. *See also* edges
mariposa lily, 182
marsh marigold, 277
mashua, 148
mayapple, 243–45
meadow blazing star, 262–63
meadowsweet, 128, 138–39
Medicago spp. (alfalfa), 187
medicinal plants, 119, 156–57, 165, 171, 190–91, 240–46, 251–55
medlar, 145
Melilotus spp. (sweet clover, yellow and white), 119
Melissa officinalis (lemon balm), 236, 238, 267
Melissodes spp. (sunflower bee), 180
Mentha spp. *See* mint (*Mentha* spp.)
Mespilus germanica (medlar), 145
mesquite, 133
microclimates, 10, 50–52, 68, 71, 81, 100, 114
microorganisms, 57, 75, 112, 203. *See also* organic material
milkweed, 182
    swamp, 138–39
Mimosa luisana, 122
mind maps, 76, 78–79, 115
miner's lettuce, 249–50
minerals, 21, 32, 34, 101–4, 108, 171. *See also* dynamic accumulators
mint (*Mentha* spp.), 31, 219–21, 223, 233–34, 236, 238, 272, 279–81
Mirabilis jalapa (four o'clocks), 129, 182
mockorange, 111
Mollison, Bill, 2, 5, 6, 7, 27, 77, 117
Monarda punctata (horsemint/spotted beebalm), 138–39
monkshood, 243, 245
monocots, 128–29
monoculture, 3, 30, 191
montados, 160–62
morning glory, 140
Morus spp. *See* mulberry (*Morus* spp.)
moss rose, 116–17
mother-in-law's tongue, 126–27
moths, 182, 184
mountain ash, 111, 145, 179
mulberry (*Morus* spp.), 28–29, 122
    M. nigra, 231, 233–34
    red (*M. rubra*), 108, 116, 119, 252–54
mulch, 13, 21, 58, 63–64, 101, 121, 137, 171
mullein, 233–34
Musa spp. (bananas), 130
mushrooms, 109, 130–31, 133, 150, 158–59, 164, 190–91
mustard, 102, 103, 119, 239
mychorrhizae, 58, 131–33. *See also* fungi

Myosotis spp. (forget-me-not), 103
Myrica spp. (bayberry), 128, 133

N

Narcissus spp. *See* daffodil (*Narcissus* spp.)
nastic growth, 141
nasturtium, 103, 121, 233, 234, 259–60
Nasturtium officinale (watercress), 103, 277–78
natural capital, 21, 23, 44, 61, 68, 69, 100
Natural Capital Plant Database, salt-tolerant plants in, 117
natural plant associates, 38
nectarine, 145–46
nectary plants, 82, 91, 137
Nepeta spp. (catnip)
    N. cataria, 149
    N. mussinii (dwarf), 277
Nephrolepis exaltata (Boston fern), 127
nettles (*Urtica dioica*), 15, 120, 127, 168, 185
niche dynamics, 47–48
niches, 32, 40, 76, 80–81, 100
    agroforestry, 191
    carrying capacity of, 42
    components of, 44–46, 67, 74
    controlled/uncontrolled, 19
    documenting conditions, 45
    enhancing number of, 28
    plant selection for, 98, 114
    pond, 56
nitrogen, 28, 32, 34, 99–100, 102–3, 137, 149, 171. *See also* nitrogen-fixing plants
nitrogen-fixing plants, 28, 58–59, 70, 81–82, 90–91, 99, 101–3, 133–35. *See also* legumes
no-till plant guilds, 178–79
nodding, 141
Nolina spp. (beargrass), 182
Nomia melanderi (alkali bee), 180
non-native species, 20
nurse plants, 42, 100, 122–23
nut trees, 99, 119, 146, 152, 191, 219–23. *See also specific trees*
nutrient cycling, 21–23, 40–42, 56–57, 69–70, 75, 91, 100
nutrients, 64, 74, 82, 102. *See also* dynamic accumulators; soil test
    biochar, absorbed by, 112
    green manures for, 101–2
    initial plants, creation by, 39
    Law of Return, 99
    over-yield of, 44
    roots and, 77
Nyssa sylvatica (sourgum), 168

O

oak (*Quercus* spp.), 18, 32, 103, 105, 109, 133, 152–64, 168, 269

black (*Q. velutina*), 158, 168
bur (*Q. macrocarpa*), 15, 109, 160, 162, 168, 190
cherrybark (*Q. falcata* var. *pagodifolia*), 168
chestnut (*Q. prinus*), 168
chinkapin (*Q. muehlenbergii*), 162, 168
coppicing and pollarding, 159–60
cork (*Q. suber*), 160, 190
fungi and, 158–59, 164
galls, 158–59
guild layout and polyculture, 163–65
holm, 160
northern pin (*Q. ellipsoidalis*), 168
northern red (*Q. rubra*), 111, 154–55, 168
Oregon white (*Q. garryana*), 111
patterns of growth, 156
pin (*Q. palustris*), 111
southern red (*Q. falcata*), 168
swamp chestnut (*Q. prinoides*), 168
swamp white (*Q. bicolor*), 162
white (*Q. alba*), 154–56, 158, 162, 168, 169
oak savannas, 65, 96, 105, 109, 160–62, 189–90, 269
oats, 102
observation and interaction, 6–7, 17, 42–45, 68, 100, 117, 204
oceanspray, 111
Ocimum basilicum (basil). *See* basil
Oenothera spp. (evening primrose), 110, 184
Olea europaea (olive), 106
olive, 106
omnivores, 193–95
onion, 129, 130, 239, 259–60, 267
    bunching, 15
    nodding wild, 138–39
open-pollinated plants, 142, 148
Opuntia spp. (prickly pear), 105
orchards, 59, 70–71, 82–91, 136, 150, 191, 203
orchid rockrose, 110
orchids, 127, 142, 145
oregano (*Origanum vulgare*), 161, 226, 230, 236, 238, 267, 272, 274, 277–78
Oregon boxwood, 110
Oregon grapeholly, 110
organic material, 18, 20, 67, 82. *See also* cover crops; decomposers/detritovores; nutrient cycling
    consumption of, 69
    cycle of, 47
    excess, 44, 68
    harvesting, 61, 204
    nutrients benefits of, 112–13
    preparing soil with, 203
    under deciduous trees, 118
Oriental poppy, 110

Origanum vulgare. See oregano *(Origanum vulgare)*
Ornithogalum umbellatum (star of Bethlehem), 243, 245
Oryza sativa (rice), 140, 147
osage orange, 169
Osmia spp. (mason bee), 137, 180–82
Ostrya virginiana (hophornbeam), 168
overstory. *See* canopy

P

Pachyrhizus erosus (jicama), 129
pachysandra, Japanese *(Pachysandra terminalis)*, 110
pale Indian plantain, 138–39
Panax spp. (ginseng), 185, 190, 251
Panicum virgatum (switchgrass), 187
Papaver orientale (Oriental poppy), 110
parsley, 185, 236, 238
parsnip, 76, 129
partial shade. *See* shade
passionflower *(Passiflora incarnata)*, 180, 231–33
patches, 19, 65–71, 83, 86, 100, 114–15, 202
patterns of growth, 140–42, 156
pawpaw *(Asimina triloba)*, 32, 120, 146, 162–63, 179, 182, 226, 229, 269–70
pawpaw delight guild, 224–30
pawpaw patch guild, 268–70
Paxistima myrtifolia (Oregon boxwood), 110
pea *(Pisum sativum)*, 28, 102, 140, 239, 266–67
    field, 102, 120–21
pea shrub, 279–81
peace lily, 127
peach, 51, 118, 140, 145–46
peanut, 124–25, 140
pear *(Pyrus* spp.), 140, 145, 147, 183, 265–66
    Asian *(P. pyrifolia)*, 219–22, 247–50
    summercrisp *(P. communis)*, 88, 212
pecan *(Carya illinoinensis)*, 125, 166, 269
Pediomelum esculentum (vine groundnut), 135, 144
peelu, 128
penstemon (beardtongue) *(Penstemon* spp.), 110, 138–39, 182
    hairy beardtongue *(P. hirsutis)*, 138–39
Peponapsis spp. (squash bee), 180
peppermint, 223, 233–34, 236, 238
peppers, 109, 140, 180, 183, 239, 260
perennial food systems, 12
perennial plants, 10, 13–15, 27, 57. *See also* perennials
    annual-perennial guild, 235–39
    as barriers, 121
    fire-resistant, 109, 110
    placement in garden of, 39, 49, 83, 86

    propagation from, 204–5
    salt-tolerant, 117
perennial polycultures, 38–40
perilla, 15
Permaculture, 1–4, 12
    ethics of, 42, 46
    garden size and, 31
    plant guilds, role of, 27, 28, 31–32
    principles of, 7, 10, 31, 42, 68–72, 100
    yield of, 32
Permaculture design. *See also* plant guild design
    components of, 61–67
    essential templates of, 7–11
    process, 5–7
    system, 4–5
permanence, scale of, 7–11
permanent agriculture, 27. *See also* perennial plants
Perovskia atriplicifolia (Russian sage), 109, 111, 249–50
persimmon *(Diospyros)*, 163
    American *(D. virginiana)*, 32, 226, 229, 271–72
    Asian *(D. kaki)*, 106, 271
    date *(D. lotus)*, 271
    Texas *(D. texana)*, 271
persimmon wood guild, 271–74
pest control, 30–31, 121, 149, 177, 184–86, 259–60, 261
    agroforestry, 191
    diversity and, 148
    preferred browse, 184
pesticides, 179, 185
Petroselinum spp. (parsley), 185
    P. crispum, 236, 238
pH, 54, 75, 103, 106–7, 116. *See also* soil test
Phalaenopsis spp. (moth orchid), 127
Phaseolus spp., 148. *See also* beans
Phellodendron amurense (Amur corktree), 252–54
phenology, plant, 38, 44, 98–99, 135
Philadelphus spp. (mockorange), 111
philodendrons, heartleaf and selloum *(Philodendron* spp.), 127
Phleum spp. (timothy), 187
phlox, creeping *(Phlox subulata)*, 110
phosphorus, 28, 102, 103
photoperiodism, 141
photosynthesis, 13, 40
Phyllostachys spp. (bamboo), 15, 129, 236, 238
Phytolacca americana (pokeberry), 243–45
phytoremediation, 108–12
pine *(Pinus* spp.), 18, 132, 134, 140, 150, 152, 164–66, 168, 179
    eastern white *(P. strobus)*, 164
    jack *(P. banksia)*, 18, 144, 164

    Korean nut *(P. koraiensis)*, 58, 122–23, 132, 251
    loblolly *(P. taeda)*, 18, 168
    lodgepole *(L. contorta)*, 164
    longleaf or slash, 18
    pinyon, 164
    ponderosa *(P. ponderosa)*, 109, 111, 164
    shortleaf *(P. echinata)*, 18, 168
    Siberian stone *(P. sibirica)*, 122
pineapple, 116
pink pussytoes, 110
pinks, 110
Pinus spp. *See* pine *(Pinus* spp.)
pioneer plants, 32, 47
Pisum sativum (pea). *See* pea *(Pisum sativum)*
plant guild design. *See also* Permaculture design
    basics, 117–28
    climate change and, 18
    flexibility of, 39
    for cooperative competition, 48–50
    from patterns to details, 69–70, 73, 100
    implementation of *(See* implementation of design)
    integration in, 70
    life cycle of, 99
    needs of garden, 57–58
    Permaculture principles applied to, 68–72
    phased planting plan for, 135–37
    process, 73, 82–91
    questions, 81, 91
    selecting plants for, 69–70, 93–150
    site context and, 42–45
    small, slow solutions, 70–71, 100
    sustainable design, components of, 58–67
plant guilds, 12–13, 31–32. *See also* polycultures
    animals, integration of, 175–99
    case studies, 215–81
    changes in, 19–18
    constructing, 75–81
    diversity of *(See* diversity)
    elevation, 29, 76
    functions and benefits of, 34–35, 99–101 *(See also* functions of plants)
    garden size and, 31
    how guilds work, 32–33
    in natural state, 32–36
    no-till, 178–79
    oak, 163–65
    overview, 26–28
    project management *(See* implementation of design)
    scientific basis for, 28–29
    stocking with plants, 142–46
    structure of, 37–91, 80–81
    trees as essence of, 151–72

INDEX

types of plants for, 118–23
plant polycultures. *See* polycultures
plant selection, 98–99
    bloom times, 137–39
    fire-resistant plants, 109–12
    for ecologically functional landscape, 70, 75
    for niches, 80–81
    for plant guilds, 69–70, 80–81, 93–150
    for resilience, 114
    for salt tolerance, 117
    nitrogen-fixing plants, 133–35
    root types, 128–30
    seasonal consideration, 135–37
    species, 69–70, 98–99
    sunlight, for available, 50, 52
Plant, The, 94
Plantago spp., 106
plantain, 106
plants. *See also* annual plants; perennial plants
    allowing for growth of, 62, 77–80
    distribution, 146–47
    diversity of (*See* diversity)
    functions of (*See* functions of plants)
    ID numbers, 84, 211
    natural range of, 146–48
    phenology of (*See* phenology, plant)
    placement in garden of, 39, 49, 59, 77–80, 83, 85–86
    propagation of, 142–47, 204–5
    rules of, 33–34
    selection (*See* plant selection)
    succession (*See* succession)
    symbiosis, 58
    tolerances of, 116–17
Plants for a Future database, 118
Platanus spp.
    P. occidentalis (sycamore), 105
    P. racemosa (western, California sycamore), 111
Pleurotis spp. (oyster mushroom), 109, 130, 158–59, 190
plum, 145, 147
    bush, 279
Poa spp., 128
Podophyllum peltatum (mayapple), 243–45
poison hemlock, 124, 241–44
poison ivy, 240–44
poison sumac, 240–44
poisonous plant guild, 240–46
pokeberry, 243–45
pollarding, oak, 159–60
pollen, 32, 137, 149, 183
pollination, 39, 137–40
pollinators, 38, 39, 91, 121, 137, 147, 149–50, 177–84. *See also* nectary plants
polycultures, 13–15, 30, 32, 38. *See also* diversity; plant guilds; redundancy; resilience
    annual, 39
    as self-supporting systems, 19
    building, strategies for, 19
    ecological functions within (*See* ecological functions of plant guilds and polycultures)
    location, effect of, 38
    oak, 163–65
    perennial, 38–40
Polygonatum biflorum. *See* Solomon's seal
pomegranate, 106–7
Pompilidae (spider wasps), 182
ponds, 56, 71
poplar, 177, 188
Populus spp.
    P. deltoides (cottonwood), 18, 119
    P. tremuloides (quaking aspen), 111, 129
    poplar, 177, 188
Portulaca spp. (moss rose, portulaca), 117
potassium, 28, 102, 103, 137
potato (*Solanum tuberosum*), 109, 118, 129, 130, 183, 239
Potentilla fruticosa (shrubby cinquefoil), 138–39
Practical Farmers of Iowa, 191
prairie, 16, 46, 108, 109, 112, 128, 144, 160, 191–92, 247. *See also* savannas
Prairie Propagations Handbook, 145
precipitation, 23–26, 74, 94–96, 137
predators, 195–96
prickly pear, 105
prince's plume, 103
prinsepia (*Prinsepia sinensis*), 281
propagation, 142–47, 204–5
Prosopis spp., 122, 133
pruning, seasonal, 204
Prunus spp., 140, 145. *See also* apricot; cherry; plum
    P. armeniaca (apricot), 146
    P. besseyi (western sandcherry), 111
    P. cerasus 'Mesabi' (Evans Bali cherry), 88, 212
    P. dulcis (hardy almond), 219–22
    P. maritima (beach plum), 281
    P. persica (peach, nectarine), 145–46
    P. virginiana (chokecherry), 111
    P. virginiana 'Schubert' (Canada red chokecherry), 111
Punica granatum (pomegranate), 106–7
purslane, 117
Pyrus spp. *See* pear (*Pyrus* spp.)

Q

Queen Anne's lace, 143, 185
Quercus spp. *See* oak (*Quercus* spp.)
quiescence, 141
quince, 145, 249–50
quinoa, 148, 150

R

rabbits, 119–20, 122
radiant heat, 50–52
radicchio, 15
radish, 121, 142, 184
raised beds, 46, 109, 185–86
ramps, 277
random plant distribution, 147
Raphanus spp. (radish), 121, 184
    R. sativus, 142
raspberry (*Rubus* spp.), 31, 118, 140, 146, 161, 192, 219–22
    hickories and, 168
    oaks and, 160
    R. arcticus, 272
    R. hayata-koidzumii, 272
    R. idaeus, 272
    R. parviflorus (thimbleberry), 272
    yellow, 231–33
Ratibida spp. (coneflower)
    R. columnifera (prairie), 110
    R. pinnata (yellow), 138–39
record-keeping, 45, 68, 117
red date. *See* jujube
red-hot poker, 110
redbud, eastern, 111, 168
redundancy, 15, 39, 70, 75, 101
reflected energy, 51–52
relationships, 36, 38, 39, 64, 81
relative permanence, scale of, 72–75
research, 17, 19, 20, 81
resilience, 66, 137
    diversity and redundancy, created by, 15, 39, 71, 101, 192
    plant functions, maintained by, 42, 99
    selecting plants for, 114, 148
resources, 18, 49, 63, 68, 69, 100
restoration of diminished systems, 69
Rhamnus spp.
    R. cathartica (buckthorn), 120, 121, 143
    R. frangula var. 'Aspleniifolia,' 'Columnaris,' 111
Rhapis excelsa (broadleaf lady palm), 127
Rheum spp. *See* rhubarb (*Rheum* spp.)
Rhizobium, 133
rhizomes, 129, 137
Rhizophora spp. (mangrove), 129
rhododendron (*Rhododendron* spp.), 106, 168
    R. macrophyllum (Pacific rhododendron), 110
    R. occidentale (western azalea), 111
rhubarb (*Rheum* spp.), 15
    Chinese (*R. officinale*), 253, 255
    R. rhabarbarum, 243, 245

Rhus spp. (sumac), 111
Ribes spp. *See* currant (*Ribes* spp.); gooseberry (*Ribes* spp.)
rice, 140, 147
Ricinus communis (castor bean), 243, 245
riparian buffer strips, 188, 191
roads. *See* access to plants
Robinia pseudoacacia. *See* locust (*Robinia pseudoacacia*)
rock cress, 109, 110
rock gardens, 120
Rocky Mountain bee plant, 149
rodents, 196
root crops, 109
root exudates, 32
root zone, 76–77
rooting, 144
roots, 12–13, 32, 39, 128–30
    clumped plant distribution from, 146
    forage species, 108
    soil types, adapted to, 106
    toxins filtered through, 108–9
    types of, 47, 52–54, 76–77, 99, 115, 128–29
rose (*Rosa* spp.), 143, 231–33, 279
    apple rose (*R. villosa*), 279, 281
    hardy shrub rose, 111
    R. moyesii, 119
    R. setigera, 119
    rugosa rose (*R. rugosa*), 88, 212, 277
    Wood's rose (*R. woodsii*), 111
rosemary (*Rosmarinus officinalis*), 106, 185, 224, 227, 230
rubber plant, 127
Rubus spp. *See* raspberry (*Rubus* spp.)
Rudbeckia spp. (blackeyed Susan), 106
Ruddock guilds, 264–81
rue, 274
Rumex acetosa. *See* French sorrel
runners, 129–30
Russian sage, 109, 111, 249–250
Ruta graveolens (rue), 274

S

St. Johnswort, 233–34
Saccharum spp. (sugarcane), 116
sage, 110, 185, 233–34
    red, 252–54
Sagittaria spp. (wapato), 117, 130
    S. latifolia (arrowhead), 276–77
saguaro cactus, 13–14, 100, 122
salal, 110
Salix spp. *See* willow (*Salix* spp.)
salts, soil, 107, 117
Salvadora oleiodes (peelu), 128
salvia, 110
Salvia spp. (salvia, sage), 110, 185

S. multiorrhiza (red sage), 252–54
S. officinalis (sage), 233–34
Sambucus nigra (elderberry). *See* elderberry (*Sambucus nigra*)
sand barrens, 105
sandcherry, western, 111
Sansevieria trifasciata (snake plant), 126–27
saprophytic fungi, 130–31
Satureja hortensis (summer savory), 236, 238
savannas, 65, 94–96, 105, 112, 156, 166, 216. *See also* oak savannas
saxifrage, 182
scale, 16–17
schisandra (magnolia vine) (*Schisandra chinensis*), 231, 233–34, 279, 281
scouring rush, 128
Scrophularia marilandica (late figwort), 138–39
Scutellaria baicalensis (baikal skullcap), 267
sea berry, 133, 279–81
sea kale, 15, 260
sea thrift, 110
seasons, 80–81, 135–37
Sechium edule (chayote), 129
sector analysis, 7–10, 43
sedges, 109, 110
sedum (*Sedum* spp.), 110, 120, 236, 238, 239
self-pollination, 140, 142, 183
self-regulation, 68, 100
self-sustaining system, 60, 63, 69
Sempervivum spp. (hens and chickens), 110
Senna hebecarpa (wild senna), 269
senna, wild, 269
serviceberry (*Amelanchier* spp.), 82, 109, 111, 119, 163, 168
    juneberry (*A. canadensis*), 277
    Saskatoon (*A. alnifolia*), 88, 212
seven-layer model, 13
sexual plant propagation, 143, 147
shade, 53, 114–15, 135–37
    dense, 115
    full, 39, 50, 114–15
    light, 114
    partial, 39, 50, 52, 114–15, 135–36
    plants for, 50, 80, 99, 115, 135–36
sheep, 22
shelterbelts, 81, 177, 191
shelters, animal. *See* animals
Sheperdia spp. (buffaloberry), 133
shrub rose. *See* rugosa rose
shrubby cinquefoil, 138–39
shrubs, 29, 39, 52, 77, 118–19, 133. *See also specific shrubs*
    agroforestry, 191
    as predator refuge, 185
    broad-leaved evergreen, 110
    deciduous, 111

fire-resistant, 109, 110–11
fruiting, 25, 29, 114
legumes, 133
placement in garden of, 49, 59, 83, 86
salt-tolerant, 117
Siberian ginseng (ciwujiaa, eleuthero), 119, 190, 269
sight lines, maintaining, 65
Silphium spp. (compass plant, cup plant), 150, 279
    S. perfoliatum (cup plant), 121, 138–39, 145, 177, 281
silvopasture (silviculture), 22, 160–61, 188–91
site assessment, 42–46, 66, 68, 78, 100
site, improvement or expansion of, 45–46
Sium sisarum (skirret), 15, 129
sketches, 45, 68, 76, 78–79, 82, 115
skirret, 15, 129
skunk cabbage, 182
slopes, 7, 10, 26, 45, 49, 52, 64. *See also* swale-and-berm systems
Smilax occidentalis (common greenbrier), 168
Smith, J. Russell, 152
snake hibernation mound, 178
snake plant, 126–27
snakeroot, 269
snow-in-summer, 110
snowberry, 111
    western, 168
soil, 13, 20, 49, 69. *See also* cover crop; nutrients; organic material; pH
    building, 82, 99, 100
    changes in, 72–75
    deteriorated, 108
    determining quality of, 52–56
    fertility, 10, 52, 54, 57, 68
    healthy, indicators of, 54
    initial plants, effect on structure of, 39
    inoculation, 58, 116, 133
    nondisturbance of, 57–58
    plant cover for, 21, 57–58
    prairies, 191
    preparation, 203
    regimes, 104–7
    salt tolerance, 107, 117
    saturated, 106, 115
    site improvement and, 45–46
    supportive resources in/on, 63
    terra preta (dark earth), 112
    toxins in, 108–9
    types, 104–7, 115–16
    water retained by, 57
soil survey map, 105
soil test, 103, 105, 109
Solanaceae, 109, 183

INDEX

Solanum tuberosum. See potato *(Solanum tuberosum)*
Solidago spp. (goldenrod), 184
    S. riddellii (Riddell's), 138–39
Solomon's seal, 135, 146, 269
Sorbus aucuparia (mountain ash), 111, 145, 179
sorghum *(Sorghum* spp.), 129
sorrel, 15
sourgum, 168
soybeans, 124–25, 184
Spathiphyllum spp. (peace lily), 127
spearmint, 221, 223, 233–34
specimen plants, 67
speedwell, 110
spicebush, 269–70
spider plant, 127
spikenard, 269
spinach *(Spinacia oleracea),* 32, 98, 141, 239
spiraling, 140
spirea
    blue-mist, 111
    bumald, 111
    western, 111
Spirea spp.
    S. × *bumalda* (bumald spirea), 111
    S. douglasii (western), 111
spring bulbs, 163, 274. *See also specific plants*
spring ephemerals, 115, 135, 161
sprouts, 76
spruce, 18, 179
squash, 76, 144, 148, 149, 180, 185, 192, 239
Stachys byzantina (lamb's ear), 110
stacking functions. *See* functions of plants
Stanleya pinnata (prince's plume), 103
star of Bethlehem, 243, 245
Stellaria media (chickweed), 103, 120
stinking gladwyn, 182
stolons, 129–30, 146
stonecrop. *See* sedum
stratification, 143
strawberry, wild strawberry *(Fragaria* spp.), 31, 82, 138–40, 147, 161–62, 192
    access to, 59
    alpine, woodland *(F. vesca),* 18, 59, 88, 118, 212, 272
    F. × *ananassa,* 249–50
    F. virginiana, 130, 138–39, 236
    fire-resistance of, 110
    in case study guilds, 238, 249–50, 266, 272
    in dwarf fruit tree guild, 118
    in oak guild, 163
    musk *(F. moschata),* 267
    nitrogen for, 99
    propagation, 145
    roots system, 106, 129–30

stress, 137, 141, 204
Stropharia spp. (wine cap mushroom), 133
    S. rugosoannulata (king), 131
structures, 51, 70, 74
subsoil plowing, 108
succession, 33, 81, 99. *See also* accelerated succession
succulent plants, 116
sugarcane, 116, 147
sumac, 111
summer savory, 236, 238
sun rose, 110
sun, aspect of land to, 40, 49, 52, 64
sun-tolerant plants, 135–36
sunchoke, 121
sunflower *(Helianthus* spp.), 108, 150, 180
    pale-leaved *(H. stumosus),* 138–39
    sunchoke *(H. tuberosus),* 121
sunlight, 40, 43, 49, 53, 80, 114–15, 191
    full sun, 50, 53, 114
    importance of, 50–52
    partitioning, 49–50, 52–53, 115
    plant needs chart, 286–87
supports, plant, 119
sustainability, 42, 58–68, 75, 99, 161
swale-and-berm systems, 23–25, 57, 68, 100, 202
sweet clover, 33
sweet potato, 129, 148, 239
sweet violet. *See* violet
sweet wormwood, 252–54
sweetflag, 277–78
sweetgum, American, 105, 111, 168
switchgrass, 187
sycamore, 105
    western or California, 111
Symphoricarpos spp. (snowberry)
    S. albus, 111
    S. occidentalis (western), 168
Symphyotrichum novae-angliae (New England aster), 88
Symphytum officinale. See comfrey *(Symphytum officinale)*
Syringa spp. (lilac), 106, 111

T

tallhedge, 111
tamarack, 118
taproots, 28, 33, 39, 47, 52–54, 115, 185
    compacted soil and, 106
    mustard plants, 103
    single- and multi-rooted plants, 128–29
    subsoil nutrients and, 101
Taraxacum officinale. see dandelion *(Taraxacum officinale)*
Taxus baccata (yew), 243, 245
technology, 69

temperatures, biomes differentiated by, 94–96
temporary (initial) plants, 39, 42, 135–37
terra preta, 112
thermal mass, 50–52
thermoperiodism, 141
thimbleberry, 272
thistle, 119, 185
Three Sisters Guild, 148–50
Thuja occidentalis L. var n*igra* (arborvitae), 280–81
thyme *(Thymus* spp.), 106, 109, 185, 267
    creeping *(T. praecox),* 59, 110, 137, 263
    T. serpyllum 'Coccineus,' 137, 263
    T. vulgaris, 236, 238
tickseed, 110
    lanceleaf, 138–39
Tilia spp. (linden), 103
    T. americana (American basswood), 168
timothy, 187
tomato *(Lycopersicon esculentum),* 98, 109, 140, 144, 183, 192, 239, 259–60
top plan grid, 76–77, 285
torch lily, 110
torric soil, 105
Toxicodendron spp.
    T. radicans (poison ivy), 240–44
    T. vernix (poison sumac), 240–44
toxins, agricultural, 108–12
Trametes versicolor (turkey tail mushroom), 131, 158–59
transitional zones. *See also* edges
transplant shock, 204
trap plants, 184
Tree Crops: A Permanent Agriculture, 152
trees, 39, 77, 118–19, 132–33. *See also* canopy; *specific trees*
    as essence of plant guild, 151–72
    as predator refuge, 185
    fire-resistant, 109
    legumes, 133
    niche occupied by, 74
    placement in garden of, 49, 59, 83, 85–86
    planting, 203
    understory, 118–19
trichloroethylene, 127
Trifolium spp. *See* clover *(Trifolium* spp.)
Trillium spp., 115, 182
Triticum spp., 140
Tropaeolum spp.
    T. majus (nasturtium), 103, 121, 233, 234, 259–60
    T. tuberosum (mashua), 148
trophic system, 23, 28–29, 41–42, 69, 99, 175, 185, 189, 199
trout lily, 161
trumpet vine, 110

Tsuga canadensis (eastern hemlock), 279–81
tulip (*Tulipa* spp.), 102, 129, 130, 135, 226, 230, 272
    T. gesneriana, 226, 230
turmeric, 129
Typha spp. (cattails), 276–78

U

udic soils, 105
Ulmus spp. (elm), 169
understory, 13, 17, 39, 50
    planning for, 136, 286–87
    plants for, 52, 118–19
    successional acceleration, 99
uniform plant distribution, 146–47
Urtica dioica. See nettles (*Urtica dioica*)
USDA
    Natural Resource Conservation Service soil survey map, 105
    planting zones, 80–81, 112–14, 288
ustic soils, 105

V

Vaccinium spp., 118
    hickories and, 168
    V. angustifolium (northblue blueberry), 256–57
    V. corymbosum (blueberry), 54, 106, 116, 118, 168, 180, 235, 236
    V. vitis-idaea (lingonberry), 118, 256–58, 279–81
Verbascum thapsus (mullein), 233–34
Verbena spp., 182
verbena, lemon, 127
Vernonia missurica (Missouri ironweed), 138–39
veronica (speedwell) (*Veronica* spp.), 110
    V. spicata, 263
Veronicastrum virginicum (Culver's root), 138–39
vertical plantings, 21, 44, 76
vetch (*Vicia* spp.), 102
    hairy, 102
    milk, 103
    wood (*V. caroliniana*), 272
    wood (*V. sylvatica*), 135
Viburnum trilobum (American, highbush cranberry), 111, 119, 163, 188
Vicia spp. *See* vetch (*Vicia* spp.)
    V. faba (fava beans), 98, 102, 160
Vigna unguiculata (cow pea), 58, 252
vine groundnut, 135, 144
vines, 77, 90, 119, 140. *See also specific vines*
    four vines guild, 231–34
Viola odorata (violet), 219, 221, 223
violet, 219, 221, 223
    dogtooth, 102, 115
viper's bugloss, 103
visually pleasing landscape, 60, 61, 64–67
Vitis spp. *See* grapes (*Vitis* spp.)

W

wake robin, 115
walls, plants as, 121
walnut (*Juglans* spp.), 106, 111, 146, 155
    black (*J. nigra*), 28–29, 120, 121
    butternut (*J. cinerea*), 269
wapato, 117, 130
wasps, 30, 39, 58, 137, 173, 178–79, 182, 195. *See also* insects
waste, 1, 68, 70. *See also* compost
    animal waste, using, 192–93
    Permaculture design and, 4–6, 32
    reusing, 37–38, 41, 46–47, 69, 100
water, 7, 32, 34, 49, 70, 191
    collection and storage, 62
    earthworks for retention of, 23–26
    in scale of permanence, 74
    maintenance of circulation and recycling, 57
    sources of, 23, 74
water chestnut, 130
water features, 52
Water for Every Farm, 26, 72
water hyacinth, 108
watercress, 103, 277–78
watershed management, 160
weather extremes, 98, 100–101, 107–8, 114
weather, timing of flowers controlled by, 137
weeds, 13, 47, 83, 108, 120–21
wet soil, 104–6, 156, 188
wet weather, plants for, 57
wheat, 32, 54, 140, 160
wild areas, 21, 38
wild ginger (*Asarum* spp.), 162, 168, 219
    A. caudatum, 135, 150, 164, 223, 226, 230
    Canadian (*A. canadense*), 106, 120, 221, 269–70
wild rice, 143, 264, 275–77
wild rice pond guild, 275–78
wild strawberry. *See* strawberry, wild strawberry (*Fragaria* spp.)
wildlife. *See* animals
willow (*Salix* spp.), 108, 111, 119, 145, 188, 279
    basket, 281
    white (*S. alba*), 119, 128
wind, 43, 52, 74, 104, 112
wind pollination, 140, 179, 183
windbreaks, 81, 112, 178, 187–88
wintergreen, 115
witchhazel, common, 168
woodlands, 46, 53, 94–96, 108, 115, 166, 190. *See also* eastern woodlands
Woods' rose, 111
worms, 22–23, 69, 171, 198–99

X

Xenoglossa spp. (gourd bee), 180
xeric (dry) soil, 105–7, 156
Xylocopa spp. (carpenter bee), 180
Xylorhiza glabriuscula (woody aster), 103

Y

yarrow (*Achillea* spp.), 90, 109, 110
    common (*A. millifolium*), 88, 213, 226, 229
Yeomans, Ken, 26
Yeomans, P. A., 7, 26, 72–75
yew, 243, 245
yield, 32, 60–61, 64, 68, 75–77, 98, 100
yucca (*Yucca filamentosa*), 110, 224, 227–30

Z

Zea mays. See corn (*Zea mays*)
Zizania spp. (wild rice), 264, 275–78
    Z. palustris (northern), 143, 277–78
    Z. texana, 275
Zizia auria (golden Alexanders), 138–39
Ziziphus jujuba (jujube, red date), 252–54
zone system, 7–10, 64, 67, 74
zones, USDA. *See* USDA planting zones

ABOUT THE AUTHORS

Wayne Weiseman is certified by the Permaculture Institute of Australia and the Worldwide Permaculture Network as an instructor of the Permaculture Design Certificate course. He is the director of Kinstone Academy of Applied Permaculture (KAAP) in Fountain City, Wisconsin, the Permaculture Project LLC, and the Permaculture Design-Build Collaborative LLC, full-service, international consulting and educational businesses promoting the ideas of eco-agriculture, renewable energy resources, and eco-construction methods. For many years he managed a land-based, self-reliant community project combining organic crop/food production, ecologically built shelters, renewable energy, and appropriate technologies.

Daniel Halsey is a certified permaculture designer and teacher for the Permaculture Research Institute. Dan travels nationally and internationally teaching permaculture and ecological design to permaculture-design-certification students, homesteaders, and landscape designers. Daniel and his wife, Ginny, manage self-sustaining forest gardens of fruiting trees, shrubs, and nut crops at SouthWoods Forest Gardens, a permaculture design, demonstration, and educational site located on a twenty-five-acre wetland savannah in Prior Lake, Minnesota.

Bryce Ruddock is certified as an instructor of permaculture teaching by the Permaculture Institute USA and the Cascadia Permaculture Institute since 2010. He authored the *Plant Guilds* e-book, a training manual used in classes by Midwest Permaculture. His interest in perennial polycultures began in 1980. Since 1984, Bryce and his partner, Debby, have been implementing permaculture-based polyculture designs at their sixth-of-an-acre urban home site in southeastern Wisconsin, where they have transformed an average suburban yard into a thriving food-and-medicinals food forest.

the politics and practice of sustainable living
CHELSEA GREEN PUBLISHING

Chelsea Green Publishing sees books as tools for effecting cultural change and seeks to empower citizens to participate in reclaiming our global commons and become its impassioned stewards. If you enjoyed *Integrated Forest Gardening*, please consider these other great books related to gardening and agriculture.

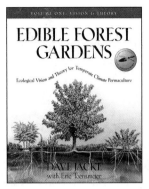
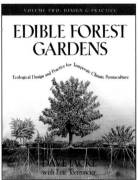

EDIBLE FOREST GARDENS (Two-Volume Set)
Ecological Vision and Theory for Temperate Climate Permaculture
DAVE JACKE and ERIC TOENSMEIER
9781890132606
Hardcover • $150.00

THE RESILIENT FARM AND HOMESTEAD
An Integrated Permaculture and Whole Systems Design Approach
BEN FALK
9781603584449
Paperback • $40.00

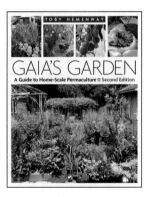

GAIA'S GARDEN, Second Edition
A Guide to Home-Scale Permaculture
TOBY HEMENWAY
9781603580298
Paperback • $29.95

PERENNIAL VEGETABLES
From Artichokes to 'Zuiki' Taro, a Gardener's Guide to over 100 Delicious and Easy-to-Grow Edibles
ERIC TOENSMEIER
9781931498401
Paperback • $35.00

For more information or to request a catalog, visit **www.chelseagreen.com** or call toll-free **(802) 295-6300**.